ESD

ESD
RF Technology and Circuits

Steven H. Voldman
Vermont, USA

John Wiley & Sons, Ltd

Other Wiley Editorial Offices

John Wiley & Sons Inc., 111 River Street, Hoboken, NJ 07030, USA

Jossey-Bass, 989 Market Street, San Francisco, CA 94103-1741, USA

Wiley-VCH Verlag GmbH, Boschstr. 12, D-69469 Weinheim, Germany

John Wiley & Sons Australia Ltd, 42 McDougall Street, Milton, Queensland 4064, Australia

John Wiley & Sons (Asia) Pte Ltd, 2 Clementi Loop #02-01, Jin Xing Distripark, Singapore 129809

John Wiley & Sons Canada Ltd, 6045 Freemont Blvd, Mississauga, Ontario, Canada L5R 4J3

Wiley also publishes its books in a variety of electronic formats. Some content that appears in print may not be
available in electronic books.

British Library Cataloguing in Publication Data

A catalogue record for this book is available from the British Library

ISBN-13 978-0-470-84755-8 (HB)
ISBN-10 0-470-84755-7 (HB)

Typeset in 10/12 pt Times by Thomson Digital.

To My People

Contents

Preface

The phenomenon of electrostatic discharge (ESD) has been known for a long time, but recently a growing interest has been observed in ESD in radio frequency (RF) technology and ESD issues in RF applications.

Why now?

Early telecommunications started with William Cooke and Charles Wheatstone in the development of the electric telegraph that became commercial in 1838. This technology was rapidly replaced by Samuel Morse, with the introduction of the "Morse Code," first introduced in 1844, which reduced the communication into dots and dashes and listening to the receiver. By 1906, Lee De Forest introduced the first three-element vacuum tube detector, opening the future to vacuum tubes for electronic applications for radio in the future. The Wireless Era began. My personal library contains some old volumes of discarded radio engineering books. An old dusty book by Herbert J. Reich on *"Theory and Applications of Electron Tubes"* is stamped on the side "RADIATION LABORATORY BLDG. 24," and superimposed is "Document Room, Research Laboratory, Mass. Inst. Technology." This is adjacent to another text, the 1947 third edition of *"Radio Engineering"* by Frederick Emmons Terman. The 1947 textbook apologizes on the first pages with a note "the quality of the materials used in the manufacture of this book is governed by continued war shortage." In the 1947 Terman text book, the new edition focuses on new issues such as television and the advancements called "radar." Adjacent to that text is my copy of *"Basic Electron Tubes"* by Donovan Geppert of General Electric Company. By 1951, the McGraw-Hill launched the "Electrical and Electronic Engineering Series" with texts such as *Fundamentals of Vacuum Tubes* by Eastman, *Vacuum Tubes* by Spangenberg, *Transmission Lines and Networks* by Johnson, *Antennas* by Kraus, and many more texts in the growing electrical engineering discipline.

In the late 1970's, I was a graduate student at Massachusetts Institute of Technology (MIT), in the Research Lab of Electronics (RLE). Our faculty were from the "Rad Lab Era" and so was the microwave equipment. The old "Rad Lab" building was still in place, and the building was filled with old machinists and the last of the glass lathe experts from the vacuum tube days. Old microwave and radio books were being discarded to make space for new texts in the MIT libraries. The Microwave Era was dying, and the faculty who brought it into existence were retiring out. The interest in "radio frequency" (RF) and microwave was limited and was not growing.

But in semiconductor military application circles, microwave semiconductor development was ongoing. Research and development of microwave semiconductor devices for high speed communications and military applications continued. Little did I know that even when I was at the University of Buffalo, in 1978, J.J. Whalen, my electrical engineering (EE) circuit's teacher, was doing research on the power-to-failure of microwave semiconductor devices and collaborating with ESD engineers in relation to ESD robustness. At this time, the Electrostatic Discharge Association was also initiated.

In the past ten to twenty years, research and development publications have been produced discussing the ESD robustness of III–V compound semiconductor devices, with a primary focus on gallium arsenide (GaAs). At the same time, new semiconductor devices have been proposed, such as silicon carbide and silicon germanium, as well as other exotic devices. But, it is only recently that these semiconductor devices have left niche markets and entered mainstream applications. In the process of leaving the niche market arena and entering the mainstream commercial marketplace, the interest in ESD has increased importance. For example, although silicon germanium transistor research began in the mid-1980s, no ESD measurements of this hetero-junction bipolar transistor were taken until approximately the year 2000. In 2000, I released the first paper on ESD robustness of the silicon germanium transistor at the International Reliability Physics Symposium (IRPS). And, ironically, it is only recently that the mainstream technology, CMOS technology, has achieved RF performance levels, making the mainstream technology players also address the same issues of achieving RF performance and maintain ESD robustness. Hence, it was at the turn of the new millennium that the interest began to accelerate as these new technologies began to leave niche markets and enter the industry in volume production.

On October 10, 2001, I presented a 3-hour tutorial, Tutorial J, entitled *"ESD Protection and RF design,"* in the Oregon Convention Center, Portland, Oregon. The response from the tutorial was very interesting; two communities came to the tutorial—the first group was of "ESD engineers" wanting to learn about RF technology and the second group was of "RF engineers" wanting to learn how to provide ESD protection to the circuits; both groups were very eager to address this growing issue. The ESD engineers were interested in learning the secrets of RF design and were keen to invent and design new circuits and structures; the RF circuit designers wanted to learn the black magic of ESD engineering. Both the RF and ESD design communities are similar in their style of work, which includes experimentation, innovation, invention, tuning, and trimming. Both groups wanted more on their areas of interest and wanted the tutorial be extended to a 4.5 hour or all-day ESD tutorial. Unfortunately at that time, in 2001, there was very little technology literature on ESD in gallium arsenide, silicon germanium, silicon germanium carbon, indium phosphide, silicon carbide, and even less circuit innovation, design techniques, experimental results, co-synthesis, or solutions.

At that time, the publisher John Wiley & Sons, anticipating the interest in this field, approached me about writing a book on the subject of ESD in RF technology. I realized at that time, although the interest was significant, the development and research were at a primitive state, and few publications, patents, and literature existed in this field and discipline. I proposed we delay the introduction of the *ESD: RF Technology and Circuits* text until the RF ESD field matured and developed further. In the meantime, I proposed an ESD book series—I proposed to write the first text *ESD: Physics and Devices,* and the second text, *ESD: Circuits and Devices* until the industry was ready for the third text on ESD in RF devices. During this period, in parallel, I initiated a special session on ESD in radio frequency components at the ESD Symposium. Additionally, in order to stimulate growth in this area, I also established an ESD RF sub-committee at the EOS/ESD Symposium. Because of the primitive state of the field, it was necessary to allow simple basic ESD RF papers to emerge that allow growth in the field. Over a 3-year period, the publications and papers went from primitive RF ESD concepts to very technical mature implementations.

The first text *ESD: Physics and Devices* has targeted the semiconductor device physicist, the circuit designer, the semiconductor process engineer, the material scientist, the chemist, the physicist, the mathematician, the semiconductor manager, and the ESD engineer. The second text *ESD: Circuits and Devices* has targeted a readership from semiconductor device physicists to circuit designers. In the third text, *ESD: RF Technology and Circuits*, a balance is established between the technology and circuits for the ESD engineer and RF circuit designer.

The first goal of the present book, *ESD: RF Technology and Circuits*, is to teach the fundamentals of a new design discipline, which we will refer to as "radio frequency electrostatic discharge (RF ESD) design." To address this new design practice, we must also address how the RF ESD design practice is different from the "ESD design practice" used for digital circuits (e.g. standard digital CMOS). Additionally, we must address how the "RF ESD design" practice is distinct from the RF design practice. An objective of this book is also to teach how the RF methodology has modified the basic ESD design practices, which involve coupling, de-coupling, buffering, ballasting, triggering, shunting, and distributing as has been discussed in the second text, *ESD: Circuits and Devices*. So, the goal is to teach a new method of design – RF ESD design – which consists of methods of substitution, cancellation, distributed loads, impedance isolation, and other techniques that utilize methods in the frequency domain as the ESD phenomenon separates from the RF application frequency. The RF ESD design practice synthesizes RF design methods and digital CMOS ESD methods leading to new structures, circuits, and innovations. Additionally, we are interested in showing how the RF design practice is modified with the requirement of having ESD protection concepts. The question is how is the RF design practice influenced with the presence of ESD networks?

The second goal is to teach a general methodology of RF ESD design without a fundamental focus on specific circuit implementations, but re-inforcing the methodologies through circuit examples. Many ESD books and publications focus on the specific circuit or device. Our goal will be to use the examples as a means to demonstrate the design practices through the RF ESD circuits. The examples of the RF ESD circuits will be from RF CMOS, RF Bipolar, and RF BiCMOS technologies.

The third goal is to demonstrate the uniqueness of the RF ESD design practice in different steps of implementation. These involve circuit conception, layout and design, design tools, characterization, implementation, testing, and failure criteria.

The fourth goal is to focus on RF technology and its ESD performance. The text will provide a wide spectrum of technologies from RF CMOS, Silicon Germanium (SiGe) technology, Silicon Germanium Carbon (SiGeC) technology, and Gallium Arsenide (GaAs) technology; it is valuable to see how the technologies influence the ESD results, circuit designs, and solutions. Additionally, the ESD-induced failure mechanisms will be highlighted.

The fifth goal is to show how to design RF ESD input networks, RF ESD rail-to-rail circuits, and RF ESD power clamps from a broad perspective. In this text, we are going to focus on issues with receivers, differential receivers, transmitters, and other circuits. The text will focus on single-ended versus differential networks ESD issues, and how the matching networks and passives change the ESD robustness of the networks. One of the objectives is to address the potential failure mechanisms in these specific circuits, and what are the possible circuit topology changes and ESD solutions to address them.

The sixth goal is to expose the reader to the prior work in the field of RF ESD. Through the early work, significant understanding can be achieved in the RF-ESD testing methodology, and power-to-failure relationships, physical models, power-to-failure electro-thermal models.

The seventh goal is to expose the reader to the patent art in the ESD field. A significant amount of activity in the ESD field can be found by reading the patent art. As a result, a number of patents are referenced, which either are first in the field and, relevant in the discussions of interest or teach methods and methodologies.

The third book in this series, *ESD: RF Technology and Circuits*, will contain the following:

Chapter 1 will introduce the reader to the fundamentals and concepts of ESD RF design. In this chapter, we will initiate the discussion of the uniqueness of this RF ESD design methodology. We will discuss concepts of substitution, cancellation, distribution, matching, and design layout practices. This chapter will review ESD pulse phenomenon and models and the relative time scales of ESD events. ESD failure mechanisms will be discussed for RF technologies—RF CMOS, Gallium Arsenide, and Silicon Germanium. RF metrics and RF ESD testing methods will also be discussed. Recent patents associated with RF ESD structures, RF ESD circuits, RF technology, and RF ESD design methodologies will be briefly discussed as a source for additional reading and reference materials.

Chapter 2 will discuss the details of RF ESD design methodology and RF ESD design synthesis. In this chapter, RF ESD design methods along with the substitution, cancellation, and impedance isolation ESD techniques will be discussed. Linearity and ESD devices will also be highlighted. In addition, the chapter will address synthesis of digital, analog, and RF circuits into a common semiconductor chip.

Chapter 3 will focus on RF CMOS ESD protection elements. A comparison of different ESD strategies from both the RF and the ESD perspectives will be given. MOSFET, shallow trench isolation defined diodes, polysilicon-bound diodes, and Silicon-controlled rectifiers will be compared from the perspectives of the RF parametrics, loading capacitance, and ESD robustness. ESD robustness and design of RF passives elements (e.g., resistors, Schottky diodes, capacitors, and inductors) will be highlighted.

Chapter 4 will discuss the RF CMOS ESD circuitry. RF ESD circuits, which utilize passive elements and co-synthesize with RF input and output-matching networks will also be shown in this chapter. It will highlight new inductor/diode networks, T-coils, distributed networks, and other RF ESD circuits; these networks will serve as examples where the RF ESD design methods discussed in Chapters 1 and 2 are utilized. ESD protection in RF LDMOS technology, and ESD design methods for RF low noise amplifier (LNA) applications will be highlighted.

Chapter 5 will focus on ESD and Bipolar technology. In this chapter, bipolar device physics of homo-junction and hetero-junctions will be discussed. This chapter will review key RF metrics and parameters of interest for ESD and RF design. Electrical stability, thermal stability, and RF stability of transistors will be shown. Electrical and thermal shunts will be discussed as well. This chapter will review the Johnson Limit, the relationship of breakdown voltages and transistor speeds, and why this is important for bipolar components,

ESD devices, and ESD circuits. Design layout of single emitter, multiple emitter designs and "ordering of the emitter, base, and collector" and its ESD implications will be also reviewed.

Chapter 6 will contain Silicon Germanium, Silicon Germanium Carbon and ESD. SiGe and SiGeC HBT device measurements from HBM to TLP in different configurations will be highlighted and compared. The chapter will discuss from TLP I-V measurements to Wunsch-Bell power-to-failure curves of Si homojunction BJTs, SiGe HBTs, and SiGeC HBTs. Usage of the SiGe HBT in emitter-base, base-collector, collector-to-emitter, and collector-to-substrate will be reviewed.

Chapter 7 will discuss Gallium Arsenide, Indium Gallium Arsenide and ESD technologies. Early GaAs MESFETs ESD measurements and failure mechanisms will be reviewed. Recent modern day GaAs HBT device HBM and TLP I–V measurements will be shown both as individual devices and devices within ESD circuits.

Chapter 8 will discuss bipolar circuits and ESD. Bipolar peripheral circuits, such as receivers and transmitters, will be shown. ESD power clamps for bipolar technology, suitable for Silicon, Silicon Germanium, Silicon Germanium Carbon, Gallium Arsenide, and Indium Phosphide technologies will be discussed. The bipolar classes of power clamps will include both forward-bias and reverse-bias breakdown trigger networks as well as capacitive-triggered networks; these include diode string trigger networks, Zener-breakdown triggered power clamps, and BV_{CEO}-breakdown triggered power clamps. Triple-well ESD power clamps will also be discussed.

Chapter 9 will discuss ESD design methodology. Whereas there are many different ESD design implementations, this chapter will focus on one system of implementation that allows for RF and ESD co-synthesis. This tool is practiced today in semiconductor design methodologies and has been found to be successful in both RF CMOS and RF Silicon Germanium technologies. By focusing on one method, the reader will have a sense of a design system that provides customers design freedom and RF-ESD co-synthesis in a mixed signal design foundry environment.

Chapter 10 will highlight non-semiconductor ESD solutions, and off-chip protection ESD design concepts. Spark gaps, air gaps, and field emission devices (FED) structures used in RF GaAs applications will be shown. In addition, mechanical shunt solutions integrated into packages will also be discussed. And finally, conductive electronic polymer surge protection concepts, applied to GaAs cell phone technology will serve as an example of off-chip protection using non-silicon and non-semiconductor solutions.

In this text, the trends and directions of RF ESD design will be shown. As with the rapid growth of this RF ESD field, devices, circuits, and design may take different directions in the future. Hopefully, the RF ESD basic concepts will fundamentally remain valid as we move from 1 to 100 GHz applications independent of the devices or specific circuit embodiments.

Enjoy the text, and enjoy the subject matter of ESD. There is still so much more to learn.

B"H
Steven H. Voldman
IEEE Fellow

Acknowledgements

I would like to thank the individuals who have helped me on the right path in my academic and professional career—to address the field of electrostatic discharge (ESD) in radio frequency (RF) technology. Faculty from the University of Buffalo, Massachusetts Institute of Technology, and University of Vermont had significant impact on my direction and interest in the area of continuum mechanics, continuum electro-mechanics, electrostatics, semiconductors, field theory, systems, and circuits as well as mathematics and physics. I am indebted to the Engineering Science, Physics Department, and Electrical Engineering curriculums at the University of Buffalo for support and interest in the thermal, mechanical, and electrical sciences; faculty includes Prof. Irving Shames, Prof. Herbert Reismann, Prof. Stephen Margolis, Prof. J.J. Whalen, Prof. R. K. Kaul, Prof. Reichert, and other faculty from The University of Buffalo – little did I know when I was J.J. Whalen's student in my first electrical engineering (EE) circuit courses at the University of Buffalo in 1978, that he was active in research on the power-to-failure and HBM testing of microwave devices with Prof. Hank Domingos of Clarkson College – some of the first work on ESD evaluation of microwave RF devices! At Massachusetts Institute of Technology (MIT), I am indebted to the Electrical Engineering (EE) Department, Physics Department, MIT Plasma Fusion Center, and MIT High Voltage Research Laboratory (HVRL) for the support in the area of plasma physics, electrodynamics, electrostatics, microwave theory, and semiconductors. It was at MIT where I had my first exposure to microwave theory, applied experimental work using microwave waveguide diagnostics in plasma physics environments, power electronics, high voltage devices. I had been surrounded by the EE faculty of the MIT Research Laboratory of Electronics (RLE) from the World War II Radar Development Era, the "Rad Lab" Building—Prof. Louis D. Smullin, Prof. James R. Melcher, and so forth—this is where I became an experimentalist and achieved my training working side-by-side with the MIT faculty. As a graduate student under Prof. Louis D. Smullin, I used to tackle experimental work as if the war did not end—no time was ever to be allowed to be wasted. As a "Melcher student", I was influenced by his analytical methodology and entranced at his academic enthusiasm. And as a student of Prof. Jin Au Kong, I was able to increase my knowledge on microwave theory, and electrodynamics from a generalist approach. I am also indebted to Prof. Markus Zahn in the MIT HVRL for guiding me in high voltage analytical and experimental work, technical writing skills, and for sharing his academic perspectives. As an MIT graduate teaching student, I was fortunate to be able to observe and participate in the teaching of semiconductor devices and circuits to the undergraduate semiconductor EE course 6.012 with Prof. Clif Fonstad, Prof. David Epstein, Prof. Wyatt, and Prof. Hank Smith. At the University of Vermont, as a student of Professor R.L. Anderson, an early hetero-junction researcher, I had my first exposure to hetero-junction bipolar transistors, III–V devices, and cryogenics, little did I know that time that I would be the first one working on ESD protection of Silicon Germanium and Silicon Germanium Carbon devices ten years later.

At IBM, I was fortunate to have many mentors and friends from IBM Burlington Vermont, IBM East Fishkill, IBM T.J. Watson Research Center, IBM San Jose, IBM Rochester, IBM Haifa Israel, and IBM RF Boston Design Center. My early years in bipolar SRAM development was influenced by Roy Flaker, Jack Gerbach, Russ Houghton, Jeffery Chu, Badih El-Kareh, John Aitken, Tak Ning, Denny Tang, George Sai-Halasz, Jack Y.-C. Sun, and Robert Dennard. In the RF CMOS and RF BiCMOS Silicon Germanium area, I would like to thank my peers who assisted in ESD work, invention, design kits, measurements, and technical support: Louis Lanzerotti, Robb Johnson, Peter Zampardi, Ephrem G. Gebreselasie, Amy Van Laecke, Stephen Ames, Susan E. Strang, Donald Jordan, C. Nicholas Perez, David S. Collins, Doug Hershberger, Alan Norris, Arnold Baizley, Bradley Orner, Michael Zierak, Robert Rassel, Peter B. Gray, Ben T. Voegeli, Q.Z. Liu, J.S. Rieh, John He, Jay Rascoe, Xue Feng Liu, Doug Coolbaugh, Natalie Feilchenfeld, Dawn Wang, J.S. Lee, Stephen St. Onge, Alvin Joseph, James Dunn, David Harame, Gary Patton, and Bernard Meyerson.

In the ESD discipline, I would like to thank for the years of support and the opportunity to provide lectures, invited talks, and tutorials on ESD and latchup; from the SEMATECH ESD group, the Electrical Overstress/Electrostatic Discharge (EOS/ESD) Symposium, the International Reliability Physics Symposium (IRPS), the Taiwan Electrostatic Discharge Conference (T-ESDC), the International Physical and Failure Analysis (IPFA), the International Conference on Electromagnetic Compatibility (ICEMAC), the Bipolar/BiCMOS Circuit and Technology Meeting (BCTM), and the International Solid State Circuit Conference (ISSCC), as well as ESD Association Education Committee, and the ESD Association Device Testing Standards Committees. I would like to thank the ESD Association office for the support in the area of publications, standards developments, and conference activities, with a special thanks to Lisa Pimpinella. In the field of ESD in RF devices, I would like to thank Prof. Elyse Rosenbaum (UIUC), Prof. Ming-Dou Ker (NCTU), Karl Heinz Bock, Eugene Worley, Corrine Richier (ST Microelectronics), Marise BaFluer (CNRS), Ann Fletcher (RF MicroDevices), Patrick Juliano (Intel), and Kathy Muhonen (RF MicroDevices), and Sami Hynoven. I would like to thank my summer students and ESD support in the RF ESD field—Patrick Juliano of UIUC, Brian Ronan of Princeton University, and Anne Watson of Penn State University for her hard work, collaboration, and participating in ESD and latchup experimentation, and discovery. A special thanks to Ephrem G. Gebreselasie for his hard work and support, in both ESD and latchup in RF CMOS and RF BiCMOS SiGe development. I would also like to thank the publisher John Wiley & Sons, Ltd and its staff in Chichester, Taiwan, Singapore, and China for taking on this first ESD book series.

And most important a special thanks to my children, Aaron Samuel Voldman and Rachel Pesha Voldman, and my wife Annie Brown Voldman. May we all fulfill what we want in our lives and walk a path of righteousness, justice, and truth. And of course, my parents, Carl and Blossom Voldman.

Baruch HaShem. . .
B''H
Dr. Steven H. Voldman
IEEE Fellow

1 RF Design and ESD

1.1 FUNDAMENTAL CONCEPTS OF ESD DESIGN

As a design discipline, the electrostatic discharge (ESD) design discipline is distinct from circuit design practices used in the development of semiconductor circuit design discipline [1,2]. Fundamental concepts and objectives exist in the ESD design of semiconductor devices, circuits, and systems in methods, layout, and design synthesis. To address the radio frequency (RF) ESD design discipline, we pose the following questions [1–3]:

- What is it that makes the ESD design discipline unique?

- How is it distinct from standard circuit design practices?

- How is RF ESD design discipline different from the RF design discipline?

- How is RF ESD design discipline different from the digital ESD design discipline?

To address the first issue of the ESD design discipline, let us first address the uniqueness of the distinction of ESD design discipline practice. Here are some of the ESD design practices [2]:

- *Device Response to External Events*: Design of devices and circuits to respond to (and not to respond to) unique current waveforms (e.g., current magnitude and time constants) associated with external environments. In ESD design, the ESD devices as well as the circuits that are to be protected can be designed to respond (and not torespond) to unique ESD current waveforms. ESD networks are typically designed to respond to specific ESD pulses. These networks are unique in that they address the current magnitude, frequency, polarity, and location of the ESD events. Hence, in ESD design, the ESD networks are designed and tuned to respond to the various ESD events. In ESD design, different stages or segments of the network can also be designed to respond to different ESD events. For example, some stages of a network can respond to human body model (HBM) and machine model (MM) events, whereas other segments respond to the charged device model (CDM)

ESD: RF Technology and Circuits Steven H. Voldman
© 2006 John Wiley & Sons, Ltd

event. These ESD events differ in current magnitude, polarity, time constant, as well as the location of the current source. Hence, the ESD circuit is optimized to respond and address different aspects of ESD events that circuits may be subjected to [2].

- *Alternate Current Loops*: Establishment of alternative current loops or current paths, that activate during high current or voltage events. A unique issue is the establishment of alternative current loops or current paths that activate during high current or voltage events. By establishing alternative current loops, or secondary paths, the ESD current can be re-directed to prevent overvoltage of sensitive circuits. In order to have an effective ESD design strategy, this current loop must respond to the ESD event and have a low impedance [2].

- *Switches*: Establishment of 'switches' that initiate during high current or voltage events. On the issue of establishment of 'switches' that initiate during high current or voltage events, the uniqueness factor is that these are at times either passive or activated by the ESD event itself. A unique feature of ESD design is that it must be active during unpowered states. Hence, the 'switches' used to sway the current into the ESD current loop are initiated passively or are initiated by the ESD event itself. Hence, the ESD event serves as the current and voltage source to initiate the circuit. These switches lead to 'current robbing' and the transfer of the majority of the current from the sensitive circuit to the alternative current loop. The ESD design discipline must use 'switches' or 'triggers' that initiate passively (e.g., a diode element) or actively (e.g., a frequency-triggered ESD network). A design objective is to provide the lowest voltage trigger allowable in the application space. Hence, a key ESD design objective is to utilize low-voltage trigger elements that serve as a means to transfer the current away from the sensitive circuit to alternative current paths. A large part of the effective ESD design discipline is the construction of these switches or trigger elements [2].

- *Decoupling of Current Paths*: Decoupling of sensitive current paths is an ESD design discipline practice. Circuit elements can be introduced that lead to the avoidance of current flow to those physical elements. The addition of 'ESD decoupling switches' can be used to decouple sensitive circuits as well as to avoid the current flow to these networks or sections of a semiconductor chip. ESD decoupling elements can be used to allow elements to undergo open or floating states during ESD events. This can be achieved within the ESD network or within the architecture of a semiconductor chip. Decoupling of sensitive elements or decoupling of current loops can be initiated by the addition of elements that allow the current loop to 'open' during ESD events. The decoupling of nodes, elements, circuits, chip subfunctions, or current loops relative to the grounded reference prevents overvoltage states in devices and eliminates undesired current paths. Decoupling elements can avoid 'pinning' of electrical nodes. Hence, integration of devices, circuits elements, or circuit functions that introduce decoupling electrical connections to ground references and power supplies references, is a unique and key ESD design practice [2].

- *Decoupling of Feedback Loops*: Decoupling of loops that initiate pinning during off condition or ESD test modes. Feedback loops can lead to unique ESD failures and lower ESD results significantly. The decoupling of nodes, elements, or current loops relative to the grounded reference prevents overvoltage states in devices and eliminates current paths initiated by the feedback elements. These decoupling elements can avoid 'pinning' of electrical nodes [2].

- *Decoupling of Power Rails*: Decoupling of electrical connections to grounded references, and power supplies [2].

- *Local and Global Distribution*: Local and global distribution of electrical and thermal phenomena in devices, circuits, and systems is a key ESD design practice and focuses on ESD development. To provide an effective ESD design strategy, the ESD design practices must focus on the local and global distribution of electrical and thermal phenomena in devices, circuits, and systems. In order to shunt the ESD current efficiently and effectively, the distribution of the current is critical in ESD design. As the current distributes, the effectiveness of the device helps improving the utilization of the total area of the ESD network or circuit element. On a circuit and system level, the distribution of the ESD current within the network or system lowers the effective impedance and lowers the voltage condition within the ESD current loop [2].

- *Usage of Parasitic Elements*: Utilization and avoidance of parasitic element is part of the ESD design practice. ESD design either utilizes or avoids activation of these parasitic elements in the ESD implementations. Utilization of parasitic elements is a common ESD design practice for ESD operation, such as utilization of parasitic lateral or vertical bipolar transistors. It is not common to use these parasitic elements in standard circuit design, whereas for ESD design it is very prevalent to utilize the parasitic devices and is part of the ESD design practice and art [2].

- *Buffering*: Utilization of current and voltage buffering of sensitive devices, circuits, or subcircuits is a key ESD design practice. In ESD design, it is also a common practice to establish current and voltage buffering of sensitive devices, circuits, subcircuits, chip level core regions, or voltage islands. A design practice is to increase the impedance in the path of the sensitive circuit either by placing of high impedance elements, establishing 'off' states of elements, voltage and current dividing networks, resistor ballasting, or by initiating elements in high impedance states [2].

- *Ballasting*: It is a standard ESD design practice to use ballasting techniques, which involves introduction of resistance to redistribute current within a single element or a plurality of elements. In digital design, ballasting is predominately achieved using resistor elements. Resistive, capacitive, or inductive ballasting can be introduced to redistribute current or voltage within a single element or a plurality of elements, circuit, or chip segment. The usage within a semiconductor device element allows for redistribution within a device to avoid electro–thermal current constriction and poor area utilization of a protection network or circuit element. The usage of ballasting allows to redistribute the source current from the ESD event to avoid thermal heating or electrical overstress within the semiconductor network or chip. Ballasting can be introduced into a semiconductor device structures achieved by semiconductor process choices, material choices, silicide film removal, introduction of discrete resistor elements, and introduction of design layout segmentation [2].

- *Usage of Unused Sections of a Semiconductor Device, Circuit, or Chip Function*: It is an ESD design practice to utilize 'unused' segments of a semiconductor device for ESD protection, which was not utilized for functional applications [2].

- *Impedance Matching between Floating and NonFloating Networks*: It is an ESD design practice to impedance match the states of floating structures. In ESD design, it is common to utilize the 'unused' segments of a semiconductor device for ESD protection and impedance match the network segments for ESD operation; this matching of conditions during ESD testing allows for current sharing during matching between networks and common triggering voltage conditions [2].

- *Unconnected Structures*: It is a common ESD design practice to address structures not containing electrical connections to the power grid or circuitry. In semiconductor chips, there are many structures that are electrically not connected to other circuitry or power grids, which are vulnerable to ESD damage. Unique ESD solutions are used to address floating or unconnected structures [2].

- *Utilization of 'Dummy Structures and Dummy Circuits'*: In the ESD design practice it is not uncommon to utilize dummy structures or dummy circuits that serve the purpose to provide better current uniformity or distribution effects; these concepts span from the usage of dummy Metal-Oxide Semiconductor Field Effect Transistor (MOSFET) poly-silicon gate fingers to dummy inverter circuits [2].

- *Nonscalable Source Events*: Another key issue is that the ESD event is a nonscalable event. Each generation, the size of devices is scaled to smaller dimensions. The ESD design practice must address the constant source input current, and the physical scaling of the structures. A unique ESD scaling theory and strategy must be initiated to address this issue [2].

1.2 FUNDAMENTAL CONCEPTS OF RF ESD DESIGN

In the ESD design discipline of radio frequency (RF) circuits, there is a fundamental difference in the focus and methods that are required, which are distinct from the ESD design practices used in ESD protection of digital circuitry [3]. This rapidly developing ESD design practice utilizes some of the ESD digital design practices when it is possible and abandons some practices, ESD circuits, and design when they are unsuitable for RF applications. In this evolution, the ESD design discipline is shifting and adapting design practices used by microwave RF circuit designers, in addition to the new, and unique RF ESD design practices that will be established to cosynthesize RF functional application needs and ESD protection [4]. The RF ESD design discipline is presently evolving as the application frequency, which continues to increase as well. The key question is what makes this new design practice unique from ESD design practice, and how does it differ from RF design practices.

The RF fundamental concepts for the ESD protection and design of RF components are as follows:

- *RF ESD Application Frequency Dependent ESD Solutions*: In RF ESD design, the solutions and methods for the ESD protection may be a function of the application frequency. Below 1 GHz, traditional digital ESD on-chip silicon ESD circuit solutions may be sufficient. Between 1 and 5 GHz, the choice of ESD device may be a function of the tradeoffs of loading and other RF parameters. Above 5–15 GHz, RF ESD cosynthesize may be a mandatory process. Above these application frequencies, off-chip protection and nontraditional ESD solutions may be necessary (Figure 1.1).

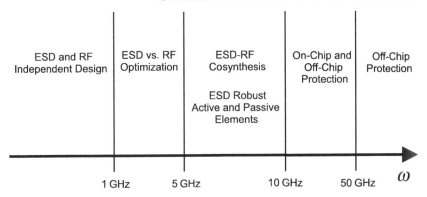

Figure 1.1   RF ESD design as a function of application frequency

- *RF Models for ESD Elements*: With RF circuits and components, d.c. and RF models are required to build RF circuits. As a result, all ESD elements must have full RF quality models. This is very different from ESD digital design practices that are not highly dependent on an ESD model. ESD design for digital design does not require a physical model. On the contrary, RF applications require some form of RF model analysis of the ESD element as it influences all of the RF functional parameters. This influences the physical design implementation.

- *RF ESD Design Methodology*: With the requirement of high-quality RF models, ESD design methodologies require full RF model support as well. As a result, the computer-aided design methodology for the ESD design methods must address this issue. As an example, it may require new computer-aided design methodologies that are not practiced in digital design and are more adaptable to the RF design environment. As an example to be discussed in a later chapter, custom fixed design sizes, growable or scalable designs, parameterized cells, and/or hierarchical parameterized cell ESD networks and methods of extraction for various size implementations may be required [5,6,7].

- *RF ESD Design Chip Subfunction Synthesis*: With RF ESD design, the synthesis of the digital, analog and RF segments may require unique structures in the substrate wafer or in the interconnect system, to isolate the electrical noise, and at the same time provide ESD protection between the various segments of the chip. This may require unique physical structures and circuits to address the circuit subfunction ESD protection. Although the same ESD networks used in digital ESD design practices are utilized, the design choices are distinct as a result of the implications on the RF application. For example, ESD diodes can be used between ground rails and between chip subfunctions. In ESD digital design, the focus may be differential voltage isolation; for the RF ESD design practice, the focus may be the capacitive coupling, and the impact on RF stability of networks [5,6].

- *RF ESD Test Methods*: In the ESD testing of RF components, unique tests need to be established on a component level and system level to evaluate the ESD degradation. Unique RF testing methods are needed that will address different d.c. and RF parametrics degradation to evaluate the pre- and post-ESD stress test conditions [8]. A distinction between digital ESD design practice and RF ESD design practice is that the digital ESD design practice focuses on d.c. voltage shifts and leakage; whereas in RF ESD design practice, the focus is on the RF parameters and what occurs first:–d.c. or RF degradation [8]. These methods may contain RF methods such as time domain reflection (TDR) and time domain transmission (TDT) methods.

- *RF ESD Failure Criteria*: In RF applications, the functional requirements are very distinct from digital applications. Unique RF parameters and ESD failure criteria require to be established on the basis of the RF parameters, d.c. parameters, and system level requirements. This is distinct from the typical digital applications, which only require d.c. leakage evaluation [8,9].

- *RF ESD Test Systems*: To address the RF ESD test methods and failure criteria, new RF ESD test systems may be required that address product evaluation. RF ESD test systems may require ESD systems that allow extraction of the RF parameters *in situ* for noise figure (NF), gain (G), output intercept third order (OIP3) harmonics as well as d.c. leakage evaluation. This may influence the direction of ESD HBM, MM, and transmission line pulse (TLP) systems. In recent times, 50 Ω-based TLP systems are compatible with 50 Ω-based RF circuits. Additionally, future TLP systems may be influenced by the needs of RF circuits.

- *ESD Frequency Spectrum Versus Functional Applications*: In advanced RF designs, the RF circuits are significantly faster than the ESD phenomenon; this allows for frequency 'bands' for the ESD phenomenon and ESD circuit element response versus the RF functional circuit operation and application frequency. As the RF application frequency exceeds 5 GHz, the application frequency will exceed the ESD CDM energy spectrum (e.g., approximately less than 5 GHz). As the application frequency exceeds the ESD phenomenon, the RF ESD design methodology allows for the utilization of the difference in the response during ESD event time scale (e.g., frequency) and application frequency.

- *ESD Frequency Domain Load Reduction Methods*: In RF ESD design, a higher focus is used to lower the loading effects by taking advantage of the frequency response of the RF networks, which are distinct from the ESD phenomena [3].

- *ESD Method of Cosynthesis of ESD and RF Circuits*: In the RF design of ESD networks, it is necessary to design the ESD device in conjunction with the RF circuit. By cosynthesis of the network, the loading effect as well as frequency modifications can be optimized to prevent the limitation of the RF ESD network [3].

- *Utilization of Shunt RF Elements as ESD Elements for the Alternative Current Loop*: In RF design, shunt elements are needed for impedance matching. These parallel shunt elements can help to provide electrical connectivity to the ESD alternative current loop. As a result, the RF shunt serves the role of providing a path to the ESD alternative current loop and must also have ESD robustness requirements to be effective.

- *ESD Method of Utilization of ESD Element as a Capacitor in RF Design*: ESD elements can serve as capacitor elements. Hence, in the cosynthesis, a method of transfer of the capacitance from the functional circuit to the ESD element to achieve the same RF performance is achievable [3].

- *ESD Method of Series to Parallel Conversion of RF Element to ESD Element*: Given a functional RF circuit that is defined as a series configuration, the representation can be modified to a parallel configuration. In the transformation from a series configuration to a parallel configuration, some portion of the element can be utilized for ESD protection using it as a parallel shunt to power or ground rails [3].

- *ESD Method of Utilization of ESD Element as a Shunt Capacitor*: Given a functional RF circuit, which is defined as a series capacitor configuration, the representation can be modified to a parallel configuration in order to establish a shunt capacitor equivalent circuit. Given a capacitor in series with a resistor element, this circuit can be transformed into an equivalent circuit of a resistor and capacitor in parallel for which it achieves the same quality factor (Q). The transformation of the network with a matched Q achieves the same circuit response. In this fashion, a series capacitor element can be substituted for a shunt ESD element, that serves as an ESD element in either diodic operation or breakdown mode of operation [3].

- *ESD Method of Parallel Susceptance Equivalent Load Compensation*: Capacitive loads that are in parallel configuration can be transformed as treating two parallel susceptances. In the implementation, the total susceptance load is the parallel configuration of the new load susceptance and the ESD susceptance. Hence, the transformation of the total load susceptance to an equivalent parallel configuration of an ESD susceptance load and a new susceptance load [3].

- *ESD Method of Series Inductive Decoupling of ESD Element Circuit*: Using inductor elements in series with the ESD element, the loading effect of the ESD element can be inductively isolated. A series inductor providing a low $L(di/dt)$ during ESD events allows for the current to flow through the ESD element to a power rail or ground. During functional RF operation, the $L(di/dt)$ allows voltage isolation of the ESD element [10,11].

- *RF ESD Method of Narrow Band Fixed Load Absorption and Resonant: Matching L-Matching Compensation Method*: ESD elements can serve as a means to provide impedance matching between the output and the load. Hence, using matching techniques, the ESD element can serve as the matching elements to provide optimum matching conditions. Using a L-match circuit, consisting of a series inductor and a shunt capacitor (e.g., ESD element), the ESD network can be used as a means to provide matching between the source and the load. The shunt capacitor must remain on a constant conductance circle on a Smith admittance chart [3].

- *RF ESD Method of Narrow Band Fixed Load ESD Absorption and Resonant Matching L-Match*: For 'absorption matching' the stray reactance are absorbed into the impedance matching network up to the maximum that are equal to the matching components. For 'resonance matching' stray reactance are resonated out with an equal and opposite reactance, providing cancellation. Hence, the stray reactance serving as an ESD element can be resonance matched to an inductor of equal reactance. An inductor element in parallel with the ESD capacitor element can null the ESD capacitance loading effect providing 'resonance matching' that hides the ESD capacitive element by matching the inductor susceptance [3].

- *ESD Method of Cancellation*: Using RF components, the loading effect of an ESD element can be hidden at the application frequency. Cancellation of the ESD loading effect can be achieved by proper loading of additional elements [12–14].

- *ESD Method of Impedance Isolation*: Using inductors in series with an ESD network, the inductors can serve as high impedance elements such that the loading effect of the ESD element is not observed at application frequencies [14–16].

- *ESD Method of Impedance Isolation using LC tank*: Using inductor and capacitor in parallel, an LC resonator tank in series with an ESD element, the loading effect of the ESD element can be reduced. The frequency of the LC tank is such that it allows operation of the ESD element but provides isolation during functional RF operation [14–16].

- *ESD Method of Lumped Versus Distributed Load*: In RF ESD design, ESD design focus on load reduction is achieved in the frequency domain by taking advantage of the distributed ESD loads instead of single component lumped elements. The 'distributed' versus 'lumped' design method can be achieved within a given element or multiple elements [17–28].

- *ESD Method of Distributed Design using Design Layout*: In RF ESD design, ESD design focus on load reduction is achieved in the frequency domain by taking advantage of the distributive nature of a single ESD element. This can be achieved through design layout by providing introducing resistance, capacitance, or inductance within a given ESD design layout. Metal interconnect design and layout distribution within diodes, MOS-FETs, and bipolar transistors can introduce distributive effects. When this effect is typically undesirable in a digital operation of the ESD elements, in an RF application, it can be intentionally utilized [29].

- *ESD Methods of Distributed Design using Multiple Circuit Element Stages*: In RF ESD design, ESD design focus on load reduction is achieved in the frequency domain by taking advantage of multiple elements. This can be achieved by multiple stage designs of equal, or variable size stages with the introduction of resistance, capacitance, or inductance within a given ESD multiple-stage design. This can be achieved through introduction of RF resistor, capacitor, or inductor components into the ESD implementation; whereas, this is typically undesirable in a digital operation of the ESD elements, in an RF application, this can be intentionally utilized [17–28,30–33].

- *ESD Method of Resistive Decoupling Using Distributed Multiple Circuit Element Stages*: In RF ESD design, ESD design focus on load reduction is achieved in the frequency domain by taking advantage of multiple elements and series resistors. This can be achieved by multiple stage designs of equal, or variable size stages with the introduction of resistors within a given ESD multiple-stage design. The introduction of resistors provides an IR voltage drop isolating the successive stages during functional operation, but not during ESD operation.

- *ESD Method of Inductive Decoupling Using Distributed Multiple Circuit Element Stages*: In RF ESD design, ESD design focus on load reduction is achieved in the frequency domain by taking advantage of multiple elements and inductors. This can be achieved by multiple stage designs of equal, or variable size stages with the introduction of on-chip or off-chip inductors within a given ESD multiple-stage design. The introduction of inductors produces an $L(di/dt)$ voltage drop isolating the successive stages during functional operation, but not during ESD operation [23,24,26–28,30–33].

- *ESD Methods of Distributed Design Using Coplanar Waveguides*: In RF ESD design, multi-stage implementations can place coplanar waveguides (CPW) to provide improvements in the power transfer, matching, and to reduce the loading effect on the input nodes [26–28,30–33].

- *ESD Method of Distributed Design for Digital Semiconductor Chip Cores*: Digital chip sectors are not inherently designed for $50\,\Omega$ matching conditions. Hence, an RF design practice is a placement of a resistive element shunt for utilization of distributed design for core chip subfunctions [26–28].

- *ESD Method of Capacitive Isolation Buffering*: Using decoupling capacitors in series with RF elements can provide impedance buffering of receiver networks, allowing operation of ESD networks. Capacitor elements can be metal–insulator–metal (MIM) capacitors, vertical parallel plate (VPP) capacitors, or metal-interlevel dielectric layer metal (M/ILD/M) capacitors. This method cannot be utilized in d.c. circuits due to the blocking of d.c. currents [34].

- *ESD Method of Architecturing an ESD circuit for Improved Linearity*: ESD networks can be designed in a fashion to eliminate linearity issues in RF design. For example, diode elements and varactor structures have capacitance variation as a function of applied voltage. Using ESD elements (e.g., double diode configuration), RF circuit linearity can be improved [35].

- *RF ESD Method of Tuning an ESD circuit for Improved Linearity*: ESD networks can be designed in a fashion to improve linearity issues in RF design by tuning. For example, diode elements and varactor structures have capacitance variation as a function of applied voltage. These variations can be modified by variable tuning by utilization of tunable ESD elements with semiconductor process or design layout techniques [35].

- *RF ESD Method of Noise and ESD Optimization*: Noise in RF circuits is a large concern. This influences the chip architecture between the digital, analog and RF circuits. Additionally, noise concerns also determine the substrate doping concentration, semiconductor profile, isolation strategy, and guard ring design. Additionally, noise may determine the acceptable ESD device type due to noise concerns. Hence, the method of cosynthesis of chip architecture, chip power grids, choice of ESD elements, and choice of ESD circuits are all influenced by the noise requirements [3].

- *RF ESD Method of Quality Factor and ESD Optimization*: The quality factor, Q, is influenced by the ESD device choice. Additionally, Q degradation can occur in RF passive elements such as resistors, inductors, and capacitors from ESD events. Hence, Q optimization and the ESD current path optimization are needed to have RF degradation mechanisms associated with ESD stress of critical circuit elements [3,12,13].

- *RF ESD Method of Stability and ESD Optimization*: In RF ESD design, circuits must be designed to achieve electrical d.c. stability, thermal stability, and RF stability. Amplifier stability is a function of the stability at both the source and the load. In RF design, these are defined as source and load stability circles. The stability of the source and load is a function of the minimum resistance requirement. With the addition of ESD ballast resistance, circuit stability can be improved. Cosynthesizing the stability requirement, ESD resistance can be integrated to improve circuit stability [3].

- *RF ESD Method of Gain Stability, Noise, Q, and ESD Optimization*: In the optimization of circuits, the gain stability, noise, quality factor, and the ESD can be cosynthesized.

An ESD circuit can be designed in such a way that the ESD elements are added to a circuit to help satisfy the Stern stability criteria. For ESD optimization, the path from source to load along the gain–noise optimum contour, which has the maximum shunt capacitance will achieve this optimized solution [3].

- *ESD Circuits Which Are Nonfrequency Triggered*: The introduction of ESD circuits with frequency-initiated trigger elements, such as RC-triggered MOSFET ESD power clamps, or RC-triggered ESD input networks, can be undesirable because of the interaction with other RF circuit response. For example, the introduction of RC-triggered networks that have inductor load can introduce undesired oscillation states and functional issues. Hence, in RF technology, nonfrequency initiated trigger networks are desirable for some RF applications (e.g., voltage-triggered networks).

- *ESD System Level and Chip Level Multistage Solutions*: At radio frequency application frequencies, ESD protection loading effects have significant impact on RF performance. ESD solutions for RF application include the combination of both on-chip and off-chip ESD solutions: spark gaps, field emission devices (FED), transient voltage suppression (TVS) devices, polymer voltage suppression (PVS) devices, mechanical shunts, and other solutions. By combining both off- and on-chip ESD protection solutions, the amount or percentage of on-chip protection solutions can be reduced.

- *On-Chip and Off-Chip Protection*: In future high speed applications, a mix of off-chip and on-chip protection may be required to reduce the capacitance loading effects. The off-chip protection can be present in the electrical cables, connectors, ceramic carriers, or on the circuit boards. As the frequency increases, or the materials change, the ESD protection may shift to only off-chip protection.

- *ESD Nonsemiconductor Devices*: Spark gaps, field emission devices (FED), transient voltage suppression (TVS) devices, polymer voltage suppression (PVS) devices, mechanical package 'crowbar' shunts, and other solutions are utilized in RF applications off-chip due loading effects, space (e.g., ESD design area), cost (e.g., cost/die), or the lack of the proper material to form ESD protection circuitry (e.g., substrate material). These solutions are not typically an option in semiconductor chips with high pin-count and packaging constraints, but for low pin-count low circuit density applications, these are an option.

1.3 KEY RF ESD CONTRIBUTIONS

In the field of electrostatic discharge (ESD) of radio frequency (RF) devices, accomplishments to advance the field are in the form of development of experimental discovery, analytical models, introduction of new semiconductor devices and circuits, test equipment, and test methods. Below is a short chronological list of key events that influenced the field of radio frequency (RF) electrostatic discharge (ESD):

- 1968 A.D. D. Wunsch and R.R. Bell introduce the power-to-failure electro-thermal model in the thermal diffusion time constant regime [36].

- 1970 A.D. D. Tasca develops the power-to-failure electro-thermal model in the adiabatic and steady-state time constant regime [37].

- 1970 A.D. Y. Anand, W. J. Moroney, G.E. Morris, V. J. Higgins, C. Cook, and G. Hall evaluate the ESD robustness of Schottky diodes for microwave mixer applications [38]. This work is significant as it demonstrates some of the first measurements and failure analysis highlighting the failure of microwave components in microsecond to nanosecond time regimes.

- 1971 A.D. Vlasov and Sinkevitch develop a physical model for electro-thermal failure of semiconductor devices [39].

- 1972 A.D. W.D. Brown evaluates semiconductor devices under high amplitude current conditions [40].

- 1979 A.D. R.L. Minear and G.A. Dodson demonstrate failure mechanisms in Silicon bipolar monolithic transistors. The significance of the work lies in the introduction of butted base contacts, known as the 'phantom emitter' to provide ESD improvements in bipolar transistors [41].

- 1979 A.D. J.J. Whalen demonstrates the power-to-failure of GaAs MESFET devices due to RF power magnitude. This study also demonstrates the relationship of absorbed energy to time-to-failure in GaAs devices [42].

- 1979 A.D. J.J. Whalen and H. Domingos evaluated the inter-relationship of the power-to-failure associated with an RF oscillating signal and ESD HBM pulse waveforms on GaAs ultra-high frequency transistors. The significance of the work is in the demonstration of the relationship between RF power stressing and ESD test stressing and between the absorbed energy and the time-to-failure models [43].

- 1981 A.D. J. Smith and W.R. Littau develop an electro-thermal model for resistors in the thermal diffusion time regime [44].

- 1981 A.D. Enlow, Alexander, Pierce, and Mason address the statistical variation of the power-to-failure of silicon bipolar transistors due to semiconductor manufacturing process, and ESD event variations [45–47].

- 1983 A.D. M. Ash evaluates the nonlinear nature of the power threshold and the temperature dependence of the physical parameters establishing the Ash relationship [9]. This study also demonstrates that Silicon and Gallium Arsenide satisfy the Ash relationship, leading to accurate prediction of the power-to-failure without addressing the temperature variation [48].

- 1983 A.D. V.I. Arkihpov, E.R. Astvatsaturyan, V.I. Godovosyn, and A.I. Rudenko derive the cylindrical nature of the electro-current constriction [49]. The significance of this work lies in the development of a physical model that quantified current constriction in high resistance wafers.

- 1989 A.D. Dwyer, Franklin, and Campbell develop a three-dimensional Green's function model for explanation of power-to-failure in Gallium Arsenide structures. The significance of the work lies in the extension of the Wunsch–Bell model to address dimensional scale length variations in the three dimensions [50].

- 1986 A.D. A.L. Rubalcava, D. Stunkard, and W.J. Roesch study the effects of ESD on Gallium Arsenide MESFETs and passive structures [51].

- 1987 A.D. F.A. Buot, W.T. Anderson, A. Christou, K.J. Sleger, and E.W. Chase study the theoretical and experimental study of subsurface burnout and ESD in GaAs FETs and HEMTs [52].

- 1990 A.D. K. Bock, K. Fricke, V. Krozer, and H.L. Hartnagel study the improvements in GaAs MESFETs using alternative metallurgy, spark gaps, and field emission devices (FEDs) [53].

- 1993 A.D. L.F. DeChairo and B.A. Unger evaluate ESD degradation in InGaAsP semiconductor lasers resulting from human body model ESD [54].

- 1997 A.D. S. Voldman develops the lateral 'polysilicon-bound ESD diode' structure. The work is significant for its invention of a device used today for RF Complementay Metal Oxide Semiconductor (CMOS) ESD applications [55].

- 1999 A.D. B. Kleveland and T. H. Lee apply distributed RF concepts to ESD protection networks. This work utilizes the concepts of distributed network to ESD structures for radio frequency applications [23,24].

- 2000 A.D. S. Voldman and P. Juliano publish the first TLP measurements of a Silicon Germanium (SuGe) heterojunction bipolar transistor (HBT). They also evaluate the SiGe power-to-failure characteristics as a function of pulse width [56].

- 2000 A.D. C. Richier, P. Salome, G. Mabboux, I. Zaza, A. Juge, and P. Mortini investigate different ESD protection strategies for 2 GHz RF CMOS applications. The significance of the work lies in the choosing of ESD elements on the basis by the ESD robustness versus RF parameter characteristics. The second significant conclusion is the effectiveness of polysilicon-bound silicon lateral diodes for RF ESD applications [35].

- 2001 A.D. S. Voldman, A. Botula, and D. Hui develop the first Silicon Germanium ESD Power Clamps [57]. This work is significant because of its utilization of the high unity current cutoff frequency/low breakdown transistor as the trigger element for the low high unity current cutoff frequency/low breakdown transistor, invoking a natural scaling relationship.

- 2001 A.D. S. Voldman, L. Lanzerotti, and R.R. Johnson evaluate the emitter-base design sensitivities of the SiGe HBT device [58]. The significant finding of the work is the implications of emitter-base scaling on the ESD robustness of SiGe HBT devices.

- 2001 A.D. P. Leroux and M. Steyart utilize on-chip inductors for a 5.25 GHz low noise amplifier (LNA) applications. This work utilizes the inductors as part of RF-ESD design cosynthesis methodologies [10].

- 2002 A.D. S. Voldman, B. Ronan, and L. Lanzerotti publish the first TLP measurements of a Silicon Germanium Carbon (SiGeC) hetero junction HBT device. This is a significant study as it provides a comparison of the SiGeC HBT and the SiGe HBT [59].

- 2002 A.D. S. Voldman, B. Ronan, S. Ames, A. Van Laecke, and J. Rascoe demonstrate testing methodologies for RF single devices, products, and systems. This study demonstrates the RF degradation typically occurs prior to ESD failure based on leakage specifications and addresses RF ESD failure criteria [8].

- 2002 A.D. S. Voldman, S. Strang, and D. Jordan demonstrate an automated ESD design system methodology that allows RF ESD cosynthesis in a RF foundry system [5–7]. The study show that the methodology achieved has the ability to provide RF-ESD cosynthesis integrated into a modern design methodology suitable for RF CMOS, and RF BiCMOS Silicon Germanium foundry environment.

- 2003 A.D. Y. Ma and G.P. Li demonstrate construction of GaAs-based ESD power clamp networks for InGaP/GaAs and GaAs technologies. The significance of the work lies in the utilization of ESD power clamps in RF GaAs applications [60].

- 2003 A.D. M.D. Ker and B.J. Kuo optimize a broadband RF performance and ESD robustness by π-model distributed ESD protection scheme. The significance of the work lies in the extension of the range of distributed networks to the 1–10 GHz RF application range [30,31].

- 2003 A.D. S. Hynoven, S. Joshi and E. Rosenbaum discuss method of resonant absorption to lower the effective loading effects of ESD, referred to as 'cancellation method.' The significance of the work is in the utilization of inductors as inductive shunts to V_{DD} or V_{SS} providing 'cancellation' of the loading effect of the diode capacitance [12,13].

- 2004 A.D. K. Shrier demonstrates ESD improvements in GaAs power amplifiers using electronic polymer off-chip protection elements [61]. This study shows the utilization of low capacitance polymer voltage suppression (PVS) devices as a solution for RF ESD protection.

- 2005 A.D. S. Hynoven and E. Rosenbaum develop single inductor/single diode ESD networks. The significance of the work is in the utilization of inductors as inductive shunts to V_{DD} or V_{SS} providing 'cancellation' of the loading effect of the diode capacitance [12,13].

1.4 KEY RF ESD PATENTS

In the past few years, a significant number of patents in the field of ESD protection of RF technologies are seen.

The patents on ESD protection of radio frequency devices address the following classes of solutions:

- RF ESD active and passive elements [61–79];

- RF ESD input circuits [80–91];

- RF ESD power clamps [92–95];

- RF ESD design integration issues [96–103];

- RF ESD design systems [7, 104–106].

1.5 ESD FAILURE MECHANISMS

ESD-induced failure mechanisms lead to both latent damage and destruction of RF semiconductor elements. ESD-induced failure mechanisms are a function of the technology

Table 1.1   ESD failure mechanisms in RF CMOS devices

RF CMOS

Device	Test type	Polarity	Failure mechanism
RF CMOS MOSFET	HBM	Positive	MOSFET Source-Drain
	CDM	Positive	MOSFET Gate Dielectric
STI-bound p^+/n-well diodes	HBM	Positive	p^+-to-well
Poly-bound P-N diodes	HBM	Positive	p^+ to n-well
	CDM		Polysilicon Gate of diode structure
Inductors	HBM	Positive or negative	Inductor Vias and metal underpass
MIM capacitors	HBM	Positve or negative	MIM dielectric
Polysilicon resistors	HBM	Positive or negative	Polysilicon film

type, device, circuit topology, and RF chip architecture. In this section, a brief tabulation of ESD failure mechanisms will be highlighted for RF CMOS, RF Silicon Germanium, RF Silicon Germanium Carbon, Gallium Arsenide, and Indium Gallium Arsenide technologies. The primary distinction between digital CMOS ESD failure mechanisms and RF failure mechanism is not the RF environment but the utilization of different technologies, circuit topologies, and extensive use of RF passive elements.

1.5.1 RF CMOS ESD Failure Mechanisms

Table 1.1 contains a brief summary of the ESD mechanisms observed in RF CMOS devices. The key distinction in this table is that in RF CMOS, there is a greater use of RF CMOS passive elements. In a RF CMOS technology that uses standard thin oxide MOSFET capacitors and RF inductors from standard CMOS film thickness, ESD failures will occur in the passive elements.

In Table 1.2, a listing of RF ESD circuit failure mechanisms is tabulated. In RF CMOS, ESD networks consist of shallow trench isolation (STI) bound diode networks,

Table 1.2   RF CMOS ESD network failure mechanisms

RF ESD Circuits

Circuit type	Test type	Polarity	Failure mechanism
RF CMOS Grounded Gate NFET	HBM	Positive	MOSFET Source-Drain
RF CMOS STI-Bound dual diodes	HBM	Positive	p^+/n-well diode
	HBM	Positive	p^+/n-well diode
RF CMOS poly-bound dual diodes	HBM	Positive	p^+/n-well
	CDM		Polysilicon Gate of diode structure
RF CMOS inductor/diode	HBM	Positive	Inductor
	HBM	Negative	Diode
RF CMOS diode/inductor	HBM	Positive	Diode
		Negative	Inductor

Table 1.3   HBM ESD failures in Silicon Germanium-based devices

Silicon Germanium

Device	Structure	ESD test	Polarity	Failure mechanism
SiGe SA npn HBT	Epitaxial SiGe film	HBM, MM	Positive, negative	Emitter–Base junction
SiGe NSA npn HBT	Epitaxial SiGe film	HBM, MM	Positive, negative	Emitter–Base junction
SiGe Raised Extrinsic Base npn HBT	Epitaxial SiGe film	HBM, MM	Positive, negative	Emitter–Base junction
SiGe Base-Collector Varactor	SiGe Base /Si Sub-collector	HBM, MM	Positive, negative	Base–Collector junction

polysilicon-bound diode networks, diode/inductor elements, impedance isolation LC tank/ diode networks, and grounded gate RF MOSFET networks. The table contains a listing of the known failure mechanisms.

1.5.2 Silicon Germanium ESD Failure Mechanisms

Table 1.3 contains a brief summary of the ESD mechanisms observed in RF Silicon Germanium devices. RF BiCMOS Silicon Germanium technologies contain both bipolar devices, CMOS devices and hybrid structures. As in RF CMOS, ESD failures will occur in the RF CMOS active and passive elements. Table 1.3 addresses the additional RF Silicon Germanium device element ESD-induced failures.

1.5.3 Silicon Germanium Carbon ESD Failure Mechanisms in Silicon Germanium Carbon Devices

Table 1.4 contains a brief summary of the ESD mechanisms observed in RF Silicon Germanium Carbon devices. RF BiCMOS Silicon Germanium Carbon technologies contain both bipolar devices, CMOS devices, and hybrid structures. As in RF CMOS, ESD failures will occur in the RF CMOS active and passive elements. Table 1.4 addresses the additional RF Silicon Germanium device element ESD-induced failures. At this point, there is no known distinction between the ESD failure mechanisms in Silicon Germanium Carbon and Silicon Germanium devices, other than the ESD failure levels.

Table 1.4   HBM ESD failures in Silicon Germanium-based devices

Silicon Germanium Carbon

Device	Structure	ESD test	Polarity	Failure mechanism
SiGeC SA *npn* HBT	Epitaxial SiGe film		Positive, negative	Emitter–Base junction
SiGeC raised extrinsic base *npn* HBT	Epitaxial SiGe film		Positive, negative	Emitter–Base junction

Table 1.5 HBM ESD failures in Gallium Arsenide-based devices

Gallium Arsenide

Device	Length (μm)	Width (μm)	HBM Failure level (V)	Failure mechanism
MESFET	0.8 (Gate)	1500	+100	Gate-to-Source
	1.0 (Gate)	2800	+350	Gate-to-Source
	2.0 (Gate)	220	+50	Gate-to-Source
HEMT	0.5 (Gate)	280	+90	Gate-to-Source
HBT		20	+120	Emitter-Base
		6 × 20	+480	Emitter-Base

1.5.4 Gallium Arsenide Technology ESD Failure Mechanisms

Gallium arsenide technology has unique ESD-induced failure mechanisms distinct from RF CMOS and RF Silicon Germanium technology because of the different materials (e.g., substrate, dopants, and metallurgy), structures, and device types. These types of devices vary from MESFETs, HEMT, to HBT devices. As a result, the ESD failure levels and failure mechanisms are very different. In Table 1.5, a brief list of GaAs ESD-induced failure mechanisms are tabulated.

1.5.5 Indium Gallium Arsenide ESD Failure Mechanisms

Indium Gallium Arsenide (InGaAs) based devices are used from lasers, photodetector to PIN diode applications. Table 1.6 shows some experimental data of ESD failure levels and failure mechanisms.

Table 1.6 HBM ESD failures in Indium Gallium Arsenide-based devices

InGaAs

Device	Structure	Width (mil)	HBM failure level (V)	ESD Failure mechanism
InGaAs Laser	GaAs substrate	10	−1500	
		20	−4000	
		30	−5500	
GaInAs/GaAlAs MODFET	GaAs substrate InGaAs Channel			Gate burnout
InGaAs/InP Photodetectors		50–80 μm	700	
InGaAs PIN	InGaAsP Substrate		−200	

Table 1.7 RF bipolar receiver circuits failure mechanisms

RF bipolar circuits

Circuit type	Test type	Polarity	Test mode	Failure mechanism
Single ended common emitter	HBM, MM	Positive or negative	Input-to-V_{SS}	Emitter–Base junction
Single ended C-E receiver with MIM series capacitor	HBM, MM	Positive or negative	Input-to- V_{SS}	Capacitor Element
Single ended C-E receiver with resistor feedback	HBM, MM	Positive or negative	Output-to-input	Feedback resistor element
Single ended C-E receiver with resistor and capacitor feedback	HBM, MM	Positive or negative	Output-to-input	Feedback resistor and capacitor element
Single ended C-E receiver with emitter resistor degeneration	HBM, MM	Positive or negative	Input-to-V_{SS}	Emitter–Base junction
Single ended C-E with Balun output	HBM, MM	Positive or negative	Output-to-output	Balun or d.c blocking capacitor failure
Differential Common-Emitter receiver	HBM, MM	Positive or negative	Input-to-input	Emitter–Base junction
Bipolar Current mirror	HBM, MM	Positive or negative	Input-to-ground	Collector-to-Base

1.5.6 RF Bipolar Circuits ESD Failure Mechanisms

RF bipolar circuits have failure mechanisms that are associated with the RF homo-junction bipolar junction transistor (BJT) or HBT, and the passive derivatives. Circuit topology plays a large role in the ESD-induced failure mechanisms observed in RF bipolar circuits. For example, failure mechanisms between single-ended mode and differential mode receiver networks are very different. Table 1.7 shows some examples of RF bipolar ESD-induced failure mechanisms.

1.6 RF BASICS

In RF ESD design, there are many RF metrics of interest. The metrics are important to understand the language and concerns of RF designers. Additionally, from an ESD perspective these RF metrics are important to understand the ESD failure mechanisms, and how they manifest themselves in RF semiconductor components and products.

Reflection Coefficient

The reflection coefficient represented as a function of characteristic impedance and load impedance,

$$\Gamma = \frac{Z_L - Z_0}{Z_L + Z_0}$$

This can also be expressed as a function of the characteristic admittance, $Y_o = Z_0$,

$$\Gamma = \frac{Y_0 + Y_L}{Y_0 + Y_L}$$

Normalized Reflection Coefficient

The normalized reflection coefficient can be expressed as normalized to the characteristic impedance. In this form it can be expressed as follows:

$$\Gamma = \frac{(Z_L/Z_0) - 1}{(Z_L/Z_0) + 1}$$

Load Relationship as a Function of Reflection Coefficient

Expressing the load as a function of the reflection coefficient, Γ,

$$Z_L = Z_0 \frac{1 + \Gamma}{1 - \Gamma}$$

Return Loss

$$RL(dB) = -10 \log |\Gamma \times \Gamma| = -20 \log |\Gamma|$$

Voltage Standing Wave Ratio

The voltage standing wave ratio (VSWR) is a measure of the forward and reflected wave conditions between the maximum voltage and the minimum voltage conditions. This can be expressed as a function of the maximum and minimum voltage, or the reflection coefficients.

$$VSWR = \frac{V_{MAX}}{V_{MIN}} = \frac{|V_f| + |V_r|}{|V_f| + |V_r|}$$

$$VSWR = \frac{1 + |\Gamma|}{1 - |\Gamma|}$$

The reflection coefficient can also be expressed as a function of the VSWR,

$$\Gamma = \frac{VSWR - 1}{VSWR + 1}$$

Mismatch Loss (ML)

Mismatch loss is associated with the loss of signal between two terminations. These terminations can be expressed as a source and a load. The mismatch loss is related to the ability to deliver power to the load.

In the general case, where the reflection coefficient on the source and load is nonzero (e.g., both the source and load are not equal to Z_0), the expression for mismatch loss can be expressed as

$$ML = \frac{|1 - \Gamma_S\Gamma_L|^2}{[1 - |\Gamma_S|^2][1 - |\Gamma_L|^2]}$$

In the case of a source with characteristic impedance, Z_0, and with a zero reflection coefficient (e.g., $\Gamma = 0$), the mismatch loss can be expressed as

$$ML = \frac{1}{1 - |\Gamma_L|^2}$$

After ESD stress, given the device-under-test (DUT) is the load, changes in the reflection coefficient in the load due to degradation can lead to changes in the mismatch loss.

Quality Factors

The quality factor, QF, or simply the 'Q' of the element is a metric associated with the ratio of the desired to the undesired electrical characteristics. In any physical element, there exists undesirable parasitics that degrade the nature of a element. For example, in a capacitor, there is resistance associated with the metal interconnects. With an inductor, there is also resistance associated with the coil wire itself. These nondesired features degrade the nature of the physical element. From an ESD perspective, the change in these ideal elements can impact circuit functionality. For example, amorphous or metal grain structure changes can lead to changes in the resistance after ESD stress. This can lead to changes in the 'Q' of the physical elements [3].

In the series expression, it is expressed as the ratio between the series reactance, X_s, and the series resistance R_s.

$$Q = \frac{X_s}{R_s}$$

For a parallel configuration, the form of the quality factor, Q, is expressed as a function of the equivalent parallel circuit expressed as susceptance, B_p, and the conductance, G_p,

$$Q = \frac{B_p}{G_p} \qquad Q = \frac{R_p}{X_p}$$

Noise Figure

The noise figure (NF) is the actual noise power to thermal noise power ratio,

$$NF(dB) = 10\log\left[\frac{P_N}{(kT)B}\right]$$

where P_N is the actual noise power, B is the bandwidth, k is the Boltzmann constant, and T is the temperature. ESD-induced degradation can change the actual noise power owing to the changes in the material properties, leading to an increase in the NF.

Output Intercept Point Third Order Harmonic

The OIP3 is the extrapolated power level where the third order inter-modulation product equals the fundamental power level (e.g., also referred to as P_{3IP}). ESD degradation in the fundamental output power can lead to a lowering in the intersection point of the fundamental power and the third order harmonics. As a result, ESD-induced degradation that degrades the fundamental power transfer curve (e.g., power output divided by power input) leads to OIP3 degradation and ESD-induced RF degradation.

Gain Compression Point

The gain compression point is a point on the output power (P_{OUT}) versus input power (P_{IN}) plot where the actual power curve deviates from the fundamental harmonic linear slope. In a linear system, the characteristic is a linear slope on the P_{OUT} versus P_{IN} plot. Power exists in the nonfundamental harmonics, which leads to an output power roll-off with increasing input power.

When the actual output power is 1 dB below the fundamental output power, a typical RF metric is 1 dB Gain Compression; this can be denoted as the P_{1dB}.

During ESD stress, the output power level of a semiconductor device or circuit can be degraded as a result of transistor degradation and impedance degradation. In a multifinger structure, or where parallel elements exist, ESD-induced interconnect failure can also lower the output power levels. This can be apparent by the 1 dB Gain Compression metric.

Dynamic Range

The dynamic range (DR) is the range of system a signal power can handle without significant noise distortion. This is evaluated by the difference between the compression gain and the output noise floor.

$$DR = P_{1dB} - (kTB + NF + G_A)$$

where DR is the dynamic range (at the output), P_{1dB} is the 1 dB gain compression point, B is the noise bandwidth, NF is the noise figure, and G_A is the available gain of the amplifier.

Spurious Free Dynamic Range

Spurious free dynamic range is the output power level where the inter-modulation product terms reach the noise floor. This can be expressed as a function of the DR and the third order intercept output power, P_{3IP}.

$$SFDR = \frac{2}{3}[P_{3IP} - DR]$$

Transmission Line Characteristic Impedance

Characteristic impedance of a lossy transmission line

$$Z_{TL} = \sqrt{\frac{R + j\omega L}{G + j\omega C}}$$

Transmission Line Electrical Length

The electrical length is the ratio of the transmission line physical length and the wavelength in the medium at a given frequency.

$$E = 360 \frac{\ell}{\lambda_G}$$

Phase constant

$$\beta = \frac{360}{\lambda_G}$$

Transmission Line Propagation constant

$$\gamma = \alpha + j\beta$$

1.7 TWO-PORT NETWORK PARAMETERS

In RF analysis, signal characterization is approached as a 'two-port network' where there are two input signals and two output signals. From an ESD perspective, ESD measurements on TLP test evaluation is also similar to this perspective. Hence, evaluation of RF ESD networks from a two-port perspective leads to a natural synergism between signal analysis and ESD TLP testing. In a two-port network, the voltage–current relationships of the terminals can be expressed where the voltages on the input and the output are expressed as a function of the currents on the input and output.

1.7.1 Z-Parameters

In a two-port network, the voltage–current relationships of the terminals can be expressed where the voltages on the input and the output are expressed as a function of the currents on the input and output. Two voltage equations are established that are coupled through the current terms, and the coefficients are the impedance terms, known as z-parameters

$$v_1 = z_{11}i_1 + z_{12}i_2$$
$$v_2 = z_{21}i_1 + z_{22}i_2.$$

This can be expressed in matrix form, where the output vector is the terminal voltages, the input vector is the terminal currents,

$$\begin{bmatrix} v_1 \\ v_2 \end{bmatrix} = \begin{bmatrix} z_{11} & z_{12} \\ z_{21} & z_{22} \end{bmatrix} \begin{bmatrix} i_1 \\ i_2 \end{bmatrix}.$$

1.7.2 *Y*-Parameters

In a two-port network, the voltage–current relationships of the terminals can be expressed where the currents on the input and the output are expressed as a function of the terminal voltages on the input and output. Two current equations are established that are coupled through the voltage terms, and the coefficients are the admittance terms, known as y-parameters

$$i_1 = y_{11}v_1 + y_{12}v_2$$
$$i_2 = y_{21}v_1 + y_{22}v_2$$

This can be expressed in matrix form, where the output vector is the terminal currents, the input vector is the terminal voltages:

$$\begin{bmatrix} i_1 \\ i_2 \end{bmatrix} = \begin{bmatrix} y_{11} & y_{12} \\ y_{21} & y_{22} \end{bmatrix} \begin{bmatrix} v_1 \\ v_2 \end{bmatrix}$$

1.7.3 *S*-Parameters

In a two-port network, the voltage–current relationships of the terminals can be expressed where the currents on the input and the output are expressed as a function of the terminal voltages on the input and output using z- and y-parameters. But, in system environments, access to the both input and output terminals are not accessible.

During ESD testing, in many cases, only two terminals are available to determine ESD degradation in RF components. Using reflected, absorbed, and transmitted signal information is of greater value. Transmission line pulse testing also is done as a 'two-pin' test. An electrical pulse is launched from the ESD test system to the DUT. The response of the device is defined by the voltage and current response of the device. The time of electrical failure and the power-to-failure can be deconvolved from the pulse response information.

Scatter parameters, also known as s-parameters, allow to quantify the reflected signal. Two equations are established in which the reflected signals, are coupled through the injected signals, and the coefficients, known as s-parameters, are the scatter terms.

$$b_1 = s_{11}a_1 + s_{12}a_2$$
$$b_2 = s_{21}a_1 + s_{22}a_2.$$

This can be expressed in matrix form, where the matrix is the scattering matrix (e.g., S-matrix),

$$\begin{bmatrix} b_1 \\ b_2 \end{bmatrix} = \begin{bmatrix} s_{11} & s_{12} \\ s_{21} & s_{22} \end{bmatrix} \begin{bmatrix} a_1 \\ a_2 \end{bmatrix}$$

where

$$S = \begin{bmatrix} s_{11} & s_{12} \\ s_{21} & s_{22} \end{bmatrix}$$

In this form, the terms in the S-matrix have physical meaning associated with reflection and transmission. The first term on the main diagonal is s_{11}, which is the input reflection coefficient, and the second term on the main diagonal is s_{22} is the output reflection coefficient. The off-diagonal terms, s_{21} and s_{12}, are the forward transmission coefficient and the reverse transmission coefficient, respectively.

S-parameters are used in the evaluation of RF semiconductor devices. Additionally, these can be used to quantify the ESD degradation after ESD stress.

1.7.4 *T*-Parameters

Another representation that is useful for RF analysis is the scattering transfer matrix. This has the advantage of relating scattering and transmission information as an output vector and an input vector.

Scatter transfer parameters, also known as t-parameters, allow to quantify the reflected signal and transmitted signals. Two equations are established

$$a_1 = T_{11}b_2 + T_{12}a_2$$
$$b_1 = T_{21}b_2 + T_{22}a_2$$

This can be expressed in matrix form, where the matrix is the scattering transfer matrix (e.g., T-matrix),

$$\begin{bmatrix} a_1 \\ b_1 \end{bmatrix} = \begin{bmatrix} T_{11} & T_{12} \\ T_{21} & T_{22} \end{bmatrix} \begin{bmatrix} b_2 \\ a_2 \end{bmatrix}$$

where

$$T = \begin{bmatrix} T_{11} & T_{12} \\ T_{21} & T_{22} \end{bmatrix}$$

To convert from S-matrix to T-matrix, we can obtain the parameters as follows:

$$T_{11} = \frac{1}{S_{21}}$$

$$T_{21} = \frac{S_{11}}{S_{21}}$$

$$T_{12} = -\frac{S_{22}}{S_{21}}$$

$$T_{22} = -\frac{\det[S]}{S_{21}}$$

and the determinant of the S-matrix is

$$\det[S] = S_{11}S_{22} - S_{12}S_{21}.$$

The conversion from the T-matrix back to an S-matrix is as follows:

$$S_{11} = \frac{T_{21}}{T_{11}}$$

$$S_{21} = \frac{1}{T_{11}}$$

$$S_{12} = -\frac{\det[T]}{T_{11}}$$

$$S_{22} = -\frac{T_{12}}{T_{11}}$$

and the determinant of the T-matrix is as follows

$$\det[T] = T_{11}T_{22} - T_{12}T_{21}$$

1.8 STABILITY: RF DESIGN STABILITY AND ESD

In RF design, circuit stability is an important design parameter. In RF ESD design, circuit stability can be influenced by ESD protection networks. A design synthesis of the RF source stability and the ESD sensitivity of the base resistance can be understood from Smith chart analysis. Different stability metrics exist for evaluation of RF circuit stability. From the Stern stability criteria, $K > 1$ achieves unconditionally stable two port networks where

$$K = \frac{1 - |s_{11}|^2 - |s_{22}|^2 + |\det S|}{2|s_{12}s_{21}|}$$

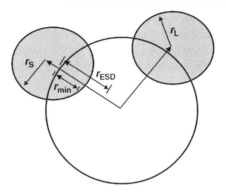

Figure 1.2 Unity radius smith chart with source and load stability circles. Region outside the Unity radius Smith chart is unstable source. Stability and ESD improvement achieved with shift along axis toward radial center

This condition initiates a specific value for the scattering parameters where stability can occur for a two-port network. This can be visualized on a Smith chart.

Source stability requirements will determine a minimum resistance, R_{min}, in order for a source connected to a bipolar base is unconditionally stable. By modification of the base resistance, it is possible to achieve both unconditionally stable circuits and ESD robust circuitry. If the additional resistance added to a bipolar device for ESD robustness improvements exceeds the minimum base resistance, both ESD and the stability criteria can be achieved. Adding resistance beyond r_{min} can exceed the minimum stability requirement with further improved ESD protection levels (Figure 1.2).

As an example, to optimize power amplifiers or other applications, this can be achieved by evaluating the unilateral constant gain and noise circles on a Smith chart. Plotting the source and load circles on the normalized chart, stability of the circuit can be evaluated.

For example, source and load stability circles can be defined by their radius and center. The radius, r, and center, C, for the source are defined as follows:

$$r_s = \frac{|s_{12}s_{21}|}{||s_{11}|^2 - |\det S|^2|}$$

and

$$C_s = \frac{s_{11}|\det S|s_{22}}{|s_{11}|^2 - |\det s|^2}$$

If the source stability circle center is contained outside the Smith chart, there are conditions that unconditional stability is not achieved.

In this case, a minimum resistance r_{min} can be defined as the one that moves the source stability circle within the boundary of the Smith chart. In this case, adding resistance R_{min} to the base of an NPN amplifier can place the amplifier into a metastable condition. Improved stability is achieved by adding a resistance R_{ESD} where

$$R_{ESD} = R_{min} + R_{ESD}$$

where the first term is for unconditional stability and the second term is additional resistor ballasting for the base region. This method can also be applied to the load stability circles to provide a stable output and identifying a G_p (min) on the constant conductance circles.

As stated above, Stern stability criteria, $K > 1$ achieves unconditionally stable two port networks where

$$K = \frac{1 - |s_{11}|^2 - |s_{22}|^2 + |\det S|}{2|s_{12}s_{21}|}$$

To achieve stability and design margin, designers should know to have $K > 1.2$. A circuit can be designed such that ESD elements are added to satisfy the Stern criteria and an additional ESD element can be added to provide the extra stability margin.

1.9 DEVICE DEGRADATION AND ESD FAILURE

1.9.1 ESD-Induced D.C. Parameter Shift and Failure Criteria

In the evaluation of ESD failure of RF devices, there are both d.c. and a.c. parameters that are important to determine the presence of latent mechanisms, device shifts, or destructive damage. The ESD failure criteria will be a function of the allowable change for the given product specification or application. For d.c. parameters, the following d.c. parameters are of interest:

- transistor emitter, base or collector leakage;

- transistor emitter-base ideality factor;

- MOSFET threshold voltage shifts;

- MOSFET gate-dielectric leakage current.

Bipolar Transistor Leakage Current

Transistor emitter, base or collector leakage current can be indicative of metallurgical junction dopant movement from thermal stress, transistor 'pipes', silicide penetration of metallurgical junction, or interconnect metallurgy poisoning of the metallurgical junction. From the ESD pulse event, the increase in temperature at the metallurgical junction can lead to a motion of the dopant materials; this will change the junction leakage current. ESD-induced transistor 'pipes' occur between the emitter and collector region, influencing the leakage characteristics. ESD-induced silicide penetration near the metallurgical junction can lead to changes in the leakage characteristics [1]. Leakage current in metallurgical junctions can increase with junction regions that use refractory metals such as titanium salicide, titanium molybdenum, and cobalt salicide. Evaluation of the d.c. shifts can be evaluated as follows:

- emitter-base reverse bias current as a function of the emitter-base voltage;

- collector-base reverse bias current as a function of the collector-base voltage;

- collector-to-emitter reverse bias current as a function of the collector-to-emitter voltage;

- collector-to-substrate reverse bias current as a function of the collector-to-substrate voltage.

Bipolar Transistor Ideality

Transistor emitter-base ideality factor can be altered as a result of ESD stress. Post-ESD stress, the transistor emitter-base ideality factor can be altered as a result of an increase in bulk or surface recombination centers in the emitter–base junction. Transistor emitter-base ideality factors can be observed from the following measurements:

- Gummel plot of base current as a function of the emitter-base voltage and extraction of the base current slope;

- Gummel plot ratio of collector-to-base current as a function of emitter-base voltage;

- extraction of the leakage current scaling with increased temperature;

- direct current bipolar gain characteristics;

- model extraction from Gummel-Poon model at low currents.

As an example, in the work of R.L. Minear and G.A. Dodson, in a CBE bipolar homojunction transistor, it was observed that the pre-ESD stress collector and base ideality factors were $n = 1.02$ and $n = 1.17$ prior to ESD stress, respectively [41]. Base and collector currents were measured for a base-emitter voltage from 0.3 to 1.0 V, with preforward bias current levels from 1×10^{-9} to 1×10^{-3} A levels. After HBM ESD stress of the emitter–base junction, the Gummel plot base characteristic (e.g., $\log I_B$ versus V_{BE}) showed a change in the slope with a new ideality factor of $n = 2.02$; no change was observed in the Gummel plot collector current characteristic [41]. In this result, the Gummel plot slope changed by $2\times$ from an ideal device (e.g., $n = 1$); this indicates that the ESD stress generated mid-band gap recombination centers in the emitter–base junction region. The recombination centers can be associated with bulk recombination centers or surface recombination centers. The recombination centers would satisfy a forward bias characteristic of

$$I_B \propto \exp\left\{\frac{qV_{BE}}{2kT}\right\}$$

MOSFET Threshold Voltage Shifts

In RF MOSFET devices, the evidence of ESD-induced degradation may be evident by an observation of the MOSFET threshold voltage shift. In a MOSFET structure, MOSFET threshold voltage shifts can occur due to MOSFET gate dielectric surface state generation, MOSFET hot-electron injection, MOSFET negative bias temperature instability (NBTI), MOSFET gate-induced drain leakage (GIDL), and MOSFET parasitic current devices. ESD-stress can lead to changes in the spatial location of fixed charge in the isolation materials (e.g., LOCOS isolation, or shallow trench isolation), or generate additional charge due to charge injection. From the d.c. MOSFET threshold measurements, ESD-induced degradation can be evaluated.

MOSFET gate dielectric leakage

Evaluation of the d.c. MOSFET gate dielectric leakage is important to determine d.c. degradation issues with RF MOSFETs. MOSFET gate dielectric leakage can be observed as 'soft breakdown' or 'hard breakdown.' MOSFET 'soft breakdown' can be induced by ESD electrical overstress. In MOSFET gate dielectric breakdown theory, evidence today demonstrates that MOSFET 'soft breakdown' is an early stage of MOSFET 'hard breakdown'. Hence, observed ESD-induced MOSFET gate leakage current may be regarded as a latent failure mechanism, which can lead to MOSFET device failure.

Additionally, it has been found that the location of the MOSFET dielectric damage can influence the d.c. characteristics of the MOSFET. MOSFET dielectric failure can occur MOSFET gate-to-drain, MOSFET gate-to-source, and MOSFET gate-to-channel; the location has an influence on the d.c. characteristics of the MOSFET device.

1.9.2 RF Parameters, ESD Degradation, and Failure Criteria

With ESD-induced degradation, the RF parameter shifts can lead to RF product failure. For example, these degradation will manifest themselves in the key RF metrics.

Reflection Coefficient ESD-induced Degradation

ESD-induced impedance changes can influence the reflection coefficient, and the power transfer into an RF component. Assuming the load is the circuit elements, the change in the reflection coefficient represented as a function of characteristic impedance and load impedance,

$$\Gamma + \Delta\Gamma = \frac{Z_L + \Delta Z_L - Z_0}{Z_L + \Delta Z_L + Z_0}$$

This can also be expressed as a function of the characteristic admittance,

$$\Gamma + \Delta\Gamma = \frac{Y_0 - (Y_L + \Delta Y_L)}{Y_0 + (Y_L + \Delta Y_L)}$$

With the ESD-induced shift in the reflection coefficient, many RF metrics will be influenced:

- load reflection coefficient;
- return loss (RL);
- mismatch loss (ML);
- voltage standing wave ratio (VSWR).

With changes in the resistance characteristics, quality factor degradation will also occur.

$$Q + \Delta Q = \frac{X_s}{R_s + \Delta R_s},$$

Changes in the RF parameters will play a role in the definition of RF product failure. In the following sections, discussion of RF ESD testing techniques and degradation in the RF parameters will be highlighted. The changes in both the d.c. and RF parameters will lead to the definition of ESD failure in the RF product.

1.10 RF ESD TESTING

Now a days, ESD testing and qualification of semiconductor components is performed using a variety of ESD models. The device level ESD testing models and methods are as follows [1,2,107–132]:

- human body model (HBM);
- machine model (MM);
- charged device model (CDM);
- charged cable model (CCM);
- cable discharge event (CDE);
- cassette model;
- transmission line pulse (TLP);
- very-fast transmission line pulse (VF-TLP).

1.10.1 ESD Testing Models

Human Body Model (HBM)

The HBM ESD pulse was intended to represent the interaction of the electrical discharge from a human being, who is charged with a component, or object [107–110]. The model assumes that the human being is initially charged. The charged human source then touches a component or object using a finger. The physical contact between the charged human being and the component or object allows for current transfer between the human being and the object. A characteristic time of the HBM model is associated with the electrical components used to emulate the human being. In the HBM standard, the circuit component to simulate the charged human being is a 100 pF capacitor in series with a 1500 Ω resistor. This network has a characteristic rise time and decay time. The characteristic decay time is associated with the time of the network

$$\tau_{HBM} = R_{HBM} C_{HBM}$$

where R_{HBM} is the series resistor and C_{HBM} is the charged capacitor. This is a characteristic time of the charged source. The HBM characteristic time constant is physically interesting as the time of the pulse is on the order of the thermal diffusion time of many materials used in the semiconductor industry. In recent times, the RC time of the HBM pulse is significantly slower than RF microwave application frequencies.

Machine Model

Another fundamental model used in the ESD industry is known as the MM pulse [111–114]. The model was intended to represent the interaction of the electrical discharge from a conductive source, which is charged with a component or object. The model assumes that the 'machine' is charged as the initial condition. The charged source then touches a component or object. In this model, an arc discharge is assumed to occur between the source and the component or object allowing for current transfer between the charged object and the component or object. In the MM industry standard, the circuit component is a 200 pF capacitor with no inherent resistor component. An arc discharge fundamentally has a resistance on the order of 10–25 Ω. The characteristic decay time is associated with the time of the network

$$\tau_{HBM} = R_{HBM} C_{HBM}$$

where R is the arc discharge resistor and C is the charged capacitor. This is a characteristic time of the charged source. The MM characteristic time scale is significantly faster than the HBM characteristic time scale due to the lack of a resistive element. The MM response is oscillatory and has significantly higher currents than the HBM ESD event, faster than the HBM ESD events, but still significantly lower than today's RF application frequencies.

Charged Device Model (CDM)

The CDM represents an electrostatic discharge interaction between a chip and a discharging means where the chip is precharged [115–117]. The charging process can be initiated by a direct charging, or field-induced charging. The discharge process initiated as contact is initiated between the charged device and the discharging means. The CDM discharge phenomenon occurs at less than 5 ns where typically the rise time of the event is on the order of 250 ps. The CDM event is the fastest of the ESD phenomenon. With a 250 ps rise time, the energy spectrum of the CDM discharge event extends to 5 GHz frequencies. For RF application frequencies below 5 GHz, this ESD event is on the same order as the RF product application. As the RF application frequencies exceed 5 GHz, the RF application frequency extends beyond the ESD phenomena of concern.

Charged Cable Model

Another fundamental model used in the ESD industry is known as the charged cable model or cable model pulse. The model was intended to represent the interaction, the electrical discharge of a charged cable, discharging to a chip, card, or system. To initiate the charging process, a transmission line or cable source is charged through a high-voltage source or tribo-electrically. The model assumes that the cable is charged as the initial condition. The charged cable source then touches a component or an object. A characteristic time of the cable model is associated with the electrical components used to emulate the discharge process. In the charged cable model, the cable acts as a capacitor element. The characteristic decay time is associated with the time of the network

$$\tau_{\text{CCM}} = R_{\text{CCM}} C_{\text{CCM}}$$

where R is the discharge resistor and C is the charged cable. The capacitance used for this model is 1000 pF. Studies are also completed using a ESD gun, which is discharged through a cable to evaluate the system level events.

Akin to this model, the CDE model is of interest [118–123]. In this case, a transmission lineis used as a source (e.g., instead of a lumped capacitor). The cable is charged, where the pulse length is a function of the length of the cable. This CDE model is very important today because of the wide usage of laptops, servers, and wireless applications. In these cases, the length of the events are significantly longer than the RF application frequencies for long cables.

Charged Cassette Model

The charged cassette model is a recent model associated with consumer electronics and game industry. In consumer electronics there are many applications where a human plugs a small cartridge or cassette into an electronic socket. These are evident in popular electronic games. In today's electronic world, there are many palm size electronic components that must be socketed into a system for nonwireless applications. To verify the electronic safety of such equipment, the cassette itself is assumed as a charged source. The 'cassette model' assumes a small capacitance and negligible resistance. This model is equivalent to a machine model type current source with a much lower capacitor component. The model assumes the resistance of an arc discharge and a capacitance of 10 pF. For RF applications that utilize charged cassettes, charged memory sticks, or other small charged items, the ESD-induced field failures may be best understood using the charged cassette model.

Transmission Line Pulse (TLP)

TLP testing has become an established methodology for the ESD design discipline [125–129]. The popularity of the TLP method is due to semiconductor and circuit engineers, interest in observing the pulsed I–V characteristic response. Historically, pulse testing is focused on the pulse waveform with the focus on the power-to-failure and energy-to-failure. In this form of ESD testing, a transmission line cable is charged using a voltage source. The TLP system discharges the pulse into the device under test (DUT). The characteristic time of the pulse is associated with the length of the cable. The pulse width of a transmission line pulse is a function of the length of the transmission line and the propagation velocity of the transmission line. The propagation velocity can be expressed relative to the speed of light, as a function of the effective permittivity and permeability of the transmission line source

$$\tau_{\text{TLP}} = \frac{2L_{\text{TLP}}}{v} = \frac{2L_{\text{TLP}}\sqrt{\mu_{\text{eff}}\varepsilon_{\text{eff}}}}{c_0}$$

As a standard practice in recent times, the TLP cable length is chosen to provide a TLP pulse width of 100 ns with less than 10 ns rise time. This choice was intended to find a pulsed waveform that was closest in total energy of a HBM pulse. As a result, the RF application frequencies today are significantly higher than this TLP method. TLP systems are designed in different configurations. TLP system configurations include current source, TDR, TDT, and time domain reflectometry and transmission (TDRT). In all configurations, the source is a transmission line whose characteristic time constant is determined by the length of the transmission line cable. The various TLP configurations influence the systems characteristic impedance, the location of the device under test, and the measurement of the transmitted or reflected signals.

An interesting synergy of TLP testing techniques and RF design is the focus on 50 Ω impedance. The TLP systems methods for TDR, TDT, and TDRT configurations are akin to the reflection and transmission methods used by RF engineers for evaluation of RF parameters. As a result, the ESD testing practice of using 50 Ω TLP systems will be similar to the RF parameter characterization needed to evaluate RF circuits and RF systems.

A second interesting point in the TLP testing method is a two-pin test (unlike the HBM and MM ESD testing methodology). As a result, the method also is closer to the two-port methodology for RF evaluation.

Very Fast Transmission line Pulse (VF-TLP)

The VF-TLP test method is similar to the TLP methodology [125–133]. The interest in the VF-TLP is driven by our understanding of the semiconductor devices in a time regime similar to the CDM time constant. The characteristic time of interest is again determined by the propagation characteristics of the transmission line cable source and the length of the transmission line cable

$$\tau_{\text{VF-TLP}} = \frac{2L_{\text{VF-TLP}}}{v} = \frac{2L_{\text{VF-TLP}}\sqrt{\mu_{\text{eff}}\varepsilon_{\text{eff}}}}{c_0}$$

The VF-TLP pulse width of interest is a pulse width of less than 5 ns and a less than 1 ns rise time. As a result, the VF-TLP characteristic time is on the same order as 1–10 GHz RF application pulses. Hence, this method has potential suitability to evaluate the power-to-failure of RF components. This is a key point. Of all the ESD test methodologies, one of the key values of the VF-TLP test is its pulse width that is closest to the RF source frequency.

Figure 1.3 shows the time constant hierarchy of the ESD phenomenon on a frequency axis. As the application frequency increases, the RF application frequency separates from the ESD pulse phenomena. A key point is that the RF application frequency extends beyond the fastest of the ESD phenomena and hence separates from the ESD pulse waveforms. As the application frequency exceeds 5 GHz, the application frequency exceedes all ESD phenomena of interest.

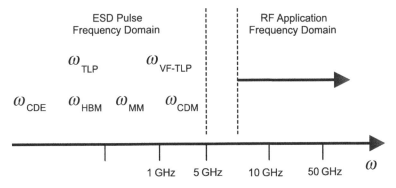

Figure 1.3   Frequency domain plot of ESD phenomena and RF application frequencies

1.10.2 RF Maximum Power-to-Failure and ESD Pulse Testing Methodology

The power-to-failure and the ability of a semiconductor device to survive electrical overstress (EOS), and ESD is important for radio frequency components. In RF components, receiver networks are subjected to electrical stimulus of different power magnitudes, and of different pulse forms and pulse widths. As a result, it is valuable to evaluate both the power-to-failure and the ESD sensitivity of RF components.

Whalen and Domingos [43] evaluated both the response of RF components to RF pulse power and ESD events. Two different independent test methodologies were utilized. A first test system addressed the RF power-to-failure using a RF pulse source. In a second test system, a square pulse test system was utilized.

Figure 1.4 shows the RF pulse system for the evaluation of the electrical overstress of a RF component. For the evaluation of the RF power-to-failure, a system was established that could evaluate the incident, reflected and transmitted power through a RF DUT. An ultra-high frequency source signal was transmitted through a 0–30 dB attenuator. The RF signal was passed through the first directional coupler for the incident power, and the second directional coupler for the reflected power evaluation. A switch followed by a load is placed after the directional couplers to allow for isolation of UHF RF source from the RF DUT. In the TDRT system, the DUT is placed in series with the incident and the transmitted signals. A directional coupler is also utilized to evaluate the transmitted power through the DUT. The three directional coupler elements are followed by crystal detectors and 50 Ω terminations, whose signals are transferred to oscilloscopes and camera sources; these diagnostics capture the incident power, reflected power, and transmitted power waveforms in time.

In this RF test system, the incident power, reflected power and transmitted power waveforms were captured. Whalen and Domingos [43] showed that at the time of failure is observable in the reflected power waveform prior to the end of the applied pulse waveform. In the case of the incident power pulse, a constant power pulse is observable. But, in the reflected power waveform, variation can be observed in the peak power as a function of time. It was noted that a transition occurs in the reflected power waveform, indicating a change in the power absorption and a reduction in the reflected power, followed by an abrupt increase in the reflected power characteristic. This was not observed in power waveforms of RF devices that did not fail at lower power levels.

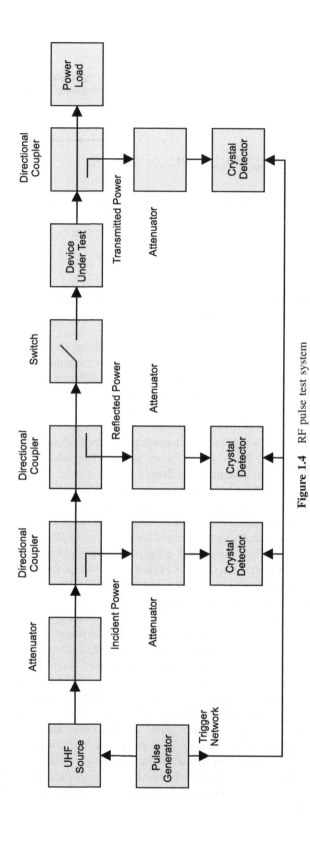

Figure 1.4 RF pulse test system

To evaluate the absorbed energy of the RF device, this can be calculated by the difference between the incident energy and the sum of the reflected and transmitted energy

$$E_A = E_I - E_R - E_T$$

where the incident energy is

$$E_I = \int P_I(t)dt$$

and the reflected energy is

$$E_R = \int P_R(t)dt$$

and the transmitted energy is

$$E_T = \int P_T(t)dt.$$

In the case of RF component failure, the integration is terminated at the time of failure, t_f. From this test methodology, the energy to failure from an RF pulse source can be determined.

Whalen and Domingos also performed a square pulse test that could vary the polarity of the pulse waveform to either positive pulses or negative pulses. Figure 1.5 shows the square pulse system utilized. As in a transmission line pulse test system, this system utilized a square pulse waveform, and was able to capture the voltage and current characteristic across the DUT. In this square pulse test system, the test system integrated a pulse source with a manual trigger, a resistor matching network, and the RF DUT. To capture the voltage characteristic across the RF DUT, a voltage probe was placed across the device, whose output was transferred to an oscilloscope. Additionally, a current probe was placed over the

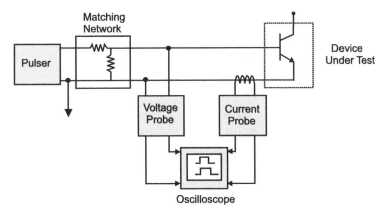

Figure 1.5   Square pulse source ESD test system

ground return line, observing the current that flowed through the RF DUT. The output of the current probe was transferred to an oscilloscope to observe the current waveform. From the voltage and current waveforms, the voltage and current across the RF DUT are evaluated. From these characteristics, the time of failure is also observable by the change in the forward bias value across the RF transistor (e.g., reduction of V_{BE}).

From the test system, the power-to-failure can be evaluated as the product of the current and voltage across the device prior to the collapse of the emitter-base voltage,

$$P_f = I_B V_{BE}$$

whereas the absorbed energy to failure is the product of the power-to-failure and time of failure,

$$E_A = P_f t_f = \{I_B V_{BE}\} t_f$$

From the test system, the voltage, current, and the time of failure are obtainable from the waveforms.

Comparison of the RF pulse system and the square pulse (e.g., TLP-like) test system shows the absorbed energy versus the time-to-failure. The absorbed energy is calculated of the power to failure up to the time of the failure. Whalen and Domingos [43] showed that during the test of RF bipolar transistors, the absorbed power-to-failure is the lowest for negative square pulses, then RF pulse events, and then positive square pulses (Figure 1.6). The results show that the forward biasing of the emitter–base junction leads to a higher absorbed energy-to-failure in comparison to any event that undergoes reverse bias of the emitter–base junction.

An RF ESD design and test methodology is as follows:

- Using a RF pulse system, electrical overstress of an RF component can be evaluated by calculating the incident, transmitted, and reflected energy. The absorbed energy-to-failure

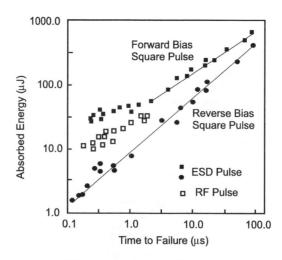

Figure 1.6 Absorbed energy-to-failure versus time-to-failure for RF pulse system and square pulse TLP system (with square pulse is for positive and negative polarity pulses)

can be calculated by the difference between the incident energy and the sum of the transmitted and reflected energy, calculated up to the time of failure.

- Using a square pulse system, electrical overstress can be evaluated in an RF component by evaluating the current and voltage across the device under test and determining the time of failure from the characteristics.

- A correlation exists between the absorbed energy-to-failure from a RF pulse system and a square pulse TLP-type system.

- The absorbed energy-to-failure versus the time-of-failure plot provides a linear plot, with the slope associated with the power-to-failure.

1.10.3 ESD-Induced RF Degradation and *S*-Parameter Evaluation Test Methodology

For the evaluation of ESD-induced degradation to RF components and system, scattering parameters (*S*-parameters) can be used as a measurement diagnostic and serve as an ESD metric for failure criteria [8]. At high frequencies, ESD-induced degradation to RF components that impact signal integrity can lead to impact the system reliability concerns. In high-speed communication systems and high-speed components, the transmission characteristics are important for the transference of signals. The transmission and reflection coefficients are the key measures of the operation and functionality of the system. As the transmission and reflection coefficients are altered, this will also be apparent from the *S*-parameter terms. Shifts in the *S*-parameters or transmission coefficients can be quantified using the TDR methodology.

An RF ESD test method can be established as follows [8]:

- *Prestress evaluation*: Evaluation of all *S*-parameter parameters terms of the RF component using a TDR methodology as a function of frequency;

- *S-parameter matrix*: Formulation of the *S*-parameter matrix terms (e.g., two-port matrix terms are S_{11}, S_{12}, S_{21}, S_{22});

- *ESD step-stress*: Apply an ESD stress between two terminals of the RF component; one of the terminals has an ESD stress applied, whereas a second terminal is the reference ground. A given ESD model waveform shape, waveform polarity, and pulse magnitude is chosen (e.g., HBM, MM, CDM, CDE, TLP, or VF-TLP);

- *Post-stress S-parameter evaluation*: Evaluation of all *S*-parameter parameters terms of the RF component using a TDR methodology as a function of frequency;

- *Post-stress S-parameter matrix formulation*: Formulation of the *S*-parameter matrix terms (e.g., two-port matrix terms are S_{11}, S_{12}, S_{21}, S_{22}) post-ESD stress;

- *ESD step-stress*: The pulse magnitude of the defined ESD stress is increased on the same RF component, and the above procedure is repeated, where the *S*-parameters are reevaluated.

This test procedure can be applied to any two ports of the RF component. Depending on the intended application of the RF component, whether a receiver element, or a transmitter

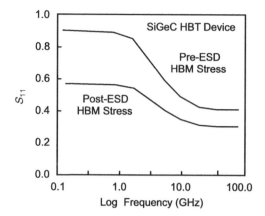

Figure 1.7 S-parameter S_{11} comparison as a function of frequency for a SiGeC HBT device before and after ESD HBM stress

element, some configuration will have more value and relevance to evaluate utilization of the component in a system environment. For RF CMOS MOSFET transistors, the ESD stress can be applied to gate-to-drain, gate-to-source, drain-to-source, and drain-to-substrate. The S-parameter values of the RF MOSFET in a given configuration are extracted, and then the ESD stress is established between any two terminals. As a RF MOSFET receiver network, ESD-induced S-parameter degradation associated with the RF MOSFET gate may be of higher interest; as a RF MOSFET transmitter the ESD-induced S-parameter degradation from drain-to-source may be of greater interest.

For a RF bipolar transistor, the ESD stress can be applied to base-to-emitter, base-to-collector, collector-to-emitter, and collector-to-substrate terminal combinations. For RF receivers, the ESD stress of the base region is of interest, whereas as a RF transmitter, the collector-to-emitter configuration S-parameter degradation may be more relevant.

Figure 1.7 shows an example of S-parameter S_{11} versus frequency before and after HBM ESD stress. The RF test structure is a RF SiGeC hetero-junction bipolar transistor placed in an S-parameter wafer-level pad set [8]. The RF SiGeC HBT device is configured in a common-emitter configuration where the two-port input terminals are between the base and the emitter, and the two-port output terminals are between the collector and the emitter.

In the test procedure, the S-parameter was extracted on a full RF functional test system and the measurements were performed on a wafer level. The HBM stress was applied using a wafer-level HBM test system. A single HBM ESD pulse was applied between the SiGeC HBT base and emitter pads. Post-ESD stress, the test wafer was reevaluated for S-parameter extraction. In Figure 1.7, the S-parameter S_{11} results are shown before and after HBM ESD stress; it can be observed that the S_{11} parameter magnitude is lower for all frequencies after ESD stress.

From this discussion, an ESD RF test method and test technique teaches the following:

- S-parameter evaluation can be used to determine the ESD-induced degradation of RF components.

- For single components, RF components or RF ESD elements are to be placed in S-parameter pad sets and accompanied with open and short pad sets for deembedding procedures and S-parameter extraction.

- ESD applied stress is established across two terminals of the two-port structure where one of the terminals has the pulse applied, and the second terminal serves as the grounded reference.

- RF functional testing and S-parameter evaluation before and after ESD stress can be performed within a single pulse or step-stress process to evaluate single pulse or cumulative degradation effects.

- S-parameters shifts and degradation can be utilized for the ESD failure criteria for a single component or system.

1.11 TIME DOMAIN REFLECTOMETRY (TDR) AND IMPEDANCE METHODOLOGY FOR ESD TESTING

Signal integrity is important to both digital systems and RF communication systems as clock frequencies and data transmission rates increase. For RF and high-data rate transmission systems, new testing techniques are needed to evaluate the impact of ESD on components. At these high-data transmission rates, ESD-induced changes to the electronics that impact signal integrity can lead to impact the system reliability. ESD damage, either permanent or latent, can lead to chip or performance issues. ESD-induced damage influences the gain, the transconductance, and the S-parameters; a new methodology and failure criterion may be needed to evaluate the ESD impacts. This can occur to circuitry on the transmission or receiving end of a system leading to unacceptable degradation levels. ESD-induced damage can include the following circuit and system issues:

- signal rise time;

- pulse width;

- timing;

- jitter;

- signal-to-noise ratio.

In high-speed communication systems and high-speed components, the transmission characteristics are important for the transference of signals. The transmission and reflection coefficients are the key measures of the operation and functionality of the system. The transmission coefficients, the S-parameters, and the impedance play a major role in the functionality of the system. Shifts in the S-parameters or transmission coefficients can be quantified using the TDR methodology.

The TDR method can be used to verify ESD failures and implications to functional system. The TDR method is a common practice in high-speed system development. The functional test requires a launching of a signal and the reflected wave is measured to evaluate the transmission, the reflection, and impedance of the port tested. Time domain reflectometry measures the reflections that occur from a signal, which is traveling through a transmission environment. The transmission environment can be a semiconductor circuit, a connector, cable, or circuit board. The TDR instrumentation launches a signal or pulse through the system to be evaluated and compares the reflections of a standard impedance to that of the unquantified transmission environment.

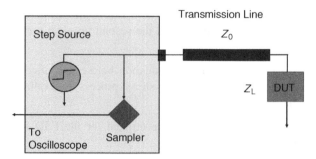

Figure 1.8   A time domain reflectometry (TDR) test methodology

Figure 1.8 shows an example of a TDR measurement system [8]. A TDR measurement sampling module consists of a step source, a 50 Ω connection, followed by a transmission line to the load. The TDR sampling module also contains a sampler circuit, which draws a signal off of a 50 Ω transmission line, whose signal is fed back to the oscilloscope. The TDR display is the voltage waveform that is reflected when a fast voltage step signal is launched down the transmission line. The waveform, which is received at the oscilloscope, is the incident step as well as the set of reflections generated from impedance mismatch and discontinuities in the transmission system.

The mathematics of the TDR method is based on the impedance ratios and a reflection coefficient ρ. The reflection coefficient ρ is equal to the ratio of the reflected pulse amplitude to the incident pulse amplitude

$$\rho = \frac{V_{\text{reflected}}}{V_{\text{incident}}}$$

The reflection coefficient can be expressed as a function of the transmission line characteristic impedance Z_0 and the fixed termination impedance Z_L. In this form, the reflection coefficient can be expressed as

$$\rho = \frac{Z_L - Z_0}{Z_L + Z_0}$$

The fast-step stimulus waveform is delivered to the DUT after propagation through the sample head, the transmission line, connectors, and the test fixture connections. The waveform, which is reflected from the device under test, is delayed by the two electrical lengths at the oscilloscope – the time of flight through all the interconnects and the return flight time. This signal is superimposed on the incident waveform at the TDR sampling head. The TDR sampling heads typically allow evaluation of the voltage waveform, the reflection coefficient, or the impedance on the TDR oscilloscope.

1.11.1 Time Domain Reflectometry (TDR) ESD Test System Evaluation

TDR is valuable for the analysis of ESD-induced failure in both components and systems. This TDR methodology can be valuable when only the input is available for analysis. As an

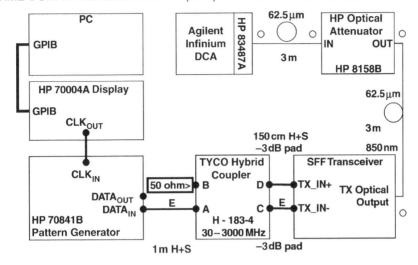

Figure 1.9 Electro-optical test system

example of application of the TDR system and ESD-induced degradation, a high-speed optical interconnect system is shown [8].

Figure 1.9 shows a high-level diagram of the optical interconnect system. The system consisted of a transmitter/receiver module, a short-wave vertical cavity surface emitting laser (VCSEL), an optical wave guide, an optical photo-detector, and an optical-to-electrical (O/E) converter. The data input signal stream is produced at approximately 2 Gb/s in a fiber channel pattern (FCPAT). The data pattern is generated by the pattern generator and converted to a differential signal via the hybrid coupler. The hybrid coupler element is connected electrically to the transceiver (TX) differential inputs. The SFF transceiver receives the data pattern and modulates the laser diode. The optical output from the SFF transceiver is connected to a multimode fiber, which conveys the signal to an optical attenuator before making its way to the optical input of the DCA. The O/E converter in the DCA filters the waveform by limiting the bandwidth and projects the displayed waveform on screen.

Optical-to-electrical conversion in the transceiver is a very precise function requiring tight control of parameters for gigabit data transfer. The electrical path from the connector to the MICC chip is a controlled 50 Ω single-ended impedance. A transceiver having this form factor has transmitter input pins exposed to ESD events. ESD events can damage the circuits and destroy the circuit integrity. Testing the O/E conversion before and after ESD events is a way to monitor the robustness of the ESD protection circuits.

The ESD test methodology for evaluation of the system comprises the following steps [8]:

- Functional characterization of the system is performed prior to ESD stress.

- The input TDR signal is evaluated in an unpowered state.

- System power is applied.

- Optical 'eye' patterns are recorded prior to ESD testing.

- ESD stress is applied to the transceiver chip input signals.

- An 'ESD gun' or ESD pulse system is directly connected to the subject pin, and the stimulus is applied.

- Each pin on the evaluation card is connected to signals, power, and grounds in the transceiver.

- Differential input TX_IN pins would be the source signal, and the return signal path was either the power or the ground pin. Both positive and negative polarities are tested.

- TDR measurements were evaluated post-ESD stress. An ESD impulse is applied to the port when the system is unpowered; the system is then retested using the TDR test methodology.

- Post-ESD stress output 'eye' test is evaluated to observe system level degradation effects.

Prior to ESD testing, the functional system is characterized. The experiment is started by observing the input TDR signal when the system is powered down. The TDR system consists of a Tektronix SD24 TDR Sampling head in a Tektronix 11801C oscilloscope. The Tektronix 20 GHz SD24 TDR sampling plug-in has a 15 ps rise time into a load and a 35 ps reflected wave rise time. The Tektronix 11801C has a 50 GHz sampling rate. After power-up the data input signal stream is initiated, and the output optical eye patterns are recorded prior to any ESD testing. The transceiver/receiver chip was first analyzed followed by full system evaluation.

The ESD test method establishes a procedure to apply ESD pulses to all pads on the transceiver via externally placed pins. An 'ESD gun' Mini-Zap 2000 is directly connected to the subject pin and the stimulus is applied in accordance to the JEDEC Standard (JESD22-A114-B). The ESD gun Mini-Zap 2000 is wired to the desired pins for source and return. Each pin on the evaluation card is connected to signals, power, and grounds in the transceiver. In this particular test, one of the TX_IN pins would be the source signal, and the return signal path was either the power or the ground pin. Note that both positive and negative ESD pulse polarities can be evaluated. The source/return path from the 'ESD gun' Mini-Zap 2000 lead is in the following order: stimulus pin, card trace, card connector, transceiver connector, transceiver trace/component, and finally MICC chip input (TX_IN). An ESD pulse stress is applied to the port when the system is unpowered; the system is then retested using the TDR test methodology. An example of the results below shows the characteristics of a system of a 1 GHz path for stress at 2000 V and at more than 2000 V.

In the test system, the semiconductor chip that may be vulnerable to ESD events is the SFF transceiver chip, which consists of RF components. In this application, a 45 GHz f_T SiGe hetero-junction bipolar transistor technology is used in the transceiver chip. Figure 1.10 shows a high-level diagram of the transceiver chip architecture. In the diagram, the differential inputs, the amplifiers, and the diode laser signals are shown.

Measurements were taken at different stress levels, evaluating the TX-pin distortion. HBM stressing was performed on the transceiver chip to determine the ESD sensitivity of the signal pins. Table 1.8 shows a table of transmission and reflection signal magnitudes at a magnitude of 2000 V HBM ESD stress. Each line in the table represents a different module placed under ESD stress. Various system level parameters of amplitude and loss are recorded

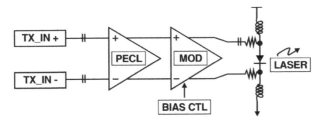

Figure 1.10 Transceiver chip architecture

for the transmitted and reflected signals. Below a 2000 V HBM level, distortions in the TX amplitude is not observed in the 1 GHz signal path. In Table 1.9, the ESD stress was increased above 2000 V HBM. From the TDR method, various parameters begin to distort as the ESD stress increases. Above 2000 V, shifts occur in the transmission port tested. Distortion of the transmitter is noted in the TX pin with a 2X increase in TX DJ (TX DJ increased from levels of 30 to 72 and 131). Hence, from this TDR methodology, the metrics are evaluated for the various signal level parameters [8].

A second means of observation is the output 'eye' test. After the input transmitter/ receiver circuit has its TDR measurement, the system is repowered and cycled at the 2 Gb/s data rate. From the output 'eye' the distortion of the output characteristics is another way of evaluation of the ESD impacts. The timing of the optical network is impacted by the ESD pulse as a result of the impact on the differential input of the input SFF chip. Experimental observations showed that, in some cases, it was hard to determine the change in the TDR input characteristic before and after ESD stress, yet observations were visible in the subtle changes in the output 'eye test' [8].

Table 1.8 TDR method results at 2000 V HBM ESD stress

Type	Serial	FCP	TX ER	TX DJ	1e-12 RX sensitivity	RX amplitude	TX fault	Rx loss
1 GHz f_c path	1505	−5.69	7.40	34.78	−18.7	756.06	OK	OK
	2735	−5.64	6.62	22.17	−19.2	761.03	OK	OK
	2779	−6.12	7.25	24.01	−19.7	743.39	OK	OK
	2750	−6.48	5.96	18.3	−18.7	717.93	OK	OK
	1502	−5.82	8.82	45.45	−19.8	749.10	OK	OK

Table 1.9 TDR method results above 2000 V HBM ESD stress

Type	Serial	FCP	TX ER	TX DJ	1e-12 RX sensitivity	RX amplitude	TX fault	Rx loss	Comments
1 GHz path	1505	−5.69	7.97	131.48	−18.8	770.85	OK	OK	TX-distorted
	2735	−5.64	6.98	24.58	−19.4	771.89	OK	OK	TX-OK
	2779	−6.12	6.75	28.85	−19.6	767.84	OKOK	OK	TX-OK
	2750	−6.48	7.25	30.57	−18.8	786.23	OK	OK	TX-OK
	1502	−5.82	7.68	72.53	−19.8	787.76	OK	OK	TX-hi crossing

Hence, using a TDR technique, and system level 'eye test,' the impact on a RF chip at the input and the output of the system can be evaluated. The TDR method, using state-of-the-art oscilloscopes with TDR sampling heads, can evaluate the reflection coefficients, the impedance, and voltage level response. Using the methodology of TDR measurements of the system input (in an unpowered state) pre- and post-ESD allow for a comparison evaluation of the stress on the TDR waveform. It was also found that the 'eye test' can also observe low-voltage degradation effects not observed from the TDR method only. With a trained eye, the subtle variations in the 'eye test' can allow a test engineer the means of observing distortions, which may impact the data transmission system.

1.11.2 ESD Degradation System Level Method – Eye Tests

In system level quantification of ESD degradation, it is not always possible to determine the ESD-induced degradation at the semiconductor chip, or at low-speed characterization. Using the 'eye test', it is possible to provide quantification of the system level impact of ESD degradation. A means of observation is the output 'eye' test. From the comparison of the output 'eye' before and after ESD degradation, a system level failure criteria can be established. From the output 'eye' the distortion of the output characteristics is another way of evaluating the ESD impacts. The eye test is a measure of the timing of signals to evaluate the worst case operational means. Figure 1.11 is a pictorial example of the 'eye' before and after ESD stress. The 'eye' is formed by the overlaying of two signals of interest where one signal is inverted. When the timing is sound, the 'eye' is open and wide. When distortion or poor timing the overlapping of the signal leads to a small eye opening. An example of an 'eye test' whose iris has been degraded as a result of ESD-induced degradation is shown in Figure 1.11.

Figure 1.12 shows an example of an 'eye test' with a system with good functional characteristics. Figure 1.13 shows the results of the output 'eye' after ESD testing. The parameters of interest from the system level from the analysis are as follows:

- rise time;
- fall time;
- jitter.

In the 'eye test' study, a range of parameter variations is shown. Prior to ESD testing, the rise time range varied from 95.6 to 97.8 ps. The fall time range varied from 113.3 to 115. 6 ps. The range of the jitter parameter is 9.4–9.9 ps.

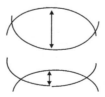

Figure 1.11 Functional eye test before and after ESD testing. The first characteristic represents a 'good eye' and the second represents a 'bad eye'

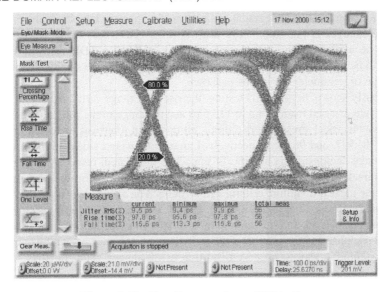

Figure 1.12   Eye diagram prior to ESD testing

Experimental observations will show that in some chip applications it will be difficult to determine the change in the semiconductor chip input characteristics, before and after ESD stress. At low current stress levels, d.c. characteristics shifts, and small degradations in the device characteristics may not appear to have significant impact on the semiconductor chip. But, on a system level these can lead to system failure. In some applications, the

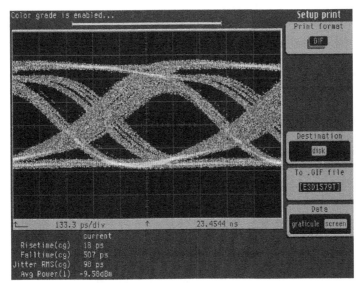

Figure 1.13   Eye test post-ESD testing

semiconductor chip level ESD degradations may not be apparent but observations are visible in the subtle changes in the output 'eye test'. At high ESD stress levels where significant change can be evident in the d.c. device characteristics, or a TDR test, the optical eye distortion will be clearly visible in the receiving output oscilloscope signals. In the 'eye test' study, a range of parameter variations is shown.

As previously shown, before ESD testing, the rise time range varied from 95.6 to 97.8 ps; the fall time range varied from 113.3 to 115.6 ps and the range of the jitter parameter was 9.4 to 9.9 ps. Post electrical stress, the rise time, fall time, and jitter was impacted to the values of range of 18, 507, and 98 ps, respectively. Large shifts were evident in the rise, fall, and jitter parameters demonstrating significant impact to the system response after ESD stress, the three parameters functional parameters (rise time, fall time and jitter) were impacted. In this system, the ideal ESD failure criteria will be defined on a system level on the basis of the timing impact (e.g., rise time and fall time), as well as in the jitter value specifications, but not on a component level. In a given system, a relationship can be established between the system level response and the internal damaged component. Hence, using the output eye test provides a qualitative way to observe distortion of the device characteristics of the input devices [8].

These results indicate that for high-speed communications systems, new criteria and test techniques may be needed to define the impact of ESD stress. Traditional ESD testing looks at the d.c. parameter shifts of the semiconductor device or input pin chip characteristics may be inadequate to quantify the system level impacts. In the future, ESD degradation may be best quantified using quantitative time-domain reflection techniques, and evaluate the impedance shifts, reflection characteristics, and transmission characteristics in determining device failure. At the same time, this can be further understood using the output 'eye' test in evaluation of the true system impact.

1.12 PRODUCT LEVEL ESD TEST AND RF FUNCTIONAL PARAMETER FAILURE

For product level evaluation, a comparison of the product level failure point anticipated by the ESD tester and the evaluation based on RF parameters is required. A key question that was being addressed is whether the ESD HBM failure levels reported from ESD HBM testers is valid for prediction of the ESD robustness of RF products. To address the question of whether ESD data from commercial HBM stress systems is valid in predicting the failure of RF products, a method was first initiated by Van Laecke, Rascoe, and Voldman [8]. In this application, a product with a 45 GHz f_T Silicon Germanium HBT technology was used for the evaluation. The ESD test method is as follows:

- A first test chip is placed on the HBM commercial test system.

- Failure criteria are defined on the basis of a d.c. shift in the $I-V$ characteristics.

- ESD step-stress is applied with pre- and post-ESD pulse leakage evaluation.

- ESD pulse step-stress is completed in 100 V HBM test increments.

- ESD failure is defined as the first pin to satisfy the 50% shift in the $I-V$ characteristic. This sets the failure level on the basis of the d.c. shift criteria, used for the second study.

- A second set of test chips are d.c. and RF characterized prior to ESD testing using a fully functional RF test system.

- A second set of test chips are stressed with a single HBM ESD pulse magnitude where the highest pulse level is the same as achieved in the prior ESD step-stress (e.g., failure level as defined by the first test). The modules are tested from zero to the highest pulse level where each module has a single pulse stress applied.

- RF characterization and full functional characterization are evaluated for the second set of modules.

- RF metrics are plotted to determine where the failure criterion is first encountered. The lowest encountered RF parameter determines the HBM stress level for 'failure.'

As an example, the HBM ESD test was applied to a first module. The HBM ESD test is done in 100 V step increments. In this application, the HBM testing of the test chip has an HBM worst-case pin-to-rail of 5000 V. This set the upper limit for the second study. In the second study, testing is completed at different ESD stress levels where the testing is stopped prior to the worst case functional failure level (of the first study). Different sample parts are tested with at a single ESD HBM stress level at 0–5000 V. Full-functional RF testing is then completed on the product chip on all the ESD stressed parts [8].

Figure 1.14 shows the OIP3 magnitude as a function of the single pulse HBM stress condition. With no ESD stress applied the OIP3 magnitude remained the same until after a stress level of 3000 V. No degradation was observed between 0 and 3000 V HBM stress levels. The modules that were stressed at 4000 V demonstrated a degraded OIP3. At an ESD stress level of 4500 V, the OIP3 is degraded with an OIP3 magnitude of 5–20. Hence, the OIP3 RF metric demonstrated degradation prior to the 5000 V HBM level reported from the commercial HBM test system (based on a 50% failure d.c. shift failure criteria).

Figure 1.15 shows the conversion gain results from the identical study. At a HBM stress level of 3000 V, the conversion gain magnitude is 8.55. As the HBM ESD stress level increased, the conversion gain characteristics decreased below 8, with a conversion gain roll-off toward zero at stress levels of 4500 V HBM. As is observed in the OIP3 metric, the RF

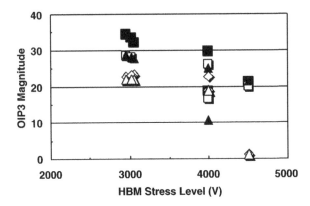

Figure 1.14 Ouput intercept point third-order harmonic (OIP3) magnitude as a function of HBM stress level

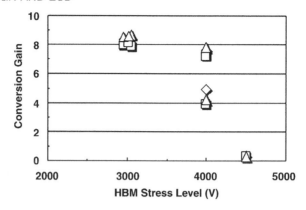

Figure 1.15 Conversion gain as a function of the HBM ESD stress

parameters degraded prior to the predicted level on the basis of the ESD HBM step-stress (e.g., the d.c. *I–V* shift failure criteria) [8].

From this product level ESD stress study, the failure criteria of RF components must address evaluation of all the RF parametrics. Using the ESD test system for d.c. characterization, this study can be used to define the upper limit of the second RF characterization study.

As was done in the single component ESD study, a relationship does exist between the d.c. characterization and the RF degradation. First, in both cases, RF degradation does occur prior to full ESD 'failure' based on a d.c. parameter. In the single component case, a clear relationship was established between transistor RF parameter degradation and the d.c. shift of the forward bias voltage (e.g., in the emitter-base study). In this product level study, the upper level of HBM failure, based on a d.c. shift criteria, was used only as an upper limit to define the stress conditions for the second full RF evaluation. In the product evaluation, there are a significant number of RF parameterics, which is dependent on the product application. Hence, the method discussed is a useful method to utilize the commercial ESD testers and establish limits for the RF ESD evaluation [8].

1.13 COMBINED RF AND ESD TLP TEST SYSTEMS

For the ESD evaluation of RF components, it is important to evaluate the RF parameters pre- and post-ESD stress in the case where there is no correlation to d.c. parameters. In ESD testing of radio frequency components, the evaluation of the RF parameters pre- and post-ESD stress can be provided by the integration of RF characterization test equipment into ESD test systems. RF parameters can be evaluated during ESD testing by the integration of RF test equipment within the testing systems. Alternatively, a second methodology would be to integrate ESD pulse sources into RF test systems. In RF test systems, the latter method would not be desirable due to potential damage or time constraints to high-cost sensitive RF test systems.

In today's commercial ESD test systems, the ability to evaluate the d.c. parameters as well as the RF parameters is not offered. For the evaluation of ESD degradation influence on

RF components, the integration of RF characterization equipment into HBM, MM, and TLP and VF-TLP test simulators is possible with providing 'switches' that allow the RF characterization between ESD pulse stress.

Hynoven, Joshi, and Rosenbaum integrated RF characterization equipment into a TLP system by applying a second single-pole double-throw (SPDT) switch to measure the RF parameters before and after application of the TLP stress [9]. Figure 1.16 shows the TLP system with integrated RF characterization.

The test system for characterization and electrical connections includes an oscilloscope, a parameter analyzer, noise figure meter, noise figure test set, signal generators, preamplifiers, relays, and SPDT switches. For the ESD pulse source, the test system includes a high-voltage generator, an ESD pulse generator source (e.g., a commercial pulse generator or a transmission line cable of predefined length) and associated electrical connections. As an example of the test equipment, the following equipments were used by Hynoven, Joshi, and Rosenbaum [9]:

- 50-Ω 40-GHz coplanar waveguide probes with ground-signal-ground (GSG) footprints.

- Bertan Series 225 High Voltage Generator.

- HP 54510B (300 MHz) Oscilloscope.

- HP 4145B Semiconductor Parametric Analyzer.

- CP Clare relays MSS41A05 (SPST) and HGJM51111K00 (SPDT).

- HP 8761A RF SPDT switch (DC-18GHz).

- HP 8970B NF Meter (1.6 GHz bandwidth).

- HP 8971B NF Test Set (mixer with control logic, up to 18 GHz).

- Agilent PSG-E8241A Signal Generator (up to 20 GHz).

- Amplifier Research LN1G11 Pre-Amplifier (<6 db NF, 27 dB gain).

- HP 8114A Pulse Generator (for relay control).

- HP E3631A Power Supply (for relay control).

- Transmission line cable.

In TLP systems, a common design are the 50 Ω impedance systems. In this RF/TLP integrated system, the SPDT relays isolate and select between the ESD pulse source (e.g., charged transmission line), the d.c. measurement equipment (e.g., parametric analyzer) and the RF test equipment. Hynoven, Joshi, and Rosenbaum [9] integrated the first relay as a conventional mercury-wetted relay used in high voltage ESD systems. The second relay utilized is a high-frequency RF switch. In both cases, the activation of the relays are controlled and initiated by programmable voltage sources. In the construction of these test systems, the RF components and cables should allow for RF characterization at the operational frequency of the semiconductor component or semiconductor chip under evaluation. Hence a key issue is that the ESD system and the RF characterization sector must be suitable to allow accurate RF characterization of the component. As a result, this may place limitation on the type of TLP test system and the impedance. TLP systems can

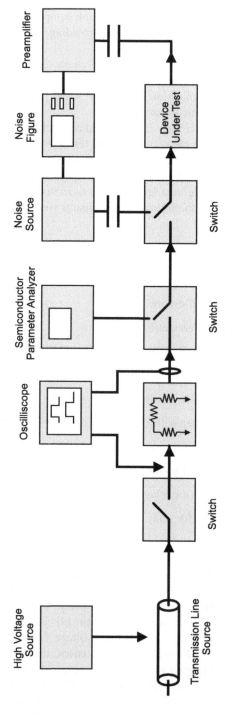

Figure 1.16 Transmission line pulse (TLP) test system with integrated RF characterization

have four classifications: TLP current source, TDR, TDT, and TDRT. For RF measurements, a 50 Ω impedance level must exist in the measurement path, limiting the classes of TLP systems that may be suitable. Additionally, with the integration of the RF characterization and the TLP measurements, the TLP pulse current level must remain under the current limit of the microwave RF probes.

It was noted by Hynoven that one must be conscious of ground loops and ground references in the combined RF/TLP test system. RF measurements require the use of coaxial cables and ground-signal-ground probes to establish a ground plane reference for the signals. Hynoven and Rosenbaum [9] noted that the RF ground establishes an undesired return path for the ESD pulse. This issue is addressed by d.c. blocking capacitors and switches. With the addition of switches and d.c. blocking capacitors, a test sequence can be established that avoids this issue.

An RF ESD test procedure for ESD TLP-stressing of an RF component can be as follows [9]:

- Initialize the state of the system.

- Initiate the charging of the transmission line cable outer conductor.

- Discharge the outer conductor potential to ground.

- Establish the state for launching TLP current pulse.

- Apply the TLP current pulse to the DUT.

- Close the TLP source switches.

- Establish switch states to initiate power to the DUT.

- Apply power to the DUT and enable RF measurement diagnostics.

- Complete RF measurements.

- Re-initiate the test sequence.

1.14 CLOSING COMMENTS AND SUMMARY

In Chapter 1, the text opens with the discussion of ESD and EOS in RF technology. The chapter discusses patents, RF parametrics, the fundamental design techniques, RF ESD metrics, and new RF ESD testing methodologies. Early contributions are reviewed to place today's efforts in the historical context and to highlight the excellent work performed in evaluation of the RF power-to-failure and ESD power-to-failure of microwave electronic components; this was intentionally done to demonstrate that the early work testing techniques, measurements, and modeling are relevant to today's rapid interest in the field of ESD in RF devices. Some things have changed since the 1970s. First, new technologies emerged. Second, the device themselves changed within a technology type. Third, the devices became faster. Fourth, testing methodologies will evolve addressing more application-specific ESD failure criteria. Fifth, there will be larger emphasis on the utilization of RF design techniques, and cosynthesizing the ESD and RF characteristics for optimization.

A plethora of cosynthesis methods of ESD and RF component design is shown for the first time to define an 'RF ESD design practice.' This RF ESD design practice is a blending of the traditional methods used in digital CMOS ESD design practices with RF design practice methods; whereas, in the case of digital ESD solutions both devices and methods may be abandoned when unsuitable for RF design. In digital design, there is a significant usage of resistors; in RF ESD design we will see the significant usage of inductors and capacitors.

Another key point is that the TLP and VF-TLP methodologies are a natural testing method for RF circuits for a few reasons. First, the TLP method is a 50 Ω-based methodology. Second, it is a two-pin test whereas RF parameter extraction are also two-port methodologies. Additionally, the device, circuit and systems can be analyzed in a common method. For future ESD test methods such as the VF-TLP, whereas its original goal was to characterize CDM failure mechanisms, the new value of this method may be for RF components where it may be the utilization for evaluation of short-pulse phenomenon akin to the RF power-to-failure methods utilized by Whalen thirty years ago. As a result, a testing synergy has begun toward the ESD testing using the TLP and VF-TLP methods, with RF characterization.

In Chapter 2, we will discuss RF codesign methods and techniques, as well as address ESD design integration in mixed signal chips that contain digital, analog, and RF subfunctions, within a common chip. In Chapter 2, we will discuss in more depth the specific methods of ESD RF cosynthesis—inductive shunts, impedance isolation, cancellation, linearity optimization, distributed loads, and matching. These practices have been only recently utilized for today's RF technologies. In Chapter 2, we will also discuss how to architecture and floor plan mixed signal chips for ESD. Additionally, some special topics on structure under pads and pad-ESD integration will be briefly discussed.

PROBLEMS

1. Given a diode element, of the width, W, the length, L, and spacing of shallow trench isolation between the $p+$ to $n+$ of spacing, W_{STI} and well sheet resistance, ρ_W, derive the series resistance of the diode element. Assuming the capacitance assuming a capacitance per unit area, C_0, develop a diode model for the series capacitor and series resistor.

2. Assume two-diode elements, where one serves as a low-impedance shunt, and the other is reverse-biased serving as a load. Assume the diode element is a low-impedance shunt of zero capacitance, whereas the second diode element is a capacitive load. Derive the figure of merit (FOM) for the 'dual diode network' using the dimensions of the prior derivation. Assume an ESD robustness per unit micron of V_0 (HBM volts per unit micron).

3. Derive the S-parameters for a diode element between an input and an output terminal. Derive the T-, Y- and Z-parameters.

4. Assuming an ideal HBM source of capacitor C and series resistor R, derive the discharge through a resistor load R_L. Assume the capacitor is pre-charged to voltage V_0, and a switch is between the series resistor and the load resistor.

5. Assuming an ideal HBM source of capacitor C, and series resistor R, derive the discharge through a capacitor load C_L. Assume the capacitor is pre charged to voltage V_0, and a switch is between the series resistor and the capacitor load.

6. Assuming an ideal HBM source of capacitor C, and series resistor R, derive the discharge through an inductor load C_L. Assume the capacitor is pre-charged to voltage V_0, and a switch is between the series resistor and the inductor load.

7. To evaluate the absorbed energy of the RF device, this can be calculated by the difference between the incident energy, and the sum of the reflected and transmitted energy.

$$E_A = E_I - E_R - E_T$$

where the incident energy is

$$E_I = \int P_I(t)\, dt$$

and the reflected energy is

$$E_R = \int P_R(t)\, dt$$

and transmitted energy is

$$E_T = \int P_T(t)\, dt$$

Assuming a transmission line pulse system, derive the power from a TLP pulse assuming a trapezoidal pulse wave form. Assume a perfect 50 Ω impedance through the system. Derive the general relation where there is an impedance mismatch at the device under test.

8. Given an inductor, ESD degradation can induced degradation in the quality factor due to resistance changes. Derive the quality factor of an inductor where a resistance degradation occurs due to an ESD events.

9. Given a capacitor, ESD degradation can induce resistance degradation due to impedance pin–holes. Derive a quality factor of a capacitor assuming a conductance variation due to an ESD event.

REFERENCES

1. Voldman S. *ESD: physics and devices*. Chichester, England: John Wiley & Sons, Ltd.; 2004.
2. Voldman S. *ESD: circuits and devices*. Chichester, England: John Wiley & Sons, Ltd.; 2005.
3. Voldman S. ESD protection and radio frequency (RF) design. *Tutorial Notes of the Electrical Overstress/Electrostatic Discharge (EOS/ESD) Symposium, Tutorial J*. Oregon, Convention Center, Portland, Oregon, September 10, 2001.
4. Voldman S. The impact of technology evolution and revolution in advanced semiconductor technologies on electrostatic discharge (ESD) protection. *Proceedings of the Taiwan Electrostatic Discharge Conference (T-ESDC)*, November 12–14, 2003. p. 1–9.
5. Voldman S, Strang S, Jordan D. An automated electrostatic discharge computer-aided design system with the incorporation of hierarchical parameterized cells in BiCMOS analog and RF

technology for mixed signal applications. *Proceedings of the Electrical Overstress/Electrostatic Discharge (EOS/ESD) Symposium*, October 2002. p. 296–305.

6. Voldman S, Strang S, Jordan S. A design system for auto-generation of ESD circuits. *Proceedings of the International Cadence Users Group*, September 2002.

7. Voldman S. Automated hierarchical parameterized ESD network design and checking system. U.S. Patent No. 5,704,179 (March 9, 2004).

8. Voldman S, Ronan B, Ames S, Van Laecke A, Rascoe J, Lanzerotti L, *et al*. Test methods, test techniques and failure criteria for evaluation of ESD degradation of analog and radio frequency (RF) Technology. *Proceedings of the Electrical Overstress/Electrostatic Discharge (EOS/ESD) Symposium*, October 2002. p. 92–100.

9. Hynoven S, Rosenbaum E. Combined TLP/RF testing system for detection of ESD failures in RF circuits. *Proceedings of the Electrical Overstress/Electrostatic Discharge (EOS/ESD) Symposium*, October 2003. p. 346–53.

10. Leroux P, Steyaert M. High performance 5.2 GHz LNA with on-chip inductor to provide ESD protection. *IEEE Electron Device Letters* 2001;**EDL-37**(7):467–69.

11. Johnson JL. Electrostatic discharge protection using inductors. U.S. Patent No. 6,624,999 (September 23, 2003).

12. Hynoven S, Joshi S, Rosenbaum E. Comprehensive ESD protection for RF inputs. *Proceedings of the Electrical Overstress/Electrostatic Discharge (EOS/ESD) Symposium*; 2003. p. 188–94.

13. Hynoven S, Joshi S, Rosenbaum E. Cancellation technique to provide ESD protection for multi-GHz inputs. *IEEE Electron Device Letters* 2003;**EDL-39**(3):284–6.

14. Ker MD, Lee CM. ESD protection design for GHz RF CMOS LNA with novel impedance isolation technique. *Proceedings of the Electrical Overstress/Electrostatic Discharge (EOS/ESD) Symposium*, 2003. p. 204–13.

15. Ker MD, Chou CI, Lee CM. A novel LC-tank ESD protection design for Giga-Hz RF circuits. *Proceedings of the IEEE Radio Frequency Integrated Circuits (RFIC) Symposium*, 2003. p. 115–8.

16. Ker MD, Lee CM, Lo WY. Electrostatic discharge protection device for giga-hertz radio frequency integrated circuits with varactor-LC tanks. U.S. Patent No. 6,885,534 (April 26, 2005).

17. Ginzton E, *et al*. Distributed amplification. *Proceedings of the IRE*, 1948. p. 956–69.

18. Sarma DG. On distributed amplification. *Proceedings of the Institute of Electrical Engineering*, 1954. p. 689–97.

19. Majidi-Ahy R, Nishimoto CK, Riaziat M, Glenn M, Silverman S, Weng S-L, Pao Y-C, Zdasiuk GA, Bandy SG, Tan ZCH. A 5–100 GHz InP coplanar waveguide MMIC distributed amplifier. *IEEE Transactions on Microwave Theory and Techniques* 1990;**38**(12):1986–94.

20. Sullivan PJ, Xavier BA, Ku WH. An integrated CMOS distributed amplifier utilizing packaging inductance. *IEEE Transactions on Microwave Theory and Techniques* 1997;**45**(10):1969–77.

21. Lee TH. *The design of CMOS radio frequency integrated circuits*. Cambridge University Press; 1998.

22. Kleveland B, Diaz CH, Vook D, Madden L, Lee TH, Wong SS. Monolithic CMOS distributed amplifier and oscillator. *Proceedings of the International Solid State Circuits Conference (ISSCC)*, 1999. p. 70–1.

23. Kleveland B, Lee TH. Distributed ESD protection device for high speed integrated circuits. U.S. Patent No. 5,969,929 (October 19, 1999).

24. Kleveland B, Maloney TJ, Morgan I, Madden L, Lee TH, Wong SS. Distributed ESD protection for high speed integrated circuits. *IEEE Transactions on Electron Device Letters* 2000;**21**(8):390–92.

25. Kleveland B, Diaz CH, Vook D, Madden L, Lee TH, Wong SS. Exploiting CMOS reverse interconnect scaling in multigigahertz amplifier and oscillator design. *IEEE Journal of Solid State Circuits* 2001;**36**:1480–8.

26. Ito C, Banerjee K, Dutton RW. Analysis and design of ESD protection circuits for high frequency/ RF applications. *IEEE International Symposium on Quality and Electronic Design (ISQED)*, 2001. p. 117–22.

27. Ito C, Banerjee K, Dutton R. Analysis and optimization of distributed ESD protection circuits for high-speed mixed-signal and RF applications. *Proceedings of the Electrical Overstress/Electrostatic Discharge (EOS/ESD) Symposium*, 2001. p. 355–63.

28. Ito C, Banerjee K, Dutton R. Analysis and design of distributed ESD protection circuits for high-speed mixed-signal and RF applications. *IEEE Transactions on Electron Devices* 2002;**49**: 1444–54.

29. Worley E, Bakulin A. Optimization of input protection diode for high speed applications. *Proceedings of the Electrical Overstress/Electrostatic discharge (EOS/ESD) Symposium*, 2002. p. 62–72.

30. Kuo BJ, Ker MD. New distributed ESD protection circuit for broadband RF ICs. *Proceedings of the Taiwan Electrostatic Discharge Conference (T-ESDC)*, November 12–13, 2003. p. 163–8.

31. Ker MD, Kuo BJ. New distributed ESD protection circuit for broadband RF ICs. *Proceedings of the Taiwan Electrostatic Discharge Conference (T-ESDC)*, 2003. p. 163–8.

32. Ker MD, Kuo BJ. Optimization of broadband RF performance and ESD robustness by π-model distributed ESD protection scheme. *Proceedings of the Electrical Overstress/Electrostatic Discharge (EOS/ESD) Symposium*, 2004. p. 32–9.

33. Hsiao YW, Kuo BJ, Ker MD. ESD protection design for a 1–10 GHz wideband distributed amplifier in CMOS technology. *Proceedings of the Taiwan Electrostatic Discharge Conference (T-ESDC)*, 2004. p. 90–4.

34. Voldman S, Williams RQ. ESD network with capacitor blocking element. U.S. Patent No. 6,433,985 (August 13, 2002).

35. Richier C. Investigation on different ESD protection strategies devoted to 3.3 V RF applications (2 GHz) in a 0.18 μm CMOS process. *Proceedings of the Electrical Overstress/Electrostatic discharge (EOS/ESD) Symposium*, 2000. p. 251–9.

36. Wunsch DC, Bell RR. Determination of threshold voltage levels of semiconductor diodes and transistors due to pulsed voltages. *IEEE Transactions on Nuclear Science* 1968;**NS-15**(6):244–59.

37. Tasca DM. Pulse power failure modes in semiconductors. *IEEE Transactions on Nuclear Science* 1970;**NS-17**(6):346–72.

38. Anand Y, Morris G, Higgins V. Electrostatic failure of X-band silicon Schottky barrier diodes. *Proceedings of the Electrical Overstress/Electrostatic Discharge (EOS/ESD) Symposium*, 1979. p. 97–103.

39. Vlasov VA, Sinkevitch VF. *Elektronnaya Technika* 1971;**4**:68–75.

40. Brown WD. Semiconductor device degradation by high amplitude current pulses. *IEEE Transactions on Nuclear Science* 1972;**NS-19**:68–75.

41. Minear RL, Dodson GA. The phantom emitter – an ESD-resistant bipolar transistor design and its applications to linear integrated circuits. *Proceedings of the Electrical Overstress/Electrostatic Discharge (EOS/ESD) Symposium*, 1979. p. 188–92.

42. Whalen JJ, Calcatera MC, Thorn M. Microwave nanosecond pulse burnout properties of GaAs MESFETs. *Proceedings of the IEEE MTT-S International Microwave Symposium*, May 1979. p. 443–5.

43. Whalen JJ, Domingos H. Square pulse and RF pulse overstressing of UHF transistors. *Proceedings of the Electrical Overstress/Electrostatic Discharge (EOS/ESD) Symposium*, 1979. p. 140–6.

44. Smith JS, Littau WR. Prediction of thin-film resistor burn-out. *Proceedings of the Electrostatic Overstress/Electrostatic Discharge (EOS/ESD) Symposium*, 1981. p. 192–7.

45. Alexander DR, Enlow EW. Predicting lower bounds on failure power distributions of silicon npn transistors. *IEEE Transactions on Nuclear Science* 1981;**NS-28**(6):145–50.

46. Enlow EN. Determining an emitter-base failure threshold density of npn transistors. *Proceedings of the Electrostatic Overstress/Electrostatic Discharge (EOS/ESD) Symposium*, 1981. p. 145–50.

47. Pierce D, Mason R. A Probabilistic estimator for bounding transistor emitter-base junction transient-induced failures. *Proceedings of the Electrostatic Overstress/Electrostatic Discharge (EOS/ESD) Symposium*, 1982. p. 82–90.

48. Ash M. Semiconductor junction non-linear failure power thresholds: Wunsch-Bell revisited. *Proceedings of the Electrostatic Overstress/Electrostatic Discharge (EOS/ESD) Symposium*, 1983. p. 122–7.

49. Arkhipov VI, Astvatsaturyan ER, Godovosyn VI, Rudenko AI. *International Journal of Electronics* 1983;**55**:395.

50. Dwyer M, Franklin AJ, Campbell DS. Thermal failure in semiconductor devices. *Solid State* 1989;**33**:553–60.

51. Rubalcava L, Stunkard D, Roesch WJ. Electrostatic discharge effects on gallium arsenide integrated circuits. *Proceedings of the Electrical Overstress/Electrostatic Discharge (EOS/ESD) Symposium*, 1986. p. 159–65.

52. Buot FA, Anderson WT, Christou A, Sleger KJ, Chase EW. Theoretical and experimental study of sub-surface burnout and ESD in GaAs FETs and HEMTs. *Proceedings of the International Reliability Physics Symposium (IRPS)*, 1987.

53. Bock K, Fricke K, Krozer V, Hartnagel HL. Improved ESD protection of GaAs FET microwave devices by new metallization strategy. *Proceedings of the Electrical Overstress/Electrostatic Discharge (EOS/ESD) Symposium*, 1990. p. 193–6.

54. DeChairo LF, Unger BA. Degradation in InGaAsP semiconductor lasers resulting from human body model ESD. *Journal of Electrostatics* 1993;**29**:227–50.

55. Voldman S. Semiconductor process and structural optimization of shallow trench isolation-defined and polysilicon-bound source/drain diodes for ESD networks. *Proceedings of the Electrical Overstress/Electrostatic Discharge (EOS/ESD) Symposium*, 1998. p. 151–60.

56. Voldman S, Juliano P, Johnson R, Schmidt N, Joseph A, Furkay S, *et al*. Electrostatic discharge and high current pulse characterization of epitaxial base silicon germanium heterojunction bipolar transistors. *Proceedings of the International Reliability Physics Symposium (IRPS)*, March 2000. p. 310–6.

57. Voldman S, Botula A, Hui D. Silicon germanium heterojunction bipolar transistor ESD power clamps and the Johnson limit. *Proceedings of the Electrical Overstress/Electrostatic Discharge (EOS/ESD) Symposium*, 2001. p. 326–36.

58. Voldman S, Lanzerotti LD, Johnson RA. Influence of process and device design on ESD sensitivity of a silicon germanium hetero-junction bipolar transistor. *Proceedings of the Electrical Overstress/ Electrostatic Discharge (EOS/ESD) Symposium*, 2001. p. 364–72.

59. Ronan B, Voldman S, Lanzerotti L, Rascoe J, Sheridan D, Rajendran K. High current transmission line pulse (TLP) and ESD characterization of a silicon germanium hetero-junction bipolar transistor with carbon incorporation. *Proceedings of the International Reliability Physics Symposium (IRPS)*, 2002. p. 175–83.

60. Ma Y, Li GP. A novel on-chip ESD protection circuit for GaAs HBT RF power amplifiers. *Proceedings of the Electrical Overstress/Electrostatic Discharge (EOS/ESD) Symposium*, 2002. p. 83–91; *Journal of Electrostatics* 2003;**59**:211–27.

61. Shrier K, Truong T, Felps J. Transmission line pulse test methods, test techniques, and characterization of low capacitance voltage suppression device for system level electrostatic discharge compliance. *Proceedings of the Electrical Overstress/Electrostatic Discharge (EOS/ESD) Symposium*, 2004. p. 88–97.

62. Voldman S, Geissler S, Nowak EJ. Semiconductor diode with silicide films and trench isolation. U.S. Patent No. 5,629,544 (May 13, 1997).

63. Lanzerotti LD, Voldman S. Self-aligned SiGe npn with improved ESD robustness using wide emitter polysilicon extensions. U.S. Patent No. 6,441,462 (August 27, 2002).

64. Voldman S. BiCMOS ESD circuit with sub-collector/trench-isolated body mosfet for mixed signal analog/digital RF applications. U.S. Patent No. 6,455,902 (September 24, 2002).

65. Voldman S. BiCMOS ESD circuit with subcollector/trench-isolated body MOSFET for mixed signal analog/digital RF applications. U.S. Patent No. 6,455,902 (September 24, 2002).

66. Brennan CJ, Voldman S. Internally ballasted silicon germanium transistor. U.S. Patent No. 6,455,919. (September 24, 2002).

67. Voldman S. ESD robust silicon germanium transistor with emitter NP-block mask extrinsic base ballasting resistor with doped facet region. U.S. Patent No. 6,465,870 (October 15, 2002).

68. Voldman S. SiGe transistor, varactor, and p-i-n velocity saturated ballasting element for BiCMOS peripheral circuits and ESD networks. U.S. Patent No. 6,552,406 (April 22, 2003).

69. Voldman S. Self-aligned silicon germanium heterojunction bipolar transistor device with electrostatic discharge crevice cover for salicide displacement. U.S. Patent No. 6,586,818 (July 1, 2003).

70. Voldman S. Method and structure for low capacitance ESD robust diodes. U.S. Patent Application No. 20030168701 (September 11, 2003).

71. Lanzerotti L, Ronan B, Voldman S. Silicon germanium heterojunction bipolar transistor with carbon incorporation. U.S. Patent No. 6,670,654 (December 30, 2003).

72. Voldman S. Dual emitter transistor with ESD protection. U.S. Patent No. 6,731,488 (May 4, 2004).

73. Lee JH. Silicon controlled rectifier for SiGe process, manufacturing method thereof and integrated circuit including the same. U.S. Patent No. 6,803,259 (October 2004).

74. Coolbaugh D, Voldman S. Carbon-modulated breakdown voltage SiGe transistor for low voltage trigger ESD applications. U.S. Patent No. 6,878,976 (April 12, 2005).

75. Voldman S. SiGe transistor, varactor and p-i-n velocity saturated ballasting element for BiCMOS peripheral circuits and ESD networks. U.S. Patent No. 6,720,637 (April 13, 2004).

76. Vashenko V, Concannon A, Hopper TJ. ESD protection methods and devices using additional terminal in the diode structures. U.S. Patent No. 6,894,881 (May 17, 2005).

77. Voldman S, Zierak M. Modulated trigger device. U.S. Patent No. 6,975,015 (December 13, 2005).

78. Voldman S. Tunable Semiconductor diodes. U.S. Patent Application No. 200515223 (July 14, 2005).

79. Voldman S. SiGe transistor, varactor, and p-i-n velocity saturated ballasting element for BiCMOS peripheral circuits and ESD networks. U.S. Patent No. 6,552,406 (April 22, 2003).

80. Brennan CJ, Hershberger DB, Lee M, Schmidt NT, Voldman S. Trench-defined silicon germanium ESD diode network. U.S. Patent No. 6,396,107 (May 28, 2002).

81. Voldman S, Ames SJ. Modified current mirror circuit for BiCMOS applications. U.S. Patent No. 6,404,275 (June 11, 2002).

82. Voldman S, Williams RQ. ESD network with capacitor blocking element. U.S. Patent No. 6,433,985 (August 13, 2002).

83. Apel TR, Bonkowski JE, Litzenberg PH. Power amplifier mismatch protection with clamping diodes in RF feedback circuit. U.S. Patent No. 6,459,340 (October 1, 2002).

84. Chang CY, Ker MD. Low substrate-noise electrostatic discharge protection circuits with bi-directional silicon diodes. U.S. Patent No. 6,617,649 (September 9, 2003).

85. Johnson JL. Electrostatic discharge protection using inductors. U.S. Patent No. 6,624,999 (September 23, 2003).

86. Lin HC, Duvvury C, Haroun B. Minimization and linearization of ESD parasitic capacitance in integrated circuits. U.S. Patent No. 6,690,066 (February 10, 2004).

87. Leete JC. Electrostatic protection circuit with impedance matching for radio frequency integrated circuits. U.S. Patent No. 6,771,475 (August 3, 2004).

88. Hatzilambrou M, Leung C, Walia R, Liang LW, Rustagi SC, Radhakrishnan MK. ESD protection system for high frequency applications. U.S. Patent No. 6,801,416 (October 5, 2004).

89. Ker MD, Lee CM, Lo WY. Electrostatic discharge protection device for giga-hertz radio frequency integrated circuits with varactor-LC tanks. U.S. Patent No. 6,885,534 (April 26, 2005).

90. Pequignot J, Sloan J, Stout D, Voldman S. Electrostatic discharge protection networks for triple well semiconductor devices. U.S. Patent No. 6,891,207 (May 10, 2005).

91. Vikram V, Alok G, Tirdad S. ESD protection circuit for use in RF CMOS IC design. U.S. Patent No. 6,894,567 (May 17, 2005).

92. Ikeda H. Negative feedback amplifier with electrostatic discharge protection circuit. U.S. Patent No. 6,900,698 (May 31, 2005).

93. Botula A, Hui DT, Voldman S. Electrostatic discharge power clamp circuit. U.S. Patent No. 6,429,489 (August 6, 2002).

94. Voldman S, Botula A, Hui DT. Electrostatic discharge power clamp circuit. U.S. Patent No. 6,549,061 (April 15, 2003).

95. Voldman S. Variable voltage threshold ESD protection. U.S. Patent No. 6,552,879 (April 22, 2003).

96. Voldman S. Electrostatic discharge input and power clamp circuit for high cutoff frequency technology radio frequency (RF) applications. Patent Application No. 20050161743 (July 2005).

97. Colclaser R. Integrated circuit with removable ESD protection. U.S. Patent No. 6,327,125 (December 4, 2001).

98. Ziers DG. Coaxial cable ESD bleed. U.S. Patent 6,217,382 (April 17, 2001).

99. Voldman S, Johnson R, Lanzerotti LD, St. Onge SA. Deep trench-buried layer array and integrated device structures for noise isolation and latch up immunity. U.S. Patent No. 6,600,199 (July 29, 2003).

100. Wang A. Bonding pad-oriented all-mode ESD protection structure. U.S. Patent No. 6,635,931 (October 21, 2003).

101. Singh R, Voldman S. Method and apparatus for providing ESD protection and/or noise reduction in an integrated circuit. U.S. Patent Application No. 20030214767, (November 20, 2003).

102. Singh R, Voldman S. Method and apparatus for providing noise suppression in an integrated circuit. U.S. Patent No. 20030214348 (November 20, 2003).

103. Ker MD, Jiang HC. Low-capacitance bonding pad for semiconductor device. U.S. Patent No. 6,717,238 (April 6, 2004).

104. Singh R, Voldman S. Method and apparatus for providing ESD protection and/or noise reduction in an integrated circuit. U.S. Patent No. 6,826,025 (November 30, 2004).

105. Perez CN, Voldman S. Method of forming guard ring parameterized cell structure in a hierarchical parameterized cell design, checking and verification system. U.S. Patent Application No. 20040268284 (December 30, 2004).

106. Collins D, Jordan D, Strang S, Voldman S. ESD design, verification, and checking system and method of use. U.S. Patent Application 20050102644 (May 12, 2005).

107. Eshun E, Voldman S. High tolerance TCR balanced high current resistor for RF CMOS and RF SiGe BiCMOS applications and Cadence-based hierarchical parameterized cell design kit with tunable TCR and ESD resistor ballasting feature. Patent Application No. 2005156281 (July 21, 2005).

108. ESD Association. ESD STM 5.1-2001: Standard test method for electrostatic discharge sensitivity testing-human body model (HBM) component level, 2001.

109. Automotive Electronics Council. Human body model electrostatic discharge test. AEC-Q100-002-Rev C, 2001.

110. JEDEC Solid State Technology Association. JEDS22-A114-B: Electrostatic discharge (ESD) sensitivity testing human body model (HBM), 2000.

111. Van Roozendaal LJ, Ameresekera EA, Bos P, Ashby P, Baelde W, Bontekoe F, *et al.* Standard ESD testing of integrated circuits. *Proceedings of the Electrical Overstress/Electrostatic Discharge (EOS/ESD) Symposium*, 1990. p. 119–30.

112. ESD Association. ESD STM 5.2-1998: Standard test method for electrostatic discharge sensitivity testing-machine model (MM) component level, 1998.

113. Automotive Electronics Council. Machine model electrostatic discharge test. AEC-Q100-003-Rev E, 2001.

114. JEDEC Solid State Technology Association. JEDS22-A115-A: Electrostatic discharge (ESD) sensitivity testing machine model (MM), 2000.

115. ESD Association. ESD TR 10-00: Machine model (MM) electrostatic discharge (ESD) investigation – reduction of pulse number and delay time, 2000.

116. ESD Association. ESD STM 5.3.1-1999: Standard test method for electrostatic discharge sensitivity testing-charged device model (CDM) component level, 1999.

117. Automotive Electronics Council. Charged device model electrostatic discharge test. AEC-Q100-011-Rev A, 2001.
118. JEDEC Solid State Technology Association. JEDS22-C101-A: Field-induced charged device model (CDM) test method for electrostatic discharge-withstand thresholds of microelectronic components, 2000.
119. Intel Corporation. Cable discharge event in local area network environment. White Paper, Order No: 249812-001, July 2001.
120. Brooks R. A simple model for the cable discharge event. IEEE802.3 Cable-Discharge Ad hoc, March 2001.
121. Telecommunications Industry Association (TIA). Category 6 Cabling: static discharge between LAN cabling and data terminal equipment. Category 6 Consortium, December 2002.
122. Deatherage J, Jones D. Multiple factors trigger discharge events in Ethernet LANs. *Electronic Design* 2000;**48**(25):111–6.
123. Pommeranke D. Charged cable event, February 23, 2001.
124. Geski H. DVI compliant ESD protection to IEC 61000-4-2 level 4 standard. *Conformity* September 2004;12–7.
125. ESD Association. ESD SP5.5-2004: Standard practice for electrostatic discharge sensitivity testing-transmission line pulse (TLP) component level, 2004.
126. Barth J, Richner J, Verhaege K, Henry LG. TLP-Calibration, correlation, standards, and new techniques. *Proceedings of the Electrical Overstress/Electrostatic Discharge (EOS/ESD) Symposium*, 2000. p. 85–96.
127. Barth J, Richner J. Correlation considerations: real HBM to TLP and HBM testers. *Proceedings of the Electrical Overstress/Electrostatic Discharge (EOS/ESD) Symposium*, 2001. p. 453–60.
128. Keppens B, de Heyn V, Natarajan Iyer M, Groeseneken G. Contribution to standardization of transmission line pulse methodology. *Proceedings of the Electrical Overstress/Electrostatic Discharge (EOS/ESD) Symposium*, 2001. p. 461–7.
129. Voldman S, Ashton R, McCaffrey B, Barth J, Bennett D, Hopkins M, *et al.* Standardization of the transmission line pulse (TLP) methodology for electrostatic discharge. *Proceedings of the Electrical Overstress/Electrostatic Discharge (EOS/ESD) Symposium*, 2003. p. 372–81.
130. Gieser H. *Verfahren zur charakterisierung von iintegrierten schaltungen mit sehr schnellen hochstromimpulsen.* Technische Universitaet Muenchen TUM, PhD Dissertation. Aachen, Germany: Shaker-Verlag; 1999.
131. Gieser H, Haunschild M. Very-fast transmission line pulsing of integrated structures and the charged device model. *Proceedings of the Electrical Overstress/Electrostatic Discharge (EOS/ESD) Symposium*, 1996. p. 85–94.
132. Juliano P, Rosenbaum E. Accurate wafer-level measurement of ESD protection device turn-on using a modified very fast transmission line pulse system. *IEEE Transactions of Device and Material Reliability (TDMR)* 2001;**1**(2):95–103.
133. ESD Association. ESD SP5.5-2004: Standard practice for electrostatic discharge sensitivity testing-very fast transmission line pulse (VF-TLP) component level. Draft D, 2006.

2 RF ESD Design

The 'RF ESD design practice' is a new design practice for ESD protection of semiconductor chips. Before 1999, few publications addressed the issue of radio frequency (RF) design and ESD design in a synthesized fashion. In the past, the focus had been on the quantification of semiconductor processing sensitivities and device characteristics of RF devices [1–3]. Between 1999 and 2005, the growth of interest in ESD protection of RF components increased rapidly [1–34]. In this chapter, the RF design and the design methods of ESD protection are synthesized to establish a new way of thinking compared to the traditional thought processes of the digital CMOS ESD design. RF ESD codesign methods and techniques, as well as ESD design integration in mixed signal chips that contain digital, analog, and RF subfunctions within a common chip will be reviewed. Specific methods of RF ESD cosynthesis—inductive shunts, impedance isolation, cancellation, linearity optimization, distributed loads, and matching opens the door to new design practices, circuits, and analyses.

2.1 ESD DESIGN METHODS: IDEAL ESD NETWORKS AND RF ESD DESIGN WINDOWS

In evaluating the effectiveness of an ESD network, an 'ESD Design Window' can be defined in both the direct current (d.c.) $I–V$ characteristics and the RF characteristics. For the d.c. $I–V$ characteristics may determine the 'on' and 'off' characteristics of the ESD network during functional operation and its ESD effectiveness as an ESD network to protect other circuitry. Additionally, the ESD device or circuit $I–V$ characteristics has certain ideal desirable features during d.c. operation of the semiconductor chip. In the case of the RF characteristics, the impedance of the element in the frequency domain determines the loading effect of the ESD network.

2.1.1 Ideal ESD Networks and the Current–Voltage d.c. Design Window

As the d.c. $I–V$ characteristics may determine the 'on' and 'off' characteristics of the ESD network during functional operation and its ESD effectiveness as an ESD network to protect

ESD: RF Technology and Circuits   Steven H. Voldman
© 2006 John Wiley & Sons, Ltd

other circuitry, an ideal ESD network has the following characteristics:

- The ESD device, circuit, or network is 'off' during the d.c. functional regime between signal levels as well as between the most negative power supply voltage and the most positive power supply (associated with the signal pin) voltage.

- The ESD network has an 'infinite resistance' when it is in the 'off' state, which can be expressed as

$$\left.\frac{dI}{dV}\right|_{off} = \frac{1}{R} = 0$$

- The ESD network is 'on' during voltage excursions that undershoot below the most negative power supply, or during voltage excursions that overshoot the most positive power supply (during ESD testing).

- The ESD network has an 'zero resistance' when

$$\left.\frac{dI}{dV}\right|_{on} = \frac{1}{R} = \infty$$

- The ESD network operation does not extend beyond the 'electrical safe-operation area' (electrical SOA) in d.c. voltage level or d.c. current level.

- The ESD network operation does not extend beyond a 'thermal safe-operation area' (thermal SOA) in d.c. voltage level or d.c current level.

ESD networks can consist of $I–V$ characteristics of the following form:

- Step function $I–V$ characteristics;

- S-type $I–V$ characteristics;

- N-type $I–V$ characteristics.

Step function $I–V$ characteristics have a single 'off' state as the structure is biased. At some voltage value, the device is 'on'. For example, a diode element has an step function $I–V$ characteristic and is suitable for ESD protection. In the case of a diode element, the ideality is a function of the on-resistance of the diode element. Figure 2.1 shows an example of a diode and an ideal ESD device in the DSD design window. The ideal ESD device has infinite resistance for voltages below the power supply voltage, and zero resistance for voltages greater then the power supply voltage. Because of the nonideality of the diode element, there is a region where the d.c. voltage of the semiconductor devices in the technology are exceeded.

S-type characteristics are semiconductor devices or circuits that have two current states for a given voltage state. For example, an n-channel MOSFET or silicon-controlled rectifier (e.g., $pnpn$ device) has an S-type $I–V$ characteristic. Figure 2.2 shows an example of an n-channel MOSFET in a MOSFET drain-to-source configuration in an ESD design window. Figure 2.2 also shows the ideal ESD device. To utilize the MOSFET as an ESD network, the MOSFET snapback must occur within the current–voltage window of the technology limits of its safe operation area (SOA) of the other structures in the technology.

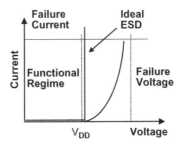

Figure 2.1 ESD design window for an ESD device (e.g., single state current–voltage characteristic)

2.1.2 Ideal ESD Networks in the Frequency Domain Design Window

From an RF ESD design perspective, the characteristics of an RF ESD design are focused on its RF characteristics at the RF application frequency. From an RF perspective, the ideal RF ESD network has the following features:

- An ideal RF ESD network would have zero impedance during an ESD pulse.

- An ideal RF ESD network has infinite impedance during RF functional applications.

- An ideal RF ESD network during RF functional operation would not be a function of the current or voltage conditions.

- An ideal RF ESD network during RF functional operation is not to be temperature dependent.

Hence from the perspective of frequency design window for RF ESD design, it is desirable to have the RF ESD network to have zero impedance at low frequencies (e.g., HBM, MM, and CDM phenomenon regime below 5 GHz), and high impedance in the application frequency (e.g., RF application frequency regime).

Figure 2.3 shows the RF ESD frequency design window and an ideal impedance characteristic imposed on the window. In a frequency regime below the CDM phenomenon, it would be desirable for the ideal ESD device to be zero impedance, and infinite impedance at the application frequency. In the figure, the capacitance and inductance as a function of

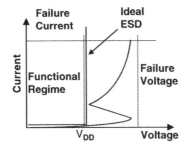

Figure 2.2 ESD design window for a S-type *I–V* characteristic ESD device

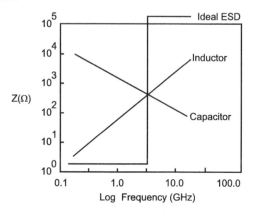

Figure 2.3   RF ESD frequency domain design window and the RF ESD device impedance

frequency is also shown. As can be observed, the capacitance dominant ESD elements has high impedance during ESD events, and low impedance at application frequencies; this is opposite of the desired device. Also note that an ESD element that has an inductor-dominated property is more consistent with the characteristics of an ideal ESD element.

2.2 RF ESD DESIGN METHODS: LINEARITY

In RF design, linearity is an important design criterion [8]. Circuits, whose capacitor, resistor, and inductor elements are not voltage dependent provide a high degree of linearity. But circuit elements whose capacitance values are voltage dependent influence the linearity of networks. Metallurgical junctions, which exist in MOSFETs, p–n diodes, and bipolar devices, are voltage dependent and hence influence the linearity of RF networks. Therefore, ESD protection networks that consist of diodes, bipolar, and MOSFET elements influence the linearity of networks at both the input and the output port of RF networks.

The effect of the ESD network on RF circuit is a function of the following:

- *Circuit Topology Symmetry*: Symmetry of the topology of the ESD network elements relative to the power supplies.

- *ESD Element Type Symmetry*: Symmetry of the ESD element used within the ESD network.

- *D.C. Voltage Condition*: The d.c. voltage state and the voltage differential within the ESD network can influence the symmetry of the voltage dependence and capacitance values of the elements.

Circuit Topology Symmetry

ESD circuit topology that contains a symmetry of element within the circuit topology can influence the voltage dependence in such a fashion to lower the linearity sensitivity. In the case where only one element exists, there is only a single voltage sensitivity function. But, in the case of multiple capacitor elements it is possible that the capacitance increases in one element and decreases in the second.

Figure 2.4 shows an example of a bipolar ESD network capacitance as a function of the applied voltage. As the voltage increases, the capacitance value changes. In the figure, this is

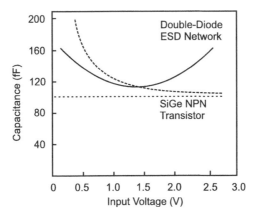

Figure 2.4   Voltage dependence an ESD network, comparing a single bipolar transistor ESD network and a double-diode ESD network

compared to an ESD double-diode network. An advantage of the double-diode network is that as the applied voltage changes, the voltage of one of the physical elements across the junction increases, while that of the other element decreases. Hence, the voltage dependence of the ESD network has been internally compensated to lessen the impact of the d.c. voltage swing [8]. Thus, in the case of a single ESD element, the topology asymmetry of the ESD network leads to a noncompensating influence, whereas in a dual-diode ESD network, which contains two elements where the capacitance value of one element decreases while that of the other increases, this leads to compensating influence.

ESD Element Type Symmetry

In many ESD networks, with a plurality of elements, the ESD elements are nonidentical. When the ESD diode elements are not identical, the built-in voltage, exponential term, and the capacitance per unit area of the capacitive elements will be different.

For example, in CMOS applications, a p^+/n-well diode is utilized for a first diode to the power supply rail, and an n-well/p-substrate diode is used to the ground potential. In this case, the two diode element have significantly different capacitance per unit area, as well as the voltage dependence. Additionally, with process variations, there will be no process-induced capacitance variation tracking.

Another implementation used in BiCMOS technology is that ESD diodes can be identical and provide ESD robustness symmetry to both power rails, as well as independence from the substrate doping concentration. In this case a single p^+/n-well diode element is used with the anode connected to the input and cathode connected to the V_{DD} power supply, and a second diode cathode connected to the input, and the cathode connected to the ground rail. In this implementation, the junction voltage dependence is matched and identical; this leads to process tracking.

ESD Voltage Bias State Symmetry

In the ESD network, the d.c. voltage state will influence the capacitance for a given voltage state. Hence, the choice of the voltage bias state can introduce symmetrical or asymmetrical capacitance variation.

Capacitance Evaluation

These conditions will be more clear by evaluating the capacitance relationships. A junction capacitance can be expressed as

$$C = AC_{jo}\left\{1 + \frac{V}{V_{bi}}\right\}^{-n}$$

where A is the junction area, C_{jo} is the junction capacitance per unit area at zero voltage bias, V_{bi} is the built-in voltage, V is the bias voltage across the junction, and n is a real number of the exponential relationship.

Capacitance Evaluation of Nonidentical Circuit Elements

Assuming a dual-diode RF ESD network, for a first ESD diode from the input pad to the V_{CC} power rail, the capacitance–voltage relationship can be expressed as [33]

$$C_1 = A_1 C_{jo1}\left\{1 + \frac{V_{CC} - V_{input}}{V_{bi}}\right\}^{-n_1}$$

For the second ESD diode element, the second diode element capacitance–voltage relationship can be expressed as follows:

$$C_2 = A_2 C_{jo2}\left\{1 + \frac{V_{input} - V_{SS}}{V_{bi2}}\right\}^{-n_2}$$

These capacitance of the first and second diode elements can be summed in order to determine the total capacitance. In this case, topology symmetry is established by an utilization of an element relative to the two different power rails

$$C_T = A_1 C_{jo1}\left\{1 + \frac{V_{DD} - V_{input}}{V_{bi1}}\right\}^{-n_1} + A_2 C_{jo2}\left\{1 + \frac{V_{input} - V_{SS}}{V_{bi2}}\right\}^{-n_2}$$

In many ESD networks, the two ESD diodes are not identical; when the ESD diode elements are not identical, the built-in voltage, exponential term, and the capacitance per unit area will be different.

Circuit Element Symmetry: When the two ESD double-diodes are identical, the area factor, the capacitance per unit area, the built-in voltage, and the power of the exponent are the identical. The distinction of the two terms is the d.c. voltage condition. As a result, the capacitance expression can be combined to form a more compact relationship. From the capacitance expression terms, these can be expressed as a series expansion about a given voltage state [33],

$$C_1 = AC_{jo}\left\{1 + \frac{V_{CC} - V_{input}}{V_{bi}}\right\}^{-n} = \sum_{i=0}^{\infty} c_i (V_{CC} - V_{input})^i$$

$$C_2 = AC_{jo}\left\{1 + \frac{V_{input} - V_{SS}}{V_{bi}}\right\}^{-n} = \sum_{i=0}^{\infty} c_i (V_{input} - V_{SS})^i$$

From this development, the terms can be combined as follows:

$$C_T = \sum_{i=0}^{\infty} c_i (V_{CC} - V_{input})^i + \sum_{i=0}^{\infty} c_i (V_{input} - V_{SS})^i$$

Note that in general, the two voltage expressions are nonidentical in value as the d.c. state can be set between any point between the V_{CC} and the V_{SS} power rails.

Voltage Bias Symmetry

Note that in the capacitance expression, as the d.c. input voltage state is equal to the midpoint between the power supply V_{CC} and the ground power supply V_{SS}, these two expressions are identical and can be combined as a function of the input voltage (e.g., additional case of d.c. voltage symmetry). Letting the input voltage state be expressed as a d.c. component and an a.c. component we have,

$$V_{input} = V_{IN} + v$$

and

$$V_{IN} = \frac{V_{CC} - V_{SS}}{2}$$

These two expressions can be combined to evaluate the total capacitance of the network as a Taylor expansion around the mid-point, where the first term is the first diode element, and the second term is the second diode element,

$$C_T(v) = \sum_{i=0}^{i=\infty} [c_i(-v)^i + c_i(v)^i]$$

Leroux and Steyaert [33] noted that in this form, the terms can be combined, and the final expression is an even function of the small signal gate voltage, or

$$C_T(v) = 2\sum_{i=0}^{i=\infty} c_{2i}(v)^{2i}$$

It was also additionally noted that the small signal current is an odd function of the input voltage [33],

$$i_D = i_{D1} + i_{D2} = svC_T(v) = 2s\sum_{i=0}^{i=\infty} c_{2i}(v)^{2i+1}$$

From this analysis, the interesting feature shows that the symmetry of the ESD elements between the power supplies can influence the capacitance, the voltage dependence, and the even and odd functional nature of the small signal response (e.g., current and voltage). The influence on the even and odd functional nature will then influence the power transfer to different RF harmonics.

An ESD RF design practice, for improved circuit linearity, is as follows:

- RF ESD networks that have topology symmetry, with multiple elements, can lead to an improved RF linearity owing to the complimentary behavior of two elements within a given network.

- RF ESD networks that utilize multiple ESD elements, which are identical, whose bias relationship is complimentary (e.g., one has increasing voltage bias when the other has a decreasing voltage bias) can lead to improved linearity and an improved statistical tracking of elements.

- RF ESD networks that utilize topology symmetry, identical elements, and d.c. bias symmetry can lead to influences on the capacitor relationship; the capacitance is an even function of the small signal voltage, with the diode current being an odd function about the bias state.

2.3 RF ESD DESIGN: PASSIVE ELEMENT QUALITY FACTORS AND FIGURES OF MERIT

Quality factors, known typically as 'Q', are important in RF applications to address the element primary function and the parasitic characteristics. In ESD RF circuit design, the ESD elements can be viewed as follows [10]:

- RF ESD elements can be treated as a passive or an active element.

- RF ESD elements can be treated as the parasitic element of a passive element that impacts the quality factor of the passive element.

- RF ESD elements can be treated as the parasitic element of an active element that impacts the quality factor of an active element.

- RF ESD elements can be cosynthesized as parallel or series elements that are additive to the parasitic component of an active or a passive element.

- RF ESD elements can be cosynthesized as parallel or series elements that are subtractive (e.g., cancellation) to the parasitic component of an active or a passive element.

Hence, from a circuit representation, the perceived nature of the circuit component or element that degrades the Q can be viewed as the ESD element itself. Hence, from an RF ESD design circuit perspective, the ESD element can be viewed as an element that leads to quality factor degradation of an active circuit component, or full circuit representation. At the same time, and ESD RF design practice for ESD RF cosynthesis, the ESD element can be viewed as an additive or a subtractive component to the active or passive element; hence it can be additive (degradation through larger component) or substractive (enhancement through cancellation component).

From the complex series impedance,

$$Z_s = R_s \pm jX_s$$

the quality factor, Q, can be defined as follows:

$$Q = \frac{X_s}{R_s}$$

From the complex parallel admittance,

$$Y_p = G_p \pm B_p$$

where the quality factor, Q, can be represented as follows:

$$Q = \frac{B_p}{G_p}$$

The quality factor for a capacitor with a series resistor element, R_s, and a series capacitor, C_s, can be expressed as follows:

$$Q_C = \frac{1}{\omega C_s R_s}$$

The quality factor for a capacitor with a parallel resistor element, R_p, and a parallel capacitor, C_p, can be expressed as follows:

$$Q_C = \omega C_p R_p$$

The quality factor for an inductor with a series resistor element, R_s, and a series inductor, L_s, can be expressed as follows:

$$Q_L = \frac{\omega L_s}{R_s}$$

whereas, a resistive shunt in parallel with a parallel inductor element has a quality factor expressed as follows:

$$Q_L = \frac{R_p}{\omega L_p}$$

In all these representations, the real (resistive) component can represent the ESD element, or the imaginary (capacitance or inductive) component can represent the ESD element, or both real and imaginary components can represent the ESD element. In another perspective, one part (either real or imaginary) can represent the ESD element and the other can represent a passive or an active circuit element.

For a circuit with both the inductor element, L, and a capacitor element, C, at the resonant frequency [27–29],

$$\left. \frac{1}{Q_{LC}} \right|_{\omega=\omega_0} = \frac{1}{Q_L} + \frac{1}{Q_C}$$

or,

$$Q_{LC}|_{\omega=\omega_0} = \frac{Q_L Q_C}{Q_L + Q_C}$$

At resonance, the impedance of the parallel inductor, capacitor, and shunt resistor element is the value of the shunt resistance,

$$Z|_{\omega=\omega_0} = R_p$$
$$Q_{LC}|_{\omega=\omega_0} = \omega_0 C R_p$$

Solving for the shunt resistance, R_p,

$$R_p = \frac{1}{\omega_0 C} \frac{Q_L Q_C}{Q_L + Q_C}$$

A common ESD design metric is the ratio of the ESD robustness and the capacitance loading. A common objective is to maximize the ESD robustness for a given amount of loading capacitance. A generalization of this ESD metric, a figure of merit can be expressed as the ratio of the ESD robustness and the shunt impedance [27–29,32],

$$\text{FOM} = Z_{\text{shunt}} V_{\text{ESD}}$$

Generalizing this, using the complex form of the series impedance,

$$\text{FOM} = (R_s \pm jX_s)_{\text{shunt}} V_{\text{ESD}}$$

The generalization of this into admittance form,

$$\text{FOM} = \frac{V_{\text{ESD}}}{Y_{\text{shunt}}}$$

or,

$$\text{FOM} = \frac{V_{\text{ESD}}}{G_p \pm jB_p}$$

This can be easily understood for the case of the capacitor element, where the ESD robustness is divided by the reactance,

$$\text{FOM} = \frac{V_{\text{ESD}}}{\omega C_{\text{ESD}}}$$

In the case of the resonant LC network at resonance, we can express the FOM as [27–29,32],

$$\text{FOM} = \frac{Q_L Q_C}{Q_L + Q_C} \left\{ \frac{V_{\text{ESD}}}{\omega_0 C} \right\}$$

2.4 RF ESD DESIGN METHODS: METHOD OF SUBSTITUTION

In RF design, passive elements exist in the peripheral circuits for biasing d.c. or a.c. isolation and matching networks. Inductors and capacitors exist in both parallel and shunt configurations. These passive elements have low-loss and high-quality factors (Q). Passive element can serve a dual role of serving as a RF matching element and an ESD protection [10].

Figure 2.5   Substitution of a capacitor passive element for an ESD element

2.4.1 Method of Substitution of Passive Element to ESD Network Element

On-chip ESD elements have capacitance values of a magnitude to serve as capacitor passive elements. On the contrary, ESD elements typically have low inductance. Hence an ESD design method be the substitution of capacitor passive elements for ESD elements. An RF ESD method of substitution of passives for ESD elements with the constraint of equivalent capacitance can be as follows [10]:

- *Full Substitution With Total Capacitance Constraint:* A passive capacitor element can be replaced by an ESD element, where in the substitution process, the size of the ESD element is equal to that of the capacitance element at the same d.c. bias condition (Figure 2.5).

- *Partial Substitution With Total Capacitance Constraint:* An ESD element is placed in parallel to the passive capacitor element, where the substitution of the new capacitor and the ESD element is equal to the initial passive capacitor element. The final capacitance at the bias voltage is equal to the original capacitance of the initial passive element (Figure 2.6).

In some application, a higher concern is the quality factor that is more important for the RF design metric. In this case, the substitution process is established under the constraint of an equivalent quality factor. Typically, ESD element quality factors are less ideal elements, hence the substitution can impact the total quality factor. Therefore, an ESD design method can be the substitution of capacitor passive elements for ESD elements where the sizes are determined on the quality factor and not on the total capacitance. An RF ESD method of

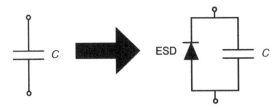

Figure 2.6   Partial substitution of a capacitor passive element for an ESD element and capacitor element

substitution of passives for ESD elements with the constraint of equivalent quality factor, Q, can be as follows [10]:

- *Full Substitution with Total Quality Factor Constraint*: A passive capacitor element can be replaced by an ESD element, where in the substitution process, the size of the ESD element size is equal to the quality factor, Q, of the initial capacitor at the same d.c. bias condition.

- *Partial Substitution with Total Quality Factor Constraint:* A passive capacitor element can be partially replaced by an ESD element, where in the substitution process, the size of the ESD element and the parallel capacitor element is determined by the quality factor, Q, of the initial capacitor at the same d.c. bias condition.

2.4.2 Substitution of ESD Network Element to Passive Element

In the ESD RF cosynthesis process, ESD networks are in series or in parallel configuration with the input and output elements. ESD element can be fully or partially substituted for passive element when the ESD elements are in a parallel configuration with a desired passive element. In this fashion, ESD elements can serve a dual role as a parallel or series passive element, RF matching element, and an ESD protection. ESD elements have both resistive and capacitive nature, and low inductance. Hence a 'method of substitution' of conversion of ESD elements serving as RF passive elements is evolved [10]

Hence, an ESD design method of substitution is the substitution of the ESD elements with capacitor passive elements. A method, which fully substituted the ESD function to an RF passive element, such as a capacitor, may lead to a significant reduction of the ESD robustness. Hence, it would be desirable to have a partial substitution of the ESD element to a passive capacitor element.

An RF ESD method of substitution of ESD elements for passive elements can be as follows [10]:

- *Full Substitution:* An ESD element can be substituted by a passive element, where in the substitution process, the size of the ESD element is equal to that of the capacitance element at the same d.c. bias condition. This method must allow for additional ESD solutions through other elements in the ESD network, or other passive elements (e.g., inductors; Figure 2.7).

- *Partial Substitution:* An ESD element is substituted for a parallel passive capacitor element, where in the substitution process, the new capacitor and the ESD element is equal to (or less than) the initial ESD element capacitance. The final capacitance at the bias voltage is equal to (or less than) the original capacitance of the initial ESD element (Figure 2.8).

In most cases, the substitution of ESD element for an ideal capacitor or inductor will lead to a higher quality factor. Hence, it is also desirable to provide an equivalent or higher quality factor in the full or partial substitution.

Figure 2.7   Substitution of an ESD element to a passive element

2.5 RF ESD DESIGN METHODS: MATCHING NETWORKS AND RF ESD NETWORKS

In RF design, the matching of the input and output of an RF circuit is critical. One of the critical problems of not cosynthesizing the ESD network with the RF circuit is the impact on the RF functional parameters. Given that the ESD network and RF circuit are not cosynthesized, the impact of the ESD element will serve as a RF degradation, or the size of the ESD element is compromised so as to avoid RF functional degradation.

An interesting synergy of design practices can be established for RF ESD design by integration of the RF matching methodology with the ESD design practices [10,13–16,33].

An ESD RF matching codesign methodology can be established as follows:

- Matching network has a predefined topology and elements, whose values are readjusted after the addition of an ESD network to provide impedance matching with the pad node.

- Matching network has a topology that uses at least one of the ESD elements in the network to become part of the desired matching topology.

- Matching network consists of all ESD elements whose ESD circuit topology is defined to provide a desired RF matching topology.

- Matching network has a predefined topology and elements, but with the addition of the ESD elements it forms a higher order or different RF matching topology (note that the magnitude of the elements can be readjusted for desired matching).

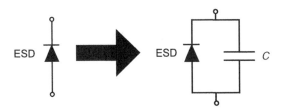

Figure 2.8   Partial substitution of an ESD element to a parallel ESD and passive element

2.5.1 RF ESD Method – Conversion of Matching Networks to ESD Networks

Some topologies of matching networks are more suitable to utilize the ESD elements as part of the RF matching network. RF matching topologies that can utilize ESD elements are the following [10,33]:

- L-matching networks;

- T-matching networks;

- Π-matching networks.

In all the matching networks, the various segments of the matching network are either passive elements (e.g., capacitor or inductor), or ESD elements serving the role of a capacitor, or shunt resistor. In these matching networks, the ESD elements can be utilized as the shunt elements between the signal and ground signals (e.g., the V_{CC} signal), and are typically not desirable in the signal path between the input and the output signals through the filter element. In all these networks, the ESD element can serve as an ideal capacitor, where the element can be in different ESD networks (e.g., MOSFET, bipolar transistor, diodes, or silicon-controlled rectifier elements).

An example of an RF-ESD cosynthesized of an L-match network, a desired element between the input and output signals of the two-port network is an inductor element (Figure 2.9). This is followed by a shunt capacitor to the ground. The shunt capacitor represents an ESD element, which is an ideal element [10]. Another implementation is to utilize an ESD shunt inductor and a series capacitor element, where the capacitor is not an ESD element but serves in the network for d.c. blocking [33].

In an example of an RF-ESD cosynthesized T-match network, the two elements in series are split by a shunt element to the ground nodes of the two-port network (Figure 2.10). The two elements in series can be inductor elements (between the input and output of the match circuit); the shunt element to ground node is a capacitor, which is an idealization of the ESD network (e.g., serving as a RF capacitor element). In the opposite configuration, the two series elements can be capacitor elements, serving as d.c. block elements, and an ESD shunt inductor element.

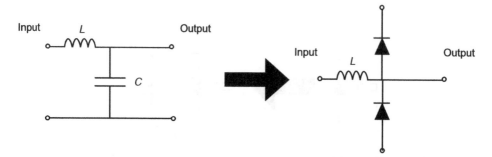

Figure 2.9 L-match networks comprising passive elements and ESD elements

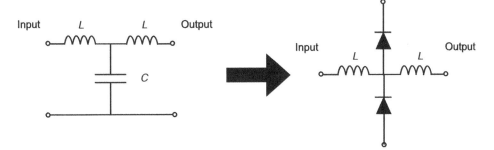

Figure 2.10   T-match networks comprising of passive elements and ESD elements

In an example of a Π-network, as in the other implementation, an inductor is used between the input and output port of the matching network (Figure 2.11). In the Π-network, the two shunt elements can be capacitor elements. In this case, the first capacitor element can be the ESD element, whereas the second element can be a passive capacitor element. In the second case, the two capacitors can be an ESD diode element forming a multi-stage ESD network with the first- and second-stage ESD network. In the complementary implementation, a series capacitor can serve as the d.c. blocking component in the match network, and the first inductor can serve as an ESD inductive shunt, and the second can be used to tune and match the complete network. In all these implementation, the ESD network and passive elements serve as elements of the matching network.

In these matching networks, when ideal passive elements are used, the ESD robustness of the network is a function of the ESD robustness of the inductor and capacitor elements. For example, the inductor design must be chosen to avoid inductor metal failure or electrical breakdown of the insulator regions. In the case of capacitor elements, the capacitor elements must be able to sustain the voltage without dielectric failure.

In RF matching network, when the size of the passive elements are on the same order as the ESD source, the RF matching network itself can influence the ESD response. For example, given an ideal shunt capacitor in the L-, T- or Π-configuration, as the capacitor value approaches the source capacitor element of the cassette model (e.g., 10 pF), or human body model (e.g., 100 pF), the response of the circuit is influenced by the RF matching

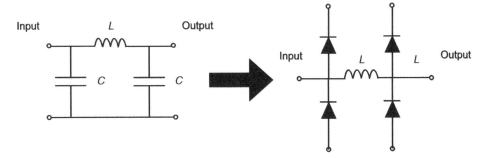

Figure 2.11   Π-match networks comprising of passive elements and ESD elements

network, and the precharged source capacitor serves to charge the RF matching network; this can lead to the second discharge process of the RF matching network into the RF circuit. For, example, given a matching filter where the RF matching inductor is 1 nH, and the capacitor is 1 pF, a 10 pF ESD capacitor source charges the 1 pF shunt capacitor element during the ESD discharge process; this is followed by a discharge of the 1 pF RF matching capacitor into the input network. Note that in the case the shunt element is an ESD forward-bias diode element, this secondary response process will not occur. Additionally, this secondary response can be limited by additional elements, such as resistor elements or higher order filter networks (e.g., additional resistor and series capacitor elements).

An RF ESD design methodology can be as follows:

- Matching networks such as L-, T-, and Π-match networks, are suitable for cosynthesis and optimization of ESD and RF characteristics, where the shunt elements in the matching network can be an ESD element whose purpose serves as a capacitor for matching, and as a shunt element for current discharge to the power rails.

- Matching networks can combine ESD and RF passive elements, where the matching network components can be pure passive elements, have at least one element from the ESD network, or the entire ESD network itself.

- Higher order matching networks can be formed by cosynthesizing the match networks and ESD circuits.

- RF matching networks using ideal capacitor elements as shunt elements can serve as to secondary discharge response to the RF circuit when the shunt capacitance is in the order of the primary ESD source capacitance; this effect can be lessened with the utilization of ESD diode elements as RF matching capacitors, or resistance limiting elements.

2.5.2 RF ESD Method: Conversion of ESD Networks into Matching Networks

As discussed in the Section 2.5.1, the ESD and matching networks components can be combined or interchanged. Hence, the ESD elements can also serve a role in the matching networks, providing RF function and topology.

2.5.2.1 Conversion of ESD networks into L-match networks

An ESD network can serve as an L-match network when the ESD elements have two elements, with at least one series element and at least one shunt element.

A method of conversion of an ESD network into an RF matching network can be achieved where the first series element allows current flow to the shunt element, and the second element serves as an ESD current shunt. For example, given the series element is an inductor, and the shunt element is a diode element (or dual-diode), an L-match network is established. During an ESD event, the ESD current must be able to flow through the first inductor element without inductor failure. Additionally, the inductor values must be chosen to appear as a 'short' at ESD event frequencies to allow the current to flow to the diode element. The ESD diodes serve the role of discharging the ESD event at the ESD frequency

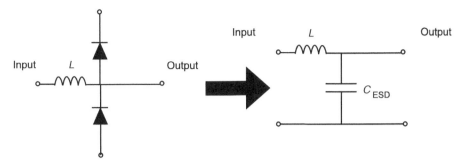

Figure 2.12   Conversion of ESD network into an L-match network

and current magnitude and serve the role of a capacitor in the L-match network at the application frequency.

As an example, Figure 2.12 shows a network that is synthesized using a first inductor and a MOSFET. During the ESD event, the MOSFET discharges the current to the V_{SS} power rails. At RF application frequency, the network forms an L-match network.

An ESD method of conversion of an ESD network into an L-match network can be established as follows:

- An ESD network is formed using at least one series element and one shunt element forming an L-type configuration.

- The shunt element within the L-type configuration serves as a ESD discharge means.

- The first series element allows the current to flow to the shunt element during ESD events and serves as an L-match element at application frequencies.

2.5.2.2 Conversion of ESD networks into Π-match networks

ESD networks can serve as matching networks. An ESD network can serve as a Π-match network when the ESD elements have two shunt elements, and a series element in series forming a 'Π' configuration. In this case, at least two shunt elements exist, separated by at least one series element.

A method of conversion of an ESD network into an RF matching network can be achieved where the first shunt element allows the current to flow to power supplies during ESD events. The series element can serve as an element that allows the current to flow to the second shunt element during ESD events or serve as a 'buffer element'. The second shunt element provides current discharge to at least one power rail during ESD events, serving as a 'second stage' ESD element. At RF application frequencies, it is desirable to have the network form a Π-match network. The ESD diodes serve the role of discharging the ESD event at the ESD frequency and current magnitude and serve the role of a capacitor in the Π-network at the application frequency.

As an example, Figure 2.13 shows a network that is synthesized using a first double-diode ESD network, and inductor, followed by a second ESD double-diode network. During the ESD event, the first and second stages of the double-diode discharges the

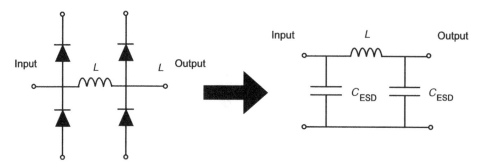

Figure 2.13 Conversion of ESD network into a Π-match network

current to the V_{DD} and V_{SS} power rails. At RF frequencies, the network forms a Π-match network. Note that in this configuration, the placement of these elements can be at different locations within a physical chip design. Also note that the series elements and second shunt element can be a 'second stage' ESD network for an improved HBM or CDM ESD protection.

An ESD method of conversion of an ESD network into a Π-match network can be established as follows:

- An ESD network is formed using at least two shunt element and a center series element forming a Π-type configuration.

- The shunt elements within the Π-type configuration serves as an ESD discharge means.

- The complete networks serve as a Π-match filter at application frequencies.

2.5.2.3 Conversion of ESD networks into T-match networks

An ESD network can serve as a T-match network when the ESD elements have two elements in series and at least one shunt element between the two series elements. A method of conversion of an ESD network into an RF matching network can be achieved where the first series element allows the current to flow to the shunt element, and the center element serves as an ESD current shunt. For example, given the two series elements are inductors, and the shunt element is a diode element (or dual-diode), a T-match network is established. During an ESD event, the ESD current must be able to flow through the first inductor element without inductor failure. Additionally, the inductor values must be chosen to appear as a 'short' at ESD event frequencies to allow the current to flow to the diode element. The ESD diodes serve the role of discharging the ESD event at the ESD frequency and current magnitude, and serve the role of a capacitor in the T-network at the application frequency. The third element (e.g., second series element) in the T-network can serve as ESD 'buffering' element at ESD event frequency or just satisfy RF application matching requirements.

As an example, Figure 2.14 shows a network that is synthesized using the first inductor, double-diode ESD devices, followed by an inductor. During the ESD event, the double-diode discharges the current to the V_{DD} and V_{SS} power rails. At RF frequencies, the network forms a T-match network.

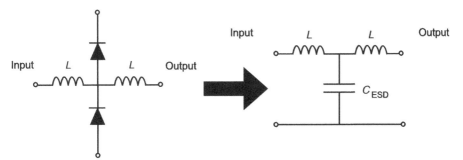

Figure 2.14   Conversion of ESD network into a T-match network

An ESD method of conversion of an ESD network into a T-match network can be established as follows:

- An ESD network is formed using at least two series element and a center shunt element forming a T-type configuration.

- The shunt element within the T-type configuration serves as an ESD discharge means.

- The first series element allows the current to flow to the shunt element during ESD events and serves as a T-match element at application frequencies.

2.6 RF ESD DESIGN METHODS: INDUCTIVE SHUNT

An ESD RF design method is to utilize both passive and active elements that can improve the ESD protection and minimize RF performance degradation. An RF ESD design method, distinct from digital methods, is the utilization of capacitor and inductor elements in the ESD protection strategy as parallel or series elements [13–16].

From circuit theory, the voltage drop across an inductor can be defined as equal to the product of the inductance and the first derivative of the current through the inductor as a function of time,

$$V = L\frac{di}{dt}$$

The current through the inductor can be obtained as

$$I(t) = \frac{1}{L}\int_{t'=0}^{t'=t} V(t')\,dt'$$

Hence for a slow-pulse phenomenon, the voltage across the inductor is large during large current transients and small during small current transients.

Using a shunt inductor between the RF pad and the ground potential, current can be transmitted between the input pad and the ground power rail. For slow current transients, the current will flow to the substrate without a significant increase in the RF pad voltage. Hence during an ESD event, it can be designed in such a fashion to transfer current from the RF pad to the ground power rail through a shunt inductor.

For HBM events, current will flow from the input pad to the ground rail. Note that the presence of the inductor element with the HBM RC source term leads to an RLC response with the HBM capacitor precharged.

For MM events, the MM capacitor source will be interactive with the pad capacitance and inductor ESD shunt element, leading to an LC response. Without the ESD shunt inductor, the response is a damped RLC response. In the presence of the ESD shunt inductor, the nature of the characteristic will change because of the ESD inductor.

For charged device model, the chip substrate serves as a capacitor. In the case of the CDM events, it is expected that the current will flow from the substrate through the ESD shunt inductor element to the RF pad.

During RF functional applications, it was suggested by Leroux and Steyart [13] that the value of the inductor chosen can be such that the RF pad capacitance and ESD inductor are defined at the resonant frequency of the LC tank element.

Figure 2.15 shows an example of a signal pad, a pad capacitance parasitic component, and an RF ESD inductor element.

The quality factor for a RF pad capacitor with a series resistor element, R_s, and a series capacitor, C_s, can be expressed as follows:

$$Q_C = \frac{1}{\omega C_s R_s}$$

The quality factor for a RF pad capacitor with a parallel resistor element, R_p, and a parallel capacitor, C_p, can be expressed as follows:

$$Q_C = \omega C_p R_p$$

In the case when the series resistor is zero, or the parallel resistance is infinite, the Q of the RF capacitor element tends to infinity.

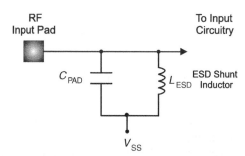

Figure 2.15 An RF input pad and a shunt ESD inductor element (with parasitic capacitance load of the RF pad)

The quality factor for an ESD inductor with a series resistor element, R_s, and a series ESD inductor, L_s, can be expressed as follows:

$$Q_L = \frac{\omega L_s}{R_s}$$

whereas, a resistive shunt in parallel with a parallel ESD inductor element has a quality factor expressed as follows:

$$Q_L = \frac{R_p}{\omega L_p}$$

For the case when the series resistance tends to zero, or the parallel resistance tends to infinity, the Q of the inductor becomes infinite.

For a circuit with both the ESD inductor element, L, and the RF pad capacitor element, C, at the resonant frequency [27–29,32],

$$\frac{1}{Q_{LC}}\bigg|_{\omega=\omega_0} = \frac{1}{Q_L} + \frac{1}{Q_C}$$

or,

$$Q_{LC}\big|_{\omega=\omega_0} = \frac{Q_L Q_C}{Q_L + Q_C}$$

At resonance, the impedance of the ESD parallel inductor, RF capacitor, and shunt resistor element is the value of the shunt resistance,

$$Z\big|_{\omega=\omega_0} = R_p$$
$$Q_{LC}\big|_{\omega=\omega_0} = \omega_0 C R_p$$

In the case where the series resistance is zero, the impedance of the LC network is zero. The resonant frequency of an ideal LC network is

$$\omega_0 = \frac{1}{\sqrt{LC}}$$

then for an application frequency ω_0, which is at a resonant frequency, an ESD inductor element can be chosen of value

$$L_{ESD} = \frac{1}{\omega_0^2 C_{pad}}$$

An RF ESD design practice can be as follows:

- Placement of an ESD inductor can serve as a low resistance ESD element between the input pad and the ground (V_{SS}) and power supply (V_{DD}) rails.

- An ESD inductor can be chosen so as to form an LC tank circuit with the input pad capacitance load.

- The value of the ESD inductor element is chosen such that the resonant frequency is at the functional application frequency,

$$L_{\text{ESD}} = \frac{1}{\omega_0^2 C_{\text{pad}}}$$

2.7 RF ESD DESIGN METHODS: CANCELLATION METHOD

2.7.1 Quality Factors and the Cancellation Method

In Section 2.2 on quality factors, it was noted that in ESD RF circuit design, the ESD elements can be viewed as follows [10,27–29]:

- RF ESD elements can be treated as passive or active elements.

- RF ESD elements can be treated as the parasitic elements of passive elements that impact the quality factor of the passive elements.

- RF ESD elements can be treated as the parasitic elements of active elements that impact the quality factor of active elements.

- RF ESD elements can be cosynthesized as parallel or series elements that are additive to the parasitic component of an active or a passive element.

- RF ESD element can be cosynthesized as a parallel or series elements that are subtractive (e.g., cancellation) to the parasitic component of an active or a passive element.

Hence, from a circuit representation, the perceived nature of the circuit component or element that degrades the Q can be viewed as the ESD element itself. Hence, from an RF ESD design circuit perspective, the ESD element can be viewed as an element that leads to quality factor degradation of an active circuit component, or full circuit representation. At the same time, from the perspective of an ESD RF design practice for ESD RF cosynthesis, the ESD element can be viewed as an additive or a subtractive component to the active or passive element; hence, it can be additive (degradation through larger component) or subtractive (enhancement through cancellation component). It was noted that from the complex series impedance,

$$Z_s = R_s \pm jX_s$$

the quality factor, Q, can be defined as

$$Q = \frac{X_s}{R_s}$$

The quality factor for a capacitor with a parallel resistor element, R_p, and a parallel capacitor, C_p, can be expressed as

$$Q_C = \omega C_p R_p$$

The quality factor for an inductor with a parallel resistor element, R_p, and a series inductor, L_p, can be expressed as follows:

$$Q_L = \frac{R_p}{\omega L_p}$$

In all these representations, the real (resistive) component can represent the ESD element, or the imaginary (capacitance or inductive) component can represent the ESD element, or both real and imaginary components can represent the ESD element. In another perspective, one part (either real or imaginary) can represent the ESD element and the other can represent a passive or an active circuit element.

2.7.2 Inductive Cancellation of Capacitance Load and Figures of Merit

Figure 2.16 shows an example of a case where there is an inductor element and pad capacitor. The inductor element can be an element of the RF circuit, and the capacitor element is the parasitic capacitance of the RF pad. Hynoven and Rosenbaum [27–29] represented the ESD network as a shunt capacitor element with a parallel resistance element. In the center of the diagram, an ESD network is represented as a capacitor element C_{ESD} and a parallel resistor element R_p.

For a circuit with both an inductor element, L, and a capacitor element, C, at the resonant frequency,

$$\frac{1}{Q_{LC}}\bigg|_{\omega=\omega_0} = \frac{1}{Q_L} + \frac{1}{Q_C}$$

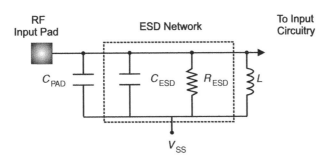

Figure 2.16   RF network comprising an RF pad, and a tuning inductor of an RF circuit, with an ESD element represented as a shunt capacitor element and shunt resistor element

or,

$$Q_{LC}\big|_{\omega=\omega_0} = \frac{Q_L Q_C}{Q_L + Q_C}$$

At resonance, the impedance of the parallel inductor, capacitor, and shunt resistor element is the value of the shunt resistance,

$$Z\big|_{\omega=\omega_0} = R_p$$
$$Q_{LC}\big|_{\omega=\omega_0} = \omega_0 C R_p$$

Solving for the shunt resistance, R_p,

$$R_p = \frac{1}{\omega_0 C} \frac{Q_L Q_C}{Q_L + Q_C}$$

A generalization of this ESD metric, a figure of merit can be expressed as the ratio of the ESD robustness and the shunt impedance,

$$\text{FOM} = Z_{\text{shunt}} V_{\text{ESD}}$$

Generalizing this, using the complex form of the series impedance,

$$\text{FOM} = (R_s \pm jX_s)_{\text{shunt}} V_{\text{ESD}}$$

This can be easily understood for the case of the capacitor element, where the ESD robustness is divided by the reactance,

$$\text{FOM} = \frac{V_{\text{ESD}}}{\omega C_{\text{ESD}}}$$

In the case of the resonant LC network, Hynoven and Rosenbaum [27–29] represented the FOM, at resonance, as follows:

$$\text{FOM} = \frac{Q_L Q_C}{Q_L + Q_C} \left\{ \frac{V_{\text{ESD}}}{\omega_0 C} \right\}.$$

From the equation, it is clear that the FOM can be optimized using two methods. Hynoven and Rosenbaum stated the two possible methods of optimization as follows:

- Maximize the term of the ESD robustness as a function of the product of application frequency and the capacitance (e.g., ESD robustness divided by the reactance).

$$\left\{ \frac{V_{\text{ESD}}}{\omega_0 C} \right\}$$

- Maximize the total Q

$$Q_{LC} = \frac{Q_L Q_C}{Q_L + Q_C}$$

Given a fixed inductor load, it is clear that the optimization can be achieved from the Q of the ESD element itself. Hence, choosing an ESD robust element, and a high Q ESD element is critical. (*Note*: in a more general development, it is clear that the FOM can be expanded as a total differential, and Lagrange multiplier constraints can be established.)

An ESD RF design practice, known as the cancellation method, is as follows:

- In an RF application, a parallel resistance element can be chosen such that the impedance at resonance of a circuit with an RF capacitance load and a tuning inductor is the resistance element, with impedance

$$Z|_{\omega=\omega_0} = R_{\mathrm{p}}$$

and parallel resistance value of R_{p},

$$R_{\mathrm{p}} = \frac{1}{\omega_0 C} \frac{Q_{\mathrm{L}} Q_{\mathrm{C}}}{Q_{\mathrm{L}} + Q_{\mathrm{C}}}$$

- Optimization of the network is achieved by obtaining the highest value of term,

$$\left\{ \frac{V_{\mathrm{ESD}}}{\omega_0 C} \right\}$$

or the term,

$$Q_{\mathrm{LC}} = \frac{Q_{\mathrm{L}} Q_{\mathrm{C}}}{Q_{\mathrm{L}} + Q_{\mathrm{C}}}$$

2.7.3 Cancellation Method and ESD Circuitry

As discussed, using resonance conditions of the LC resonator circuits, the capacitance load or the inductor load of a circuit can be tuned out so as to reduce the load on the RF input or output circuit. Hence, the ESD elements can be the following:

- ESD element can be an ESD capacitive load, tuning out an inductance of the circuit whose capacitance value is chosen for a desired resonance frequency.
- ESD element can be an ESD inductive load, tuning out a capacitance of the circuit whose ESD inductor value is chosen for a desired resonance frequency.
- ESD elements can be both the inductive and the capacitance load whose values are chosen for a desired resonant frequency.

Figure 2.17 shows an example of an ESD RF circuit with an ESD double-diode network. ESD double diode elements are designed as low resistance structure. The ESD double-diode network can be represented as a capacitor element. In the a.c. analysis, the two ESD diode

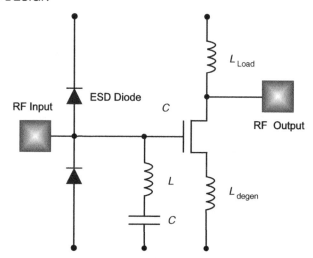

Figure 2.17   ESD double-diode network, cancellation ESD shunt inductor element, and a capacitor element

elements are in parallel configuration between the input pad and the a.c. ground reference. This capacitive element is also in parallel with the input pad capacitance. In the case of ESD double-diode networks, both the capacitor elements are voltage dependent (e.g., capacitance variation is a function of the metallurgical junction). Hence, the d.c. bias state of the ESD double-diode circuit alters the capacitance load condition.

Hynoven *et al.* [27–29] noted that a cancellation method can use an ESD inductor element to cancel the loading effect of the ESD double-diode element. The inductor element value is chosen in such a way that during the cancellation, the combined loading effect of the ESD double-diode and inductor is minimized at the functional application frequency. It was also noted that direct electrical connection of the ESD inductor to a ground reference is not desirable. Hence, placement of a capacitor element in series with the inductor is desirable (as shown in Figure 2.17).

Figure 2.18 shows an RF ESD double-diode network, an ESD shunt capacitor, and a series capacitor element, as well as a d.c. blocking capacitor in series with the input of the RF circuit. Hynoven *et al.* [27–29] noted that in this case, the addition of a d.c. blocking capacitor in series with the input node of the RF circuit allows biasing of the ESD double-diode/inductor network to a d.c. potential to tune out the capacitance of the RF input pad, and the ESD diode elements.

As an alternative ESD network, the inductor can be substituted for one of the ESD diode elements (Figure 2.19); In the first case, the ESD inductor element is placed between the input node and the V_{DD} power supply node. The advantages of the diode/inductor implementation are as follows:

- Elimination of the ESD diode (e.g., diode to V_{DD}) reduces the total capacitance load and design area.

- Elimination of ESD diode capacitance requires larger shunt impedance with reduced impact on RF performance.

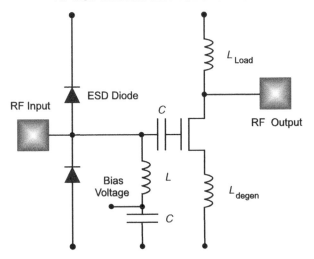

Figure 2.18 ESD double-diode network, cancellation ESD shunt inductor element with series capacitor, and an additional RF input series d.c. blocking capacitor element. Note the addition of the d.c. bias electrical connection

- The d.c. bias condition of the single diode element reduces the capacitance load of the diode element.

The disadvantages of this network are as follows:

- Inductor area;
- ESD robustness limited to the inductor coil ESD robustness;

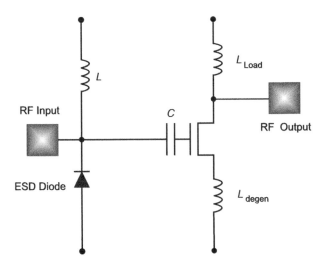

Figure 2.19 ESD diode/inductor network with series d.c. blocking capacitor element for RF inputs (inductor to V_{DD} power supply)

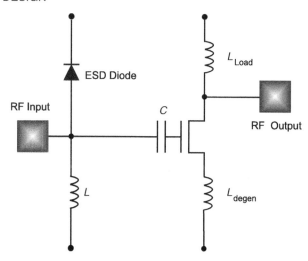

Figure 2.20 ESD diode/inductor network with series d.c. blocking capacitor element for RF inputs (inductor to ground reference)

- Input voltage transient $L(di/dt)$ response to HBM and MM pulse;
- CDM transient response (precharged V_{DD} power rail) leads to voltage overshoot and resonance oscillation of cancellation network.

As a second alternative ESD network, the inductor can be substituted for the ESD diode to the ground reference (Figure 2.20); In the first case, the ESD inductor element is placed between the input node and the V_{DD} power supply node. The advantages of the diode/inductor implementation are as follows [27–29]:

- Elimination of the ESD diode (e.g., diode to V_{SS}) reducing the total capacitance load and design area.
- Elimination of ESD diode capacitance requires larger shunt impedance with reduced impact on RF performance.

The disadvantages of this network are as follows:

- Inductor area;
- ESD robustness limited to the inductor coil ESD robustness;
- Input voltage transient $L(di/dt)$ response to HBM and MM pulse;
- CDM transient response (precharged V_{SS} power rail) initiates voltage undershoot and resonance oscillation.

An ESD RF design practice, known as the cancellation method, is as follows:

- In an RF application, an ESD inductor element can be chosen such that the combined capacitance load of an input pad and the ESD diodes form an LC resonance cancellation circuit.

- The ESD inductor element value is chosen such that the load is minimized during RF functional applications.

- An alternative ESD implementation includes the usage of a single inductor and single diode element to form the LC resonance cancellation circuit.

- RF ESD optimization factors between the ESD double-diode and ESD inductor-diode networks include capacitance load, design area, ESD robustness of diode and inductor elements, and response to ESD events.

2.8 RF ESD DESIGN METHODS: IMPEDANCE ISOLATION TECHNIQUE USING LC RESONATOR

A method to reduce the effective loading in a circuit can utilize resonant conditions of LC resonator circuits. With the placement of a LC resonator circuit in series with an ESD network, the impedance of the network will approach infinity at the LC resonator resonant frequency. Hence, by a placement of an LC resonator network in series with ESD elements, the loading effect of the complete network approaches infinity (e.g. open circuit) at the resonant state. By choosing the resonance condition of the LC network to be equal to that of the application frequency, the ESD loading does not impact the circuit performance. This RF ESD technique of 'impedance isolation' was demonstrated by Ker and Lee [22]. Ker and Lee demonstrated two types of impedance isolated RF ESD networks.

An RF ESD network can be constructed as follows:

- An ESD element connected to a signal pad in series with an LC network connected to a power rail (whose LC resonant frequency equals the application frequency).

- An LC network (whose resonant frequency equals the application frequency) connected to a signal pad in series with an ESD element connected to a power rail.

In this fashion, four types of ESD networks are possible for a two power supply systems (e.g., V_{DD} and V_{SS}). Ker and Lee demonstrated two types of networks [22]: the first network where the LC resonator is connected to the RF input pad, followed by a series p–n diode element (e.g. LC-D); the second network of a diode element connected to the RF input pad, followed by a series LC resonator (e.g. D-LC).

Figure 2.21 shows an example of a standard double-diode ESD network, and an impedance isolation ESD network with the LC resonator connected to the RF input pad (e.g., LC-D). Figure 2.22 shows again the standard digital CMOS double-diode ESD network and an impedance isolation ESD network with the diode connected to the RF input pad followed by the LC resonator network connected to the power rails (e.g. D-LC).

An ESD RF design practice, known as the impedance isolation method, is as follows [22]:

- In an RF application, an ESD inductor element and ESD capacitor element are placed in a parallel configuration to establish an LC resonator whose resonant frequency equals the application frequency.

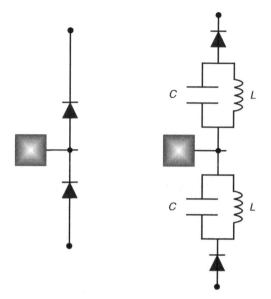

Figure 2.21   Standard digital ESD double-diode network and the RF LC resonator–diode ESD network

- The LC-resonator is placed in series with an ESD element between the signal pad and the power supply voltage such that at the application frequency.
- The series configuration of the LC resonator and ESD element's impedance approaches infinity at the application frequency.

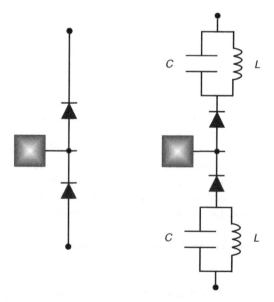

Figure 2.22   Standard digital ESD double-diode network and the RF diode–LC resonator ESD network

2.9 RF ESD DESIGN METHODS: LUMPED VERSUS DISTRIBUTED LOADS

In designing protection circuits for RF applications, the loading capacitance will be a concern as the reactance continues to decrease. Given a scaling condition of a constant reactance, where reactance is

$$X_C = 1/(2\pi f C_{ESD})$$

as the application frequency increases, the ESD loading capacitance requires to be reducing. In the case of a narrow-band application, an ESD resonant cancellation technique can be utilized to eliminate the loading effect (e.g., adding a parallel inductor element). Hence, for a single narrow-band application, any inductive element can be utilized to circumvent the impact of the capacitor element (e.g., at a given d.c. voltage state). This resonance cancellation technique is not acceptable for broadband applications.

Hence, an RF ESD protection device is needed that has the following characteristics:

- low impedance during ESD events;
- high impedance during RF functional application;
- low capacitance load during RF functional application;
- low impact on bandwidth.

A common RF design technique for broadband applications has been to use distributed circuit networks for amplifier and oscillator networks [4–9,11,12].

Using a distributed RF ESD network, a segmented transmission line is placed between successive stages of an ESD network. The RF ESD elements are distributed along the transmission line in such a manner that they match with the external signal line. The transmission line can consist of the following elements:

- micro-strip transmission line;
- coplanar waveguide (CPW);
- coplanar strip-line;
- bond wire;
- inductor element.

As an explanation, Figure 2.23 shows an example of a 50 Ω source, a series resistor, an ESD capacitance load, and a circuit load. The series resistance represents the interconnect wiring, pad, and ESD element resistance. The ESD capacitance load is the total capacitance coupling to the ground or power supply rails of the ESD element and other parasitic capacitance elements. In this case, the series resistor element and the ESD element are chosen to provide the optimum ESD protection scheme. In this case, the power loss occurs because of the mismatch between the input source and the effective load condition. An additional issue is that the voltage variation of the capacitor element can lead to poor linearity. ESD implementations

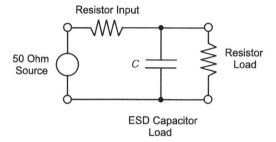

Figure 2.23 Example of a 50 Ω source, a series resistor, a ESD equivalent capacitance load, and a 50 Ω circuit load

that utilize metallurgical junctions can initiate poor linearity in RF circuits; this is true for RF GG MOSFET, single diode, dual-diode, and bipolar ESD networks.

2.9.1 RF ESD Distributed Load with Coplanar Wave Guides

Coplanar waveguides have the advantage of forming circles on a Smith chart, allowing the ability to branch to different impedance states utilizing ESD shunts at selective locations. Figure 2.24 shows an example of a coplanar waveguide segment replacing the series resistance element. In this fashion, the effective ESD network comprises an LC component with a capacitor shunt element. The LC transmission line can provide matching conditions for the 50 Ω input source. Additionally, the linearity of the circuit can be improved as the LC transmission line is not a voltage dependent capacitor element.

Figure 2.25 shows an example of the subdivision of the same ESD capacitor element into multiple segments. In this manner, the coplanar waveguide splits the load of the ESD element into the first segment, followed by another segment. As more segments are added and the size of the network load is distributed more, the network becomes a more ideal transmission line, and the broadband matching of the 50 Ω input source and the 50 Ω output load improves [4–9,11,12].

An RF ESD design practice can be as follows:

• With a distributed RF ESD network, the deleterious loading effects of an ESD network can be reduced (e.g., impedance mismatch, reflection, and power transfer).

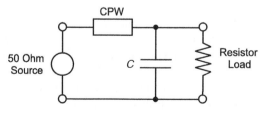

Figure 2.24 Example of a 50 Ω source, a coplanar waveguide (CPW), an ESD equivalent capacitance load, and a 50 Ω circuit load

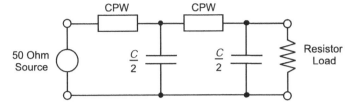

Figure 2.25 Example of a 50 Ω source, a dual segment coplanar waveguide (CPW), and an ESD equivalent capacitance load

- With a distributed RF ESD network, S-parameter degradation of reflection parameter S_{11}, and transmission parameter S_{21} is minimized (compared with single lumped RF ESD network).

- With the use of transmission lines in a distributed RF ESD network, circuit linearity can be improved because of the voltage independent nature of LC transmission lines.

2.9.2 RF ESD Distribution Coplanar Waveguides Analysis Using ABCD Matrices

With the utilization of lossless transmission line elements, such as coplanar waveguide structures, the loading effect of ESD networks can be minimized. From the reflection coefficient relationship,

$$\Gamma_i = \frac{Z_1 - Z_{CPW}}{Z_1 + Z_{CPW}}$$

From the reflection coefficient relationship, the input impedance can be expressed as a function of the reflection coefficient and the coplanar waveguide term,

$$Z_i = \left\{ \frac{1 + \Gamma_i}{1 - \Gamma_i} \right\} Z_{CPW}$$

The length of the coplanar waveguide can be expressed as a function of the wavelength,

$$L_{CPW} = \frac{\phi - \pi}{2\pi} \frac{\lambda}{2}$$

As an RF metric, as proposed by Ito et al. [11,12], the matching of the input impedance can be evaluated by quantifying the standing wave ratio (SWR),

$$SWR = \frac{1 + \left| \dfrac{Z_i - Z_0}{Z_i + Z_0} \right|}{1 - \left| \dfrac{Z_i - Z_0}{Z_i + Z_0} \right|}$$

As the impedance is matched, the value of SWR approaches unity. From the ABCD matrix form of the two-port matrix of the coplanar waveguide,

$$
\begin{bmatrix} A & B \\ C & D \end{bmatrix} = \begin{bmatrix} \cos \dfrac{2\pi l}{\lambda} & j\dfrac{Z_{ol}}{Z_o} \sin \dfrac{2\pi l}{\lambda} \\ j\dfrac{Z_o}{Z_{ol}} \sin \dfrac{2\pi l}{\lambda} & \cos \dfrac{2\pi l}{\lambda} \end{bmatrix}.
$$

The ABCD matrix for the input impedance can be expressed as

$$
\begin{bmatrix} A & B \\ C & D \end{bmatrix} = \begin{bmatrix} 1 & 0 \\ Z_0/Z & 1 \end{bmatrix}.
$$

Given the impedance is an RC load, and Z_0 is a 50 Ω impedance, then

$$
\begin{bmatrix} A & B \\ C & D \end{bmatrix} = \begin{bmatrix} 1 & 0 \\ 50\left\{\dfrac{1}{R} + j\omega C\right\} & 1 \end{bmatrix}.
$$

The total matrix can be obtained by matrix multiplication, where the first matrix is for the coplanar waveguide, and the second is for the RC elements

$$
\begin{bmatrix} A_i & B_i \\ C_i & D_i \end{bmatrix} = \begin{bmatrix} \cos \dfrac{2\pi l}{\lambda} & j\dfrac{Z_{ol}}{Z_o} \sin \dfrac{2\pi l}{\lambda} \\ j\dfrac{Z_o}{Z_{ol}} \sin \dfrac{2\pi l}{\lambda} & \cos \dfrac{2\pi l}{\lambda} \end{bmatrix} \begin{bmatrix} 1 & 0 \\ 50\left\{\dfrac{1}{R} + j\omega C\right\} & 1 \end{bmatrix}.
$$

In this ESD RF methodology, the size of the ESD structure must be determined without failure to maintain ESD robustness. Ito *et al.* [11,12] propose the length of the coplanar waveguide that is determined from the matrix algebra. The length of the coplanar waveguide can be solved and expressed as follows:

$$
\frac{l}{\lambda} = \tan^{-1} \frac{2\omega C}{\dfrac{Z_{ol}}{Z_o^2} + Z_{ol}(\omega C)^2 + \dfrac{1}{Z_{ol}}}
$$

Using the length of the coplanar waveguide, Ito *et al.* [11,12] solved the input impedance of the network. The input impedance can be obtained using the expression,

$$
R_i = \cos^2\left\{\frac{2\pi l}{\lambda}\right\} + \left[\frac{Z_o}{Z_{ol}} \sin\left\{\frac{2\pi l}{\lambda}\right\} + \omega C Z_o \cos\left\{\frac{2\pi l}{\lambda}\right\}\right]^2
$$

2.10 ESD RF DESIGN SYNTHESIS AND FLOOR PLANNING: RF, ANALOG, AND DIGITAL INTEGRATION

In floor planning a RF semiconductor chip, an ESD architecture and methodology must be established to provide a good ESD protection in all pins and chip sectors. The 'whole-chip' ESD design addresses not only the individual circuit elements and circuits but also the design synthesis and integration of the higher order subfunctions, power bus architecture, and floor plan. Moreover, an ESD methodology must be established that addresses the power rail placement and placement of the ESD network, as well as solutions between the subfunctions. In essence, the ESD design synthesis must incorporate ESD solutions that address both electrical and spatial connectivity. With high-level integration of RF applications, monolithic microwave integrated circuits (MMIC) semiconductor chips may contain the following subblocks:

- digital circuit block;

- analog circuit block;

- RF circuit block.

In today's system-on-chip (SOC) and network-on-chip (NOC) applications, these three subblocks exist as independent sectors of a chip design, or core designs. The digital, analog, and RF subblocks are segmented with their own corresponding physical and electrical subdomains to avoid noise and coupling issues.

Electrical decoupling between the different subfunctions is established by the following means:

- independent power bus rails;

- independent ground bus rails;

- high resistance substrate regions;

- triple well isolated epitaxial regions with separate electrical nodes.

Spatial decoupling is achieved as follows:

- spatial separation of the digital, analog, and RF subfunctions on a common substrate (by floor plan design);

- guard ring structures between the subfunctions;

- deep trench and isolation troughs between the subfunctions;

- placement of the digital, analog, and RF subfunctions on separate chips interconnected on a carrier structure.

With the spatial and electrical separation of different subfunctions, noise decoupling is achieved but the ESD protection is compromised. As a result, the digital, analog, and RF subfunctions must be coupled electrically in order to provide ESD protection. This is achieved through the ESD design strategy implementation and floor planning of the digital, analog and RF domains with ESD protection in mind.

With the introduction of separate power rails, regulators, predesigned modular design, and the isolation of the digital, analog and RF circuitry, it is more difficult to achieve the above good ESD protection results. The ESD design practice objectives are achievable with the following:

- V_{DD} to V_{SS} core ESD power clamps;

- bidirectional V_{DD} digital core to V_{DD} digital I/O ESD power clamps;

- bidirectional V_{SS} digital core to V_{SS} I/O ESD circuitry;

- V_{DD} to V_{SS} analog ESD power clamps;

- bidirectional V_{DD} digital core to analog V_{DD} ESD circuitry;

- bidirectional V_{SS} digital core to analog V_{SS} ESD circuitry;

- V_{DD} to V_{SS} RF ESD power clamps;

- bidirectional V_{DD} digital core to RF V_{DD} ESD circuitry;

- bidirectional V_{DD} digital core to RF V_{DD} ESD circuitry.

2.10.1 ESD Power Clamp Placement Within a Domain

ESD power clamps play a key role in the ESD protection of semiconductor chip [34–41]. Because of the bus wiring resistance, the placement of these power clamps plays a significant role in the pin-to-pin ESD variation within a semiconductor chip and the effectiveness of the ESD power clamps.

As the frequency of the power clamp placement increases, the effectiveness of the whole chip solution improves. The advantages of the increased frequency of power clamp placement is as follows:

- *Reduction of the electrical bus resistance between each pin and power clamp* (e.g., lowering the electrical impedance through the ESD current loop improving overall whole chip effectiveness).

- *Reduction of the pin-to-pin distribution along the power bus* (e.g., the ESD robustness of the pin-to-pin variation along the power bus is decreased.)

- *Reduction of the time response between the pin and power clamp* (e.g., time scale length of the power bus is reduced).

- *Current distribution in the interconnects* (e.g., lowering the current density in the interconnects and substrate).

- *Parallelism of the ESD power clamp networks* (e.g., increase the number of parallel current paths).

In the alternative current loop established by the ESD protection networks, the current flows through the ESD network, the power bussing, and the ESD power clamp. By increasing the frequency of placement of the power bus, the system becomes a distributed system where the incremental segment of the distributed system is a resistance–conductance $(R–G)$ transmission line. The power clamp can be modeled as a conductance and as either a frequency dependent switch or a voltage dependent power supply. During conduction, the ESD power clamp can be treated as a conductance term. As the successive power clamps are initiated, the distributed system forms a $R–G$ transmission line response.

As the ESD power clamps become more distributed along the power rail, pin-to-pin variation of the ESD decreases. In many chip designs, the ESD results improve when the I/O pin or signal pin is placed near the ESD power clamp; this continues to a worst-case pin condition. Eventually, the results begin to improve as the pin comes adjacent to the next power clamp. In an evaluation of the spatial variation, the ESD pin distribution will have a hyperbolic nature (e.g., or at times, half-sine wave nature) along the power rail; the peak-to-valley pin-to-pin variation is a function of the frequency of the power clamps, and the height is a function of the resistance voltage drop along the power rail.

Additional to the pin-to-pin variation between the power clamps spatially, the time response of the network also comes into play. From the voltage-diffusion equation, the voltage 'diffuses' along the system associated with a characteristic time constant. The bus resistance plays a role in deciding how fast the ESD current distributes along the distributed RC network. From the perspective of a RCG distributed system, the bus resistance and power clamp placement influence the temporal response (e.g., the resistance element and the conductance element). As the resistance decreases, a larger part of the semiconductor chip participates in the distribution in time. With the participation of the distributed network, the ESD current propagates and distributes through more current paths. The current density is also lowered, decreasing the likelihood of interconnect wiring and metal bus failures.

2.10.2 Power Bus Architecture and ESD Design Synthesis

Power bus architecture and the ESD design must be cosynthesized to achieve the digital, analog, and RF functional objectives, as well as ESD objectives. One of the primary reasons that ESD protection is not achieved is because the functional objectives and the ESD objectives are in dichotomy. In SOC applications, power grids are separated for the following issues:

- mixed voltage interface;

- ΔI noise from digital I/O circuitry;

- regulated internal power supplies;

- modular construction of floor plan of semiconductor chips;

- power management.

In mixed voltage interfaces, the peripheral circuits must interface with high voltage conditions above the native voltage of a semiconductor chip. In this case, the exterior power rail must be isolated from the internal chip circuitry. As a result, the I/O or peripheral circuit power rail (e.g., I/O V_{DD}) may be at a higher voltage than the internal core voltage (V_{DD}).

Coincident switching of peripheral circuitry and digital circuitry can disturb the voltage states of the internal circuitry and analog applications. Hence, semiconductor chips are structured to have the peripheral rail isolated from the internal, as well as segmenting the digital, analog, and RF chip sectors.

2.10.3 V_{DD}-to-V_{SS} Power Rail Protection

It is a common ESD designing practice to implement ESD power clamp networks between the power supplies in CMOS digital semiconductor chips. ESD power clamps provide multiple objectives in ESD design:

- *Alternative Current Loop*: ESD power clamps are placed between the V_{DD} and V_{SS} power rails to establish the alternative ESD current loop in parallel with the current loop through the input pad circuitry.

- *Lower Impedance:* ESD power clamps are placed between the V_{DD} and V_{SS} power rails to lower the effective impedance between the two power rails.

- *Input Pin ESD Protection:* ESD V_{DD} and V_{SS} power clamps provide an improved ESD protection at the input pin.

- *Power Supply Pin Protection*: ESD V_{DD} and V_{SS} power clamps provide ESD protection between the V_{DD} and V_{SS} power rails.

In a CMOS or RF-CMOS digital or analog subfunction, CMOS-based ESD power clamps are preferred from MOSFET elements, or CMOS parasitic elements. A CMOS or RF-CMOS digital or analog subfunction can consist of the following ESD power clamp solutions:

- RC-triggered MOSFET ESD power clamps;

- voltage-triggered MOSFETs ESD power clamp;

- gate-modulated MOSFETs ESD power clamp;

- silicon-controlled rectifiers (SCR) ESD power clamp.

It has been noted that circuit designers prefer to utilize the CMOS ESD power clamps for the CMOS digital circuitry chip segments in a mixed signal application. The advantages of using the CMOS circuitry is circuit compatibility, suitability, and self-tracking with the CMOS digital logic networks.

In a bipolar analog subfunction, or a bipolar RF subfunction, ESD power clamps using bipolar element is preferred. Bipolar-based ESD power clamps are preferred for the following reasons:

- low capacitance;

- size;

- high current drive capability;

- low $1/f$ noise;

- circuit stability criteria;

- high breakdown voltage;

- negative power supply utilization;

- manufacturing tracking with the bipolar functional networks.

Circuit designers that construct BiCMOS and mixed signal applications do not desire to use CMOS ESD power clamps because of the breakdown voltage incompatibility, area, and MOSFET noise generation. As a result, in the design process of a mixed signal semiconductor chip, the CMOS-based ESD power clamps are used on the CMOS digital logic cores and networks, and bipolar-based ESD power clamps are used on the bipolar analog and RF chip sectors. The RF subfunction can utilize the following types of circuits:

- bipolar grounded base common-emitter single stage *npn* bipolar element;

- bipolar multistage Darlington-configured common-emitter ESD power clamps;

- RC-triggered Darlington-configured common-emitter ESD power clamps.

2.10.4 V_{DD}-to-Analog V_{DD} and V_{DD}-to-RF V_{CC} Power Rail Protection

In the integration of RF, analog, and digital networks, it is a common practice to isolate the analog function from the core logic sector to reduce the noise injection from the simultaneous switching noise of the digital circuitry. Additionally, it is a common practice to isolate the RF V_{CC} power rail from the digital core V_{DD} supply. Yet, complete isolation can introduce ESD failure of the analog sector or RF sector when no electrical current path exists. It is not desirable to have direct electrical connections between the V_{DD}-to-Analog V_{DD} and V_{DD}-to-RF V_{CC} power rails using ESD power clamps. A preferred choice is to utilize ESD power clamps in its own power domain, and to provide bidirectional connectivity through the substrate. Typical ESD protection can consist of the following ESD solutions:

- symmetric back-to-back diode strings;

- asymmetric back-to-back diode strings.

In digital circuit applications, the number of diode elements is chosen to allow voltage differential between the ground rails. In the case of RF applications, the focus is on the capacitive coupling between the ground rails. The placement of ESD networks between ground rails influence both ESD and noise.

An ESD design practice in the integration of RF, analog, and digital chip sectors is as follows:

- Utilize ESD power clamps between the power and ground rail of each chip sector.

- Utilize ESD power clamps that are common to the circuit type (e.g., use MOSFET-based ESD power clamps in the CMOS chip sectors, and bipolar-based ESD power clamps in the analog and RF Bipolar chip sectors).

- Utilize bidirectional elements ESD networks between the ground rails of the RF, analog, and digital sectors.

- Choose the number of diodes in series on the basis of both the necessary voltage differential and the capacitive coupling requirements.

2.10.5 Interdomain ESD Protection Networks

In mixed signal applications between the power domains, signal lines traverse the separated digital, analog, and RF functional blocks. During functional operation, the transmitted and receiving circuits are designed to prevent electrical overstress from the incoming signal. But during ESD events, the receiving functional block receiver can get electrically overstressed by ESD events. There are multiple design environments that lead to ESD failure in these internal networks [42,43]:

- There is no ESD protection network between the two electrically separated power domains and the only electrical connection is through the internal signal line (e.g., receiver).

- There is an ESD protection network between the two electrically separated power domains, but the ESD device turn-on voltage between the two chip segments is such that the electrical overstress of the internal signal line occurs first.

- There is an ESD protection network between the two electrically separated power domains, the ESD device turn-on voltage between the two chip segments is low, but the bus resistance voltage drops are such that electrical overstress of the internal signal line occurs first.

- There are multiple paths between the transmitting signal of the first power domain to a high fan-out of receiver networks in a plurality of different power domains, where at least one overstress condition occurs in one of the domains.

- There are functional blocks that are spatially adjacent functional blocks and those that are spatially separated functional blocks, where the substrate spacing is significant.

This concern can be addressed by interposing an additional functional block that provides ESD protection between the transmitting and the receiving functional blocks. An ESD design practice can be as follows:

- 'Intervening' functional logic blocks or a 'third party' block can be interposed between the two separated power domains that transmits signal from one to the other and that can absorb the voltage stress and ESD stress conditions (e.g., two inverter stages in series configuration).

- Intervening functional block has ESD power rail protection for both V_{DD} and V_{SS} power rails; between the first transmitting function block power supply V_{DD} rail and the intervening functional block V_{DD} power rail for the V_{DD}-to-V_{DD} power rail, and between the first transmitting function block power supply V_{SS} rail and the intervening functional block V_{SS} power rail for the V_{SS}-to-V_{SS} power rail.

- Intervening functional block has a ESD V_{DD}-to-V_{SS} power clamp between the power rails.

An alternative design strategy to address internal cross-domain ESD events is to provide local ESD networks on the internal nodes as is utilized on external nodes. Traditionally, on internal receivers, there is no need for 'internal' ESD protection. But in this interdomain power environment, the concept of using an ESD network on an internal logic node may be needed to address CDM events. An example of an internal node protection concern can be solved using a series resistor between the transmitter and the receiver network, as well as a local double-diode ESD network at the internal receiver node [42,43].

2.11 ESD CIRCUITS AND RF BOND PAD INTEGRATION

For RF applications, the wire bond pad can influence the RF circuit performance. Both the inductance and capacitance of the bond pad influence the impedance of the RF network. The wire bond itself introduces a series inductance. The bond pad introduces an inductive shunt component and a capacitor shunt to the chip substrate. The inductance of the RF bond pad can be reduced by the introduction of thick interlevel dielectric (ILD) films, conductive field shields, and insulators regions in the chip substrate. With the construction of inductors, thick ILD films are introduced to provide high quality factor (Q) inductors. Using semiconductor interconnect metal films, slotted field plates are placed under RF bond pads to reduce eddy currents in the chip substrate. Additionally, trench structures [e.g., shallow trench isolation (STI), trench isolation (TI), and deep trench (DT isolation)] are formed in a mesh to prevent the flow of substrate eddy currents.

For the RF network, the capacitance of the RF bond pad introduces a shunt capacitance term. With the introduction of a RF ESD network, such as an ESD diode network, or a grounded gate n-channel MOSFET, the capacitance of the RF bond pad and the ESD network are an additive shunt capacitance term to the RF circuit. The capacitance load of the components can be expressed as follows:

$$C_{load} = C_{wirebond} + C_{bondpad} + C_{ESD} + C_{ckt}$$

As the capacitance load of the bond pad and the ESD network are additive, in order to reduce the total capacitance load on the RF network, it is advantageous to reduce the capacitance of the bond pad structure. A solution to reduce the bond pad structure without an additional semiconductor processing cost is to use a smaller bond pad for RF pins compared to the other analog and digital signal pins. Hence, it is common to introduce octagonal pads by forming an octagon shape within the square shape of the standard pad used in a given technology.

Figure 2.26 shows an example of an octagonal pad structure. The octagon is effectively formed by the removal of the corners of the square pad structure of the standard analog and digital pad structures. Note that in some applications, all the pads can be identically designed to assume the same shape whether RF pins, analog pins, or digital pins. The removal of the corners of the square pad introduces a reduction in the pad capacitance load; this capacitance reduction allows for either higher RF performance, or allows for a larger acceptable load capacitance for the RF ESD element.

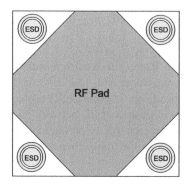

Figure 2.26   RF octagonal pad structure with circular ESD elements

With the removal of the corners of the square pad structure, in RF applications, this additional area can be utilized for additional circuit function or ESD networks. Figure 2.26 show an example of the utilization of circular ESD structures on the corners of the octagonal pad structure. Note that the octagonal and circular ESD designs are designed so as to fit within the normal square pad foot print; this imposes a limit of the area and size of the ESD elements (e.g., the ESD device diameter). In the case of mixed signal applications, with the analog, digital, and RF pins all using octagonal pad structures, different ESD design practices are established. The implementations can be as follows:

- *Identical Elements*: Each element is an independent ESD diode element where, all are identically designed. In this case, the anode and cathode elements can be interchanged and provide 4× variations in the load capacitance and ESD structure size.

- *Nonidentical ESD Elements*: Two of the four corner elements are independent ESD diode elements that are used for diodes between the input signal and the power supply (e.g. 'up' diodes), and the remaining two are the diodes used between the input signal and the ground rail (e.g. 'down' diodes). In this fashion, a 2× variation in the load capacitance and ESD structure size is possible.

A common mixed-signal and RF ESD design practice is to utilize the identical structures for analog and RF pins. But for the analog pins and digital circuits, the area utilized is double, and for the RF pins, the load capacitance is reduced by not utilizing all the four corners of the given signal pin.

An RF ESD design practice is as follows:

- Octagonal pads are used for RF pins, where octagonal or circular ESD structures are placed on the corner regions.

- Identical or nonidentical ESD elements are placed on the corners of the octagonal pad structure.

- For analog and digital pins, ESD diode networks are formed using all four of the ESD diode elements on the corners of the octagonal pad structures, when octagonal pads are used on a given design for all signal and power pins.

- For RF pins, only one or two of the ESD diode elements are used on the corners of the octagonal pad structure (e.g., one-fourth or one-half of the capacitance load).

2.12 ESD STRUCTURES UNDER WIRE BOND PADS

Applications of the RF technology extend from small single transistor chips (e.g., GaAs power amplifiers), to system-on-chip applications. In the wireless marketplace, the small die size allows for low cost components. In many of the small die size chips, a few active and passive elements are connected. In these RF applications, the number of circuits on a chip is such that the total bond pad area is a significant percentage of the total chip area. Additionally, in small chips (e.g., 1 to 4 mm^2), the competitive environment forces a high yield low-cost product solution to maximize the number of chips per wafer. Hence, structures under bond pads reduces chip size and cost. The ability to provide RF structure under pads for RF products (e.g., RF CMOS, SiGe, GaAs, and InP) to lower chip costs is important. To save space, RF ESD structures can be placed under bond pads [44–48]. The following concerns exist for the placement of structures under bond pads in RF applications:

- changes in the semiconductor device characteristics from the mechanical strain;
- failure of the semiconductor component (e.g., gate dielectric failure and dislocation);
- metal deformation and ILD cracking;
- mechanical failure of the bond pad (e.g., separation from the semiconductor component);
- change in the RF characteristics and models.

In designing RF ESD structures under bond pads, one must address the issue of metal wiring design to avoid ILD film cracking due to the mechanical pressure. The ability of the placement of the ESD structures under the pads is a function of the following:

- dielectric material (e.g., low-k or high-k materials);
- dielectric thickness;
- number of metal levels;
- metal material (e.g., aluminum or copper);
- ESD metal wiring design;
- insulator topography;
- bond pad stress test (e.g., pull or de-lamination test).

Adding ESD structure under bond pads has been demonstrated since the early 1990s without mechanical stress issues, by a proper design of the metallurgy [44–47]. In 1993, ESD protection structures under wire bond pads were demonstrated in a 0.5 μm LOCOS and STI CMOS technology in a silicon dioxide ILD and titanium/aluminum/titanium interconnect system [44]. LOCOS and STI defined diode structures were placed under bond pads, where only first level metal (M1) was used with a plurality of thin metal connections (e.g., no large rectangular shapes were used). No metal levels or contacts were placed over the structures. No active gated structures were placed under the bond pad regions. Active devices, such as MOSFETs, were successfully constructed under bond pads by Chittipeddi *et al.* [45] in both an aluminum and dual-damascene copper interconnect process.

RF technologies have the advantage of thick ILD film stack because of the need to create high quality factor passive elements, such as high Q inductors, and metal–insulator–metal (MIM) capacitors. The high quality factor inductor structures are achieved by the placement of the inductor far from the substrate region and the thick metal films. The thick metal films lead to a significant increase in the insulator film stack, reducing the concern for dielectric film cracking from solder ball placement and wire bonding. Hence, RF CMOS and RF BiCMOS SiGe technologies have an advantage for the placement of active elements and ESD structures under wire bond pads.

In RF technology, Gebreselasie *et al.* [48] demonstrated the ability to place RF passive elements and components under bond pads in RF CMOS and RF BiCMOS Silicon Germanium technology. Additionally, a family of SiGe-based derivative devices, CMOS devices (e.g., hyper-abrupt varactor, MOS-varactor, isolated triple-well MOSFETs), RF passive elements, and ESD elements and circuits (e.g., p^+/n-well diodes, MOSFETs, diode strings, back-to-back diode string networks, SiGe Darlington ESD power clamps, and RC-triggered MOSFET ESD power clamps) were constructed and tested under bond pad environments [48]. As part of the evaluation, both the functional d.c. characterization, RF testing and ESD transmission line pulse (TLP) testing were evaluated. For the evaluation of the RF bipolar elements, silicon germanium heterojunction bipolar transistors (HBT) devices were placed under wire bond pads and solder balls. In the study, the breakdown voltages (e.g., SiGe HBT BV_{EBO}, and BV_{CBO} breakdown voltages) for three different transistors: a high performance HBT, a medium performance HBT, and a high breakdown HBT were evaluated. Comparisons were evaluated to determine if the mechanical stress from the solder balls modulated the breakdown voltage or lead to an early failure. Experimental results showed no substantial difference in the breakdown voltages between the case of a solder ball/pad versus the control case. For the RF CMOS, both RF MOSFETs and triple-well MOSFET structures were placed under wire bond pad. RF MOSFETs and triple-well isolated MOSFET $I–V$ current characteristics were evaluated with and without solder balls, with no apparent issue with the active RF structures under the pads. From all of the testing (e.g., the d.c., a.c. and ESD stress), there was no significant concern.

An additional practical issue is that the placement of RF components under wire bond pads can influence the RF model. Figure 2.27 and 2.28 show the evaluation of the S-parameters as a function of frequency from 0 to 50 GHz. The parameters S_{11} and S_{22} both show that as the frequency increases, the difference of separation between the structure with wire bond pad and the structure without the wire bond pad increases. For the case of S_{11}, at low frequency, S_{11} is near a unity value. As the frequency increases, S_{11} also increases. E. Gebreselasie, and A. Das Gupta showed at approximately 20 GHz, the case of with and without the wire bond separate, and below 20 GHz, there was little observed differences. Note that for both S-parameters, the separation between the two cases occurs after 30 GHz [48].

In the electrical connections of an RF ESD structure under a wire bond pad, there is an interesting tradeoff between the requirements that will be needed to avoid mechanical cracking under a wire bond pad, the optimum wire connections to minimize capacitance

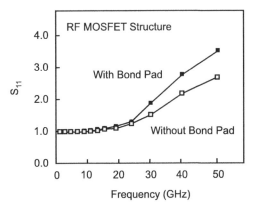

Figure 2.27   RF MOSFET scattering parameter S_{11} with and without wire bond pad

loading, and prevent ESD induced interconnect failure. Mechanical cracking concerns will be reduced using metal design levels near the silicon surface and by avoid the higher level metallization levels. At the same time, it is undesirable to have ESD currents to flow in the lower level metals owing to the metal film thickness. And, additionally, the capacitance loading of the RF component can be reduced by avoiding the metal levels near the surface. Hence, the metal design of an RF ESD circuit under a wire bond pad must address the issue of mechanical stability to failure, the ESD-induced metallization failure and the RF capacitance load from the interconnects.

On the issue of RF ESD structures under pads, an RF ESD design practice is as follows:

- Significant area reduction in semiconductor chip size for RF components can be achieved in low pin count pad-limited chip design by the placement of RF ESD elements under pads (e.g., RF input ESD networks, and RF ESD power clamps).

- RF CMOS and RF BiCMOS technologies have lower risks of insulator mechanical failure because of the thicker interlevel dielectric films utilized for high quality factor inductor elements and passives (compared to CMOS technology).

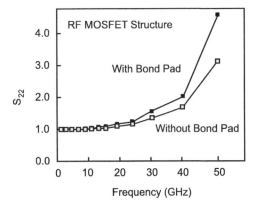

Figure 2.28   RF MOSFET scattering parameter S_{22} with and without wire bond pad

- RF ESD structures under bond pads must address the tradeoff between the mechanical cracking issues (which prefers metal connections close to the silicon surface) and capacitance loading issues of the interconnects (which prefers metal connections far from the silicon surface).

- Placement of RF ESD structures under wire bond pads must evaluate the d.c., RF, and ESD measurements with and without the bond pad.

- Placement of RF ESD structures under wire bond pads must evaluate the change in the RF circuit model to evaluate the influence on the ESD elements, and total ESD circuit.

2.13 SUMMARY AND CLOSING COMMENTS

In this chapter, we have discussed RF codesign methods and techniques, as well as address ESD design integration in mixed signal chips that contain digital, analog, and RF subfunctions, within a common chip. We have also discussed in more depth the specific methods of ESD RF cosynthesis—inductive shunts, impedance isolation, cancellation, linearity optimization, distributed loads, and matching. Using inductor shunts, direct discharge can be established to ground. Additionally, the use of inductors in combination with diode elements allows for both rectification, current discharge and 'tuning out' the capacitive loading effect of the semiconductor diodes. Using the inductor in parallel with the ESD element, the loading effect of the ESD element is 'cancelled' from an impedance loading perspective. An alternative modification, using LC tanks in series with ESD elements, it was shown that the loading effect of the series element is reduced at LC tank resonance frequencies. Additionally, using two elements where one has an increasing bias voltage and the second has a decreasing bias voltage, linearity improvements can be realized. In addition, using distributed loads, the ESD load at application frequencies can be reduced. And finally, with the integration of the ESD element and RF input and output matching circuits, the ESD function and the matching function elements can be mixed and matched; where sometimes the function is for ESD protection, matching or both, and sometimes the passive element is being used for ESD protection, matching or both.

In this chapter, we have discussed the important concepts of integration between digital, analog, and RF chip sectors; the discussion is related to RF issues where there is a higher concern of capacitive coupling induced noise. Integration of octagonal RF pads and octagonal ESD elements as well as structures under RF pads, and the implications are discussed.

In Chapter 3, we will discuss the ESD protection of RF CMOS. We will also discuss the type of ESD element choice, RF ESD design layout, RF ESD metrics, RF ESD passive elements, and RF ESD degradation issues in active and passive elements.

PROBLEMS

1. Plot the impedance of a capacitor on an impedance versus log frequency plot for a given capacitor of size, C_0 at zero frequency. Plot a constant impedance on the plot. Show the dif- ference between the ideal frequency independent load and the frequency dependent load.

2. Plot the impedance of an inductor on an impedance versus log frequency plot for a given inductor of size, L_0, at zero frequency. Plot a constant impedance on the plot. Show the difference between the ideal frequency independent load and the frequency dependent load.

3. Given an LC tank with an inductor and a capacitor ideal element. Derive the impedance of the LC tank circuit. Derive the reactance of the LC tank network. Derive the quality factor of the LC tank. Plot it as a function of frequency.

4. Given a design constraint of a constant reactance, derive the required capacitance as the frequency increases. Assume the capacitance is an ESD diode. How does the diode size decrease with the application frequency increase to maintain constant reactance? What is the ESD robustness required per unit micron as the frequency increases to maintain a constant ESD robustness?

5. Given a design constraint of a constant reactance, derive the required capacitance as the frequency increases. Assume that the capacitance is an ESD diode. Assume a constant ESD robustness must also be maintained as the frequency of the RF application increases. What is the ESD robustness required per unit micron as the frequency increases to maintain a constant ESD robustness?

6. Draw all possible two component matching filters for L-match topology, using resistor, capacitor and inductor elements. Which matching filters would not be suitable for ESD events? Why?

7. Draw all possible two component matching filters for T-match topology, using resistor, capacitor and inductor elements. Which matching filters would not be suitable for ESD events? Why?

8. Draw all possible two component matching filters for Π-match topology, using resistor, capacitor, and inductor elements. Which matching filters would not be suitable for ESD events? Why?

9. Show examples of the three different topologies, where in each case an inductive shunt is used. Assume the series element is a series capacitor. How will the circuit respond to CDM events?

10. As in the above problem, assume the series element is a resistor. How will the circuit respond to ESD events on the input node? How will the circuit respond to CDM events?

11. In technology scaling, the substrate resistance increases to reduce the capacitance loading and to improve noise isolation between circuits. Assuming a substrate resistance scaling parameter, α, derive how this influences the diode ideality of a n-well to substrate diode?

12. In technology scaling, the substrate resistance increases to reduce the capacitance loading and to improve noise isolation between circuits. Assuming a substrate resistance scaling parameter, α, derive the effect on the self-heating (e.g., temperature is equal to the power times the thermal resistance $T = PR_{TH}$).

13. Given a square signal pad, form an octagonal pad contained within the pad. What is the dimension of the side of the octagonal pad? What is the ratio of the capacitance of the

octagonal pad compared to the square pad? What is the maximum diameter of the ESD element? Given an ESD capacitance per unit area of C_{ESD}, and a pad capacitance of capacitance per unit area, C_{pad}, derive the loading relationship for the cases of a square pad, a octagonal pad, and for all cases of the number of diodes (e.g., $N = 0$, 1, 2, 3 and 4).

14. Assuming the reactance is a constant, derive a required RF ESD scaling relationship as frequency is increased according to a scaling parameter, α.

REFERENCES

1. Voldman S. ESD: *Circuits and Devices*. Chichester, England: John Wiley & Sons, Ltd; 2005.
2. Voldman S. *ESD: Physics and Devices*. Chichester, England: John Wiley & Sons, Ltd; 2004.
3. Voldman S. The state of the art of electrostatic discharge protection: physics, technology, circuits, design, simulation, and scaling. *IEEE Journal of Solid-State Circuits* 1999;**34**(9):1272–82.
4. Lee TH. *The design of CMOS radio frequency integrated circuits*. Cambridge University Press; 1998.
5. Kleveland B, Diaz CH, Vook D, Madden L, Lee TH, Wong SS. Monolithic CMOS distributed amplifier and oscillator. *Proceedings of the International Solid State Circuits Conference (ISSCC)*, 1999. p. 70–1.
6. Kleveland B, Lee TH. Distributed ESD protection device for high speed integrated circuits. U.S. Patent No. 5,969,929 (October 19, 1999).
7. Kleveland B, Maloney TJ, Morgan I, Madden L, Lee TH, Wong SS. Distributed ESD protection for high speed integrated circuits. *IEEE Transactions Electron Device Letters* 2000; **21**(8):390–2.
8. Richier C, Salome P, Mabboux G, Zaza I, Juge A, Mortini P. Investigations on different ESD protection strategies devoted to 3.3 V RF applications (2 GHz) in a 0.18 μm CMOS process. *Proceedings of the Electrical Overstress/Electrostatic Discharge (EOS/ESD) Symposium*, 2000. p. 251–60.
9. Kleveland B, Diaz CH, Vook D, Madden L, Lee TH, Wong SS. Exploiting CMOS reverse interconnect scaling in multi-gigahertz amplifier and oscillator design. *IEEE Journal of Solid-State Circuits* 2001;**36**:1480–8.
10. Voldman S. ESD Protection and RF Design, *Tutorial J, Tutorial Notes of the Electrical Overstress/ Electrostatic Discharge (EOS/ESD) Symposium*, 2001.
11. Ito C, Banerjee K, Dutton RW. Analysis and design of ESD protection circuits for high frequency/ RF applications. *IEEE International Symposium on Quality and Electronic Design (ISQED)*, 2001. p. 117–22.
12. Ito C, Banerjee K, Dutton R. Analysis and optimization of distributed ESD protection circuits for high-speed mixed-signal and RF applications. *Proceedings of the Electrical Overstress/Electro-static Discharge (EOS/ESD) Symposium*, 2001. p. 355–63.
13. Leroux P, Steyart M. High performance 5.25 GHz LNA with on-chip inductor to provide ESD protection. *IEEE Transactions Electron Device Letters* 2001;**37**(7):467–69.
14. Janssens J. *Deep submicron CMOS cellular receiver front-ends*. PhD thesis, K.U. Leuvens, Belgium, July 2001.
15. Leroux P, Janssens J, Steyaert M. A 0.8 dB NF ESD-protected 9 mW CMOS LNA. *Proceedings of the IEEE International Solid State Circuits Conference (ISSCC)*, 2001. p. 410–1.
16. Leroux P, Steyaert M. A high performance 5.25 GHz LNA with an on-chip inductor to provide ESD protection. *IEEE Electron Device Letters* 2001;**37**(5):467–9.
17. Worley ER, Bakulin A. Optimization of input protection for high speed applications. *Proceedings of the Electrical Overstress/Electrostatic Discharge (EOS/ESD) Symposium*, 2002. p. 62–72.

18. Ker MD, Lo WY, Lee CM, Chen CP, Kao HS. ESD protection design for 900 MHz RF receiver with 8 kV HBM ESD robustness. *Proceedings of the IEEE Radio Frequency Integrated Circuit (RFIC) Symposium*, 2002. p. 427–30.

19. Lee CM, Ker MD. Investigation of RF performance of diodes for ESD protection in GHz RF circuits. *Proceedings of the Taiwan Electrostatic Discharge Conference (T-ESDC)*, 2002. p. 45–50.

20. Leroux P, Janssens J, Steyaert M. A new ESD protection topology for high frequency CMOS Low noise amplifiers. *Proceedings of the IEEE International Symposium on Electromagnetic Compatibility*, September 2002. p. 129–33.

21. Leroux P, Vassilev V, Steyaert M, Maes H. A 6 mW 1.5 dB NF CMOS LNA for GPS with 3 kV HBM protection. *Proceedings of the Electrical Overstress/Electrostatic Discharge (EOS/ESD) Symposium*, 2002. p. 18–25.

22. Ker MD, Lee CM. ESD protection design for GHz RF CMOS LNA with novel impedance isolation technique. *Proceedings of the Electrical Overstress/Electrostatic Discharge (EOS/ESD) Symposium*, 2003. p. 204–13.

23. Leroux P, Steyaert M. RF-ESD co-design for high performance CMOS LNAs. *Proceedings of the Workshop on Advances in Analog Circuit Design*, Graz, Austria, 2003.

24. Leroux P, Vassilev V, Steyaert M, Maes H. High performance, low power CMOS LNA for GPS applications. *Journal of Electrostatics* 2003;**59**(3–4):179–92.

25. Vassilev V, Thijs S, Segura PL, Leroux P, Wambacq P, Groseneken G, *et al.* Co-design methodology to provide high ESD protection levels in the advanced RF circuits. *Proceedings of the Electrical Overstress/Electrostatic Discharge (EOS/ESD) Symposium*, 2003. p. 195–203.

26. Ker MD, Kuo BJ. New distributed ESD protection circuit for broadband RF ICs. *Proceedings of the Taiwan Electrostatic Discharge Conference (T-ESDC)*, 2003. p. 163–8.

27. Hynoven S, Joshi S, Rosenbaum E. Comprehensive ESD protection for RF inputs. *Proceedings of the Electrical Overstress/Electrostatic Discharge (EOS/ESD) Symposium*, 2003. p. 188–94.

28. Hynoven S, Joshi S, Rosenbaum E. Cancellation technique to provide ESD protection for multi-GHz RF inputs. *IEEE Transactions Electron Device Letters* 2003;**39**(3):284–6.

29. Hynoven S, Rosenbaum E. Diode-based tuned ESD protection for 5.25 GHz CMOS LNAs. *Proceedings of the Electrical Overstress/Electrostatic Discharge (EOS/ESD) Symposium*, 2003. p. 188–94.

30. Ker MD, Kuo BJ. Optimization of broadband RF performance and ESD robustness by π-model distributed ESD protection scheme. *Proceedings of the Electrical Overstress/Electrostatic Discharge (EOS/ESD) Symposium*, 2004. p. 32–9.

31. Hsiao YW, Kuo BJ, Ker MD. ESD protection design for a 1–10 GHz wideband distributed amplifier in CMOS technology. *Proceedings of the Taiwan Electrostatic Discharge Conference (T-ESDC)*, 2004. p. 90–4.

32. Rosenbaum E. ESD protection for multi-GHz I/Os. *Proceedings of the Taiwan Electrostatic Discharge Conference (T-ESDC)*, 2004. p. 2–7.

33. Leroux P, Steyaert M. *LNA-ESD Co-design for fully integrated CMOS wireless receivers*. Dordecht, The Netherlands: Springer; 2005.

34. Merril R, Issaq E. ESD design methodology. *Proceedings of the Electrical Overstress/Electrostatic Discharge (EOS/ESD) Symposium*, 1993. p. 233–8.

35. Dabral S, Aslett R, Maloney TJ. Designing on-chip power supply coupling diodes for ESD protection and noise immunity. *Proceedings of the Electrical Overstress/Electrostatic Discharge (EOS/ESD) Symposium*, 1993. p. 239–50.

36. Dabral S, Aslett R, Maloney TJ. Core clamps for low voltage technologies. *Proceedings of the Electrical Overstress/Electrostatic Discharge (EOS/ESD) Symposium*, 1994. p. 141–9.

37. Maloney TJ, Dabral S. Novel clamp circuits for power supply protection. *Proceedings of the Electrical Overstress/Electrostatic Discharge (EOS/ESD) Symposium*, 1995. p. 1–12.

38. Ker MD, Liu SC. Whole-chip ESD protection design for submicron CMOS technology. *Proceedings of the International Electron Device and Materials Symposium*, Taiwan, 1996. p. 55–8.

39. Ker MD. Whole chip ESD protection design with efficient VDD to VSS ESD clamp circuit for submicron CMOS VLSI. *IEEE Transactions of Electron Devices* 1999;**ED-46**(1):173–83.

40. Torres C, Miller JW, Stockinger M, Akers MD, Khazhinsky MG, Weldon JC. Modular, portable, and easily simulated ESD protection networks for advanced CMOS technologies. *Proceedings of the Electrical Overstress/Electrostatic Discharge (EOS/ESD) Symposium*, 2001. p. 82–95.

41. Stockinger M, Miller JW, Khazhinsky MG, Torres CA, Weldon JC, Preble BD, *et al.* Boosted and distributed rail clamp networks for ESD protection in advanced CMOS technologies. *Proceedings of the Electrical Overstress/Electrostatic Discharge (EOS/ESD) Symposium*, 2003. p. 17–26.

42. Worley E. Distributed gate ESD network architecture for inter-domain signals. *Proceedings of the Electrical Overstress/Electrostatic Discharge (EOS/ESD) Symposium*, 2004. p. 238–47.

43. Huh Y, Bendix P, Min K, Chen JW, Narayan R, Johnson LD, *et al.* ESD-induced internal core device failure: new failure modes in system-on-chip (SOC) designs. *Proceeding of the 2005 International Workshop on System on Chip (IWSOC)*, July 20–24, 2005, Banff, Alberta, Canada.

44. Countryman R, Gerosa G, Mendez H. Electrostatic discharge protection device, U.S. Patent No. 5,514,892 (May 7, 1996).

45. Chittipeddi S, Cochran W, Smooha Y. Integrated circuit with active devices under bond pads, U.S. Patent No. 5,751,065 (May 12, 1998).

46. Anderson W. ESD protection under wire bond pads. *Proceedings of the Electrical Overstress/Electrostatic Discharge (EOS/ESD) Symposium*, 1999. p. 88–94.

47. Chittipeddi S, Cochran W, Smooha Y. Process for forming a dual damascene bond pad structure over active circuitry. U.S. Patent No. 6,417,087 (July 9, 2002).

48. Gebreselasie EG, Sauter W, St. Onge S, Voldman S. ESD structures and circuits under bond pads for RF BiCMOS silicon germanium and RF CMOS technology. *Proceedings of the Taiwan Electrostatic Discharge Conference (T-ESDC)*, 2005. p. 73–8.

3 RF CMOS and ESD

In CMOS technology, through the 1990s to 2000, the ESD focus was to understand how the evolutionary and revolutionary changes in CMOS influenced ESD robustness and scaling [1–18]. In that time frame, technology transitions included high resistance substrates, retrograde implanted wells, shallow trench isolation, silicides, tungsten stud contacts, and copper interconnects to low-k interlevel dielectrics. As discussed in the last two items on decoupling, establishing a means of decoupling from a power supply rail prevents "pinning" of nodes to the power rail during ESD testing and undesirable current paths.

With the migration to RF CMOS, few revolutionary changes are occurring in CMOS technology. The primary change is MOSFET scaling to thin dielectrics and small MOSFET channel lengths. With these issues, the MOSFET performance has increased to allow RF applications to be acceptable in technology.

In RF CMOS, the ESD strategy that is taking place is a function of the application frequency. As the application frequencies are increasing from 1 to 10 GHz, the decisions and choices of usage of ESD, cosynthesis, and the utilization of RF design techniques are taking a greater role.

With future technology generations and the power supply scaling, noise reduction between digital, analog, and RF circuits become more critical. To reduce noise and junction capacitance, the substrate wafer resistance is scaled to higher substrate doping concentrations. Figure 3.1 shows the scaling of the substrate resistivity with time as a function of application frequencies. Figure 3.1 shows that as the RF CMOS is scaled, the direction is to higher substrate resistance. Substrate resistance will influence the RF CMOS ESD capacitance, the choice of designs and technology ESD robustness.

3.1 RF CMOS: ESD DEVICE COMPARISONS

Choosing the correct ESD device or ESD network for a given RF CMOS application is important as RF CMOS becomes an important technology in RF communication systems [19]. RF CMOS has an advantage in the RF marketplace because of the high density, low cost, integration advantages, and global proliferation of CMOS technology foundries. RF

ESD: RF Technology and Circuits   Steven H. Voldman
© 2006 John Wiley & Sons, Ltd

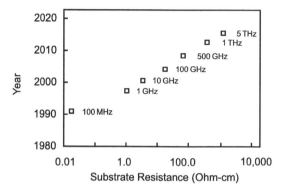

Figure 3.1   Scaling of substrate wafer resistivity as a function of time and application frequency

CMOS can be utilized in RF communication systems and optical interconnect systems. In advanced optical communication systems, the systems contain CMOS, RF CMOS, silicon germanium (SiGe), and gallium arsenide (GaAs); each plays a unique role in the achievement of performance objectives. With MOSFET scaling, as the MOSFET effective channel length decreases, the unity current gain cutoff frequency (f_T) increases. As a result, RF CMOS can be utilized in high-speed applications.

An additional advantage is that RF CMOS can be integrated with standard CMOS, SiGe, and silicon germanium carbon (SiGeC) technologies. Today, RF CMOS technologies can be established by conversion of CMOS technology by the addition of passive elements (e.g., high quality factor inductors, resistors, and capacitors), or spin-off from BiCMOS silicon germanium technology by the removal of the silicon germanium bipolar transistor. Additionally, RF CMOS can be integrated coincidently with CMOS and SiGe technology on a common system-on-chip (SOC) or network-on-chip (NOC).

For a RF CMOS technology, the choice of an ESD network is a function of the following:

- ESD structure robustness (e.g., power-to-failure, or critical current to failure);

- ESD structure area (power-to-failure per unit area, or critical current per unit area);

- capacitance loading (e.g., total capacitance load on the input circuit);

- impedance (e.g., influence on the matching characteristics);

- trigger voltage;

- noise factor.

In RF CMOS, the following ESD elements can be utilized [19]:

- grounded gate n-channel MOSFET (GGNMOS);

- shallow trench isolation (STI) defined diodes;

- polysilicon-defined diodes;

- silicon controlled rectifiers (SCRs).

For evaluation, critical ESD metrics can be established by the ratio of the various parameters (e.g., the ratio of ESD robustness to capacitance load). Richier *et al.* [19] provided a study of comparison of the various ESD options for RF CMOS applications where there were two fundamental criteria.

- The loading capacitance was required to be below a given critical capacitance value.

- An ESD metric was established as the ratio of ESD robustness per unit capacitance (e.g., ESD HBM robustness [kV]/capacitance [fF]).

From this metric, ESD design curves can be constructed that evaluates the relative quality of different ESD devices. Establishing an ESD design curve, where the y-axis is the loading capacitance of the ESD network, and the x-axis is the ESD HBM failure level, the different devices can be compared for the aforementioned metrics. The slope of the experimental results can be expressed at any size device, as

$$\frac{\Delta C_{ESD}}{\Delta V_{ESD}}$$

The inverse of the slope can also be expressed as a FOM_{ESD}

$$FOM_{ESD} = \frac{\Delta V_{ESD}}{\Delta C_{ESD}}$$

The ESD results can be expressed as a linear relationship, where the first term is the intercept of the ESD results at x-axis intercept (e.g., load capacitance is zero).

$$V_{HBM} = \{V_{HBM}\}_0 + FOM_{ESD}C_{ESD}$$

As a first example, a plot of the loading capacitance versus the ESD HBM failure level of a grounded gate MOSFET is analyzed, on the basis of the results of Richier *et al.* [19]. Figure 3.2 shows the plot of a grounded gate MOSFET loading capacitance vs. ESD failure voltage level with and without MOSFET drain extension. To improve the MOSFET ESD

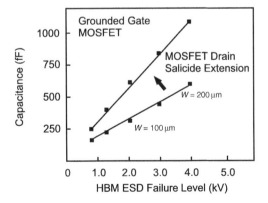

Figure 3.2 RF CMOS ESD device comparison of a grounded gate n-channel MOSFET with and without drain extension

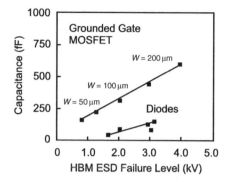

Figure 3.3 RF CMOS ESD device comparison of a grounded gate n-channel MOSFET and STI diode structure

robustness, a standard practice is the extension of the drain region as a resistor ballasting element. With the increase in the drain area, the capacitance load of the ESD protection device increases. As can be observed from the figure, the slope of the $C_{ESD}-V_{ESD}$ plot increases with the increase in the drain extension; this degrades the figure of merit, FOM_{ESD}.

Figure 3.3 shows a plot of the comparison of a grounded gate n-channel MOSFET and a STI diode structure [3–7,11]. Experimental results show that in the case of the MOSFET structure, for a given ESD failure level, the load capacitance is higher than the STI diode structure. For an HBM failure level of 2000 V, the MOSFET loading capacitance exceeds 200 fF, whereas the diode structure remains under 200 fF, even up to 3000 V HBM levels. Additionally, the ESD figure of merit (e.g., kV/fF) is higher for the STI diode structure. Hence, as these structures are scaled to lower capacitance objectives, the diode structure has an advantage over the grounded gate MOSFET structure.

In a standard CMOS technology, a four region SCR can be formed using the p^+ diffusion, the n-well region, the p^- substrate, and an n^+ diffusion [15–17]. SCR structures have the advantage of regenerative feedback to provide a low voltage–high current state providing good current conduction. Figure 3.4 shows the comparison of a GGNMOS ESD structure and a SCR ESD structure on the $C_{ESD}-V_{ESD}$ design curves. From the plot, the first point of interest is that the SCR can achieve over 2000 V HBM ESD failure levels for capacitance

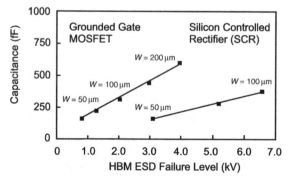

Figure 3.4 RF CMOS ESD device comparison of a grounded gate n-channel MOSFET and SCR structure

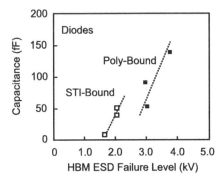

Figure 3.5   STI-defined and poly-silicon gate-defined ESD diode structures

loads under 200 fF. Second, the FOM_{ESD} is significantly greater using a SCR compared to the GGNMOS ESD device. Hence, from these results, it is clear that it is possible to construct SCR ESD networks with low load capacitance and a high FOM_{ESD} as structures are scaled in future RF technologies.

For RF CMOS applications, different type of diode structures can be utilized. Additional to the STI-defined, a poly-silicon-bordered diode structure (also referred to as a poly-silicon-bound diode structure) can be utilized for a low turn-on voltage, low capacitance ESD structure [12]. Voldman first introduced the poly-silicon-bound ESD diode structure for bulk CMOS and SOI technology to address STI pull-down effects and to integrate with SOI technology [12]. For bulk CMOS technology, the poly-silicon-bound diode structure was used to avoid STI process integration and salicide issues. For SOI technology, it was necessary to eliminate STI regions to provide lateral current flow. Figure 3.5 shows the STI-defined and poly-silicon gate-defined ESD diode structures on the $C_{\text{ESD}} - V_{\text{ESD}}$ design plot. From the work of Richier *et al.*, both structures can achieve 2000 V HBM levels well below 200 fF capacitance loads. Second, the slope on the design plots showed equivalent results, but the x-intercept for the poly-silicon-gated diode structure was superior. The significant outcome from these experimental studies demonstrate that both the STI-bound ESD diode structure and the poly-silicon-bound ESD diode structure can be utilized for RF CMOS applications [19].

In the RF CMOS ESD design, the following RF ESD design practice can be applied:

- RF ESD design plots of the capacitance load versus the ESD robustness can be utilized for determination of the intercept and slope for evaluation of the equation

$$V_{\text{HBM}} = \{V_{\text{HBM}}\}_0 + \text{FOM}_{\text{ESD}} C_{\text{ESD}}$$

where

$$\text{FOM}_{\text{ESD}} = \frac{\Delta V_{\text{ESD}}}{\Delta C_{\text{ESD}}}$$

- The intercept and the slope of the RF ESD design plot can be utilized for evaluation of acceptable loading capacitance and scaling implications.

3.2 CIRCULAR RF ESD DEVICES

Low capacitance diode designs are that have a high figure of merit of ESD robustness per unit area is desirable for radio frequency (RF) applications. Circular ESD diode designs are desirable in RF applications because of the following issues [20–22]:

- small physical area [23–25];

- elimination of isolation and salicide issues [12];

- elimination of corner effects [12];

- elimination of wire distribution impact on ESD robustness (e.g., parallel and antiparallel wire distribution issues [23–25]);

- current density symmetry;

- integration with RF octagonal bond pad structures.

Circular diodes can be placed in small physical areas under bond pads, whether square or octagonal pad structures. Additionally, the small diode structures can be placed in the center or corners of RF octagonal pads.

Isolation issues and corner issues can be eliminated using circular ESD structures because of the enclosed nature of the anode or cathode structures. Figure 3.6 shows an example of an RF ESD diode structure with the p^+ anode in the center area; this is separated by an isolation region, and an n^+ cathode ring structure. As a result, there are no corners in the anode structure which can lead to current concentrations or three-dimensional current distribution effects. It has been shown that the corners of linear p^+ diode structures can form ESD failure in shallow trench defined diode structures [3,4]; this was evident from optical microscope image, atomic force microscope images, and emission microscope (EMMI) tool emissions. Three-dimensional electro-thermal simulation also indicated this as a region of peak temperature during ESD HBM stress [7]. Hence circular diodes have the advantage of avoidance of these geometrical issues. In linear diode structures, metal distribution can play a role in the nonuniform current distribution; nonuniform current distribution impacts the ESD FOM of the ratio of the ESD robustness to capacitance load. Additionally, due to the

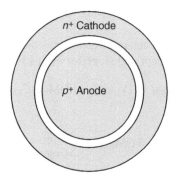

Figure 3.6   RF circular ESD diode pad structure with a p^+ anode in the center region enclosed by an n^+ cathode

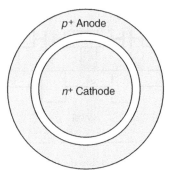

Figure 3.7   RF circular ESD diode pad structure with an n^+ cathode in the center region enclosed by a p^+ anode

physical symmetry, there is a natural design symmetry and no preferred directionality. Additionally, a geometrical advantage of a circular diode is the radial current flow; the geometrical factor of the $1/r$ current distribution leads to a radially decreasing current density from the center.

Figure 3.7 shows a RF circular ESD diode element with the n^+ cathode in the center region, enclosed by a p^+ anode region; this structure can be utilized for negative polarity ESD events.

An additional advantage can be established by the reduction of the capacitance associated with the metal wiring [23–25]. With the circular geometry, the capacitance load of the anode-to-cathode metal capacitance can be optimized. In the case of the multi-finger linear diode, the spacing is established by the wire-to-wire pitch of the anode and cathode element.

The disadvantage of the RF ESD circular element are as follows:

- design width and area limitations;

- computer-aided design (CAD) issues with nonrectangular shapes;

- CAD automation convergence and process time;

- CAD-parameterized cells for nonrectangular shapes is not always possible preventing variation in the size of the RF ESD element for different circuits or applications;

- lithography, etching, and polishing issues;

- limitation on the wire density.

An RF ESD design practice is as follows:

- Utilization of circular ESD networks for RF applications have the advantage of providing improved current uniformity, elimination of isolation effects, elimination of geometric, and semiconductor process corner effects.

- Design integration can be established by utilization in the corner areas of octagonal bond pads.

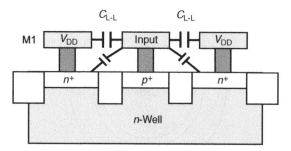

Figure 3.8 An RF ESD diode structure highlighting the interconnect capacitance terms for a structure whose input wiring and output wiring are on the same film level (e.g., input and output on M1 metal level)

3.3 RF ESD DESIGN—ESD WIRING DESIGN

Low capacitance diode designs are that have a high figure of merit of ESD robustness per unit area is desirable for RF applications. The capacitance load components of an ESD diode network is associated with the metallurgical junction capacitance of the diode elements and the interconnect wiring capacitance.

In CMOS technology, the scaling of the metal lines was a fundamental ESD scaling issue in high pin count applications [26–29]. At that time, the metal lines were narrow for wiring density requirements, performance, and reduce capacitance loading effect. Signals were lowered from top level of metal through the lower levels through a 'via stack'; at the time, this was primarily performed to avoid wire track blocking of lower level signal lines.

With the introduction of RF CMOS, the issue is changed purely to loading capacitance as the primary concern. In RF CMOS, the issue is with the scaling of the diode sizes, the metal capacitance is becoming a substantial part of the loading capacitance [30]. Figure 3.8 is an example of an STI defined diode structure highlighting the capacitance elements in the interconnect region. The example shows the input and the output (e.g., V_{DD}) on the same metallization level. In this representation, the capacitance can be represented as a capacitor between the metal input to the silicon surface. A capacitor is formed between the input wire and the silicon cathode region. Additionally, a second capacitor element, a line-to-line capacitor, C_{L-L}, is formed between the two metallurgy films. The first capacitor term, the input line to silicon capacitor element will be a function of the edge of the metal film relative to the cathode implant edge and total metal perimeter. The line-to-line capacitor, C_{L-L}, is a function of the metal film height, the spacing between the input metal film and the V_{DD} metal film, the total perimeter of the diode structure, and the dielectric constant for the dielectric between the two films. As higher metal levels are used, the metal interconnect-to-silicon surface capacitance will decrease. Concurrently, for the same spacing the metal thickness increases, the line-to-line capacitance term becomes a more significant part of the total interconnect capacitance. In advanced semiconductors, the line-to-line capacitance can be reduced with the use of low-k materials.

To reduce the interconnect capacitance, it is a common practice in high performance digital applications to transfer the input signal from the top metal pad level through a via

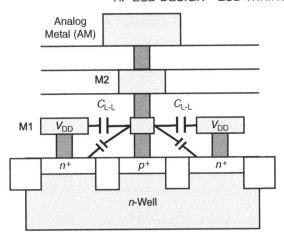

Figure 3.9   An RF ESD diode structure highlighting the interconnect capacitance terms for a structure whose input wiring and output wiring are on different metal film levels (e.g., input on AM level and output on M1 level)

stack to the silicon surface. This serves a first purpose of providing a low resistance path through the interconnect system. Second, this prevents the need to widen the metal film in off-chip driver and receiver circuits as the signal is transmitted from the metal pad level to the silicon surface (note: with the thinner metal films near the surface, interconnect ESD robustness would decrease due to smaller film cross-sectional area). A third advantage is the avoidance of wire channel blockage in densely wired designs. It was pointed out by Hynoven and Rosenbaum [30] that this metal design practice also has an advantage in RF ESD design.

Figure 3.9 shows an example of an RF ESD diode structure where the input signal is transferred from the analog metal (AM) level to the silicon surface, where the output wiring is preserved on the lower metal level. In the figure, the input signal film is at the AM design level, and the output V_{DD} wiring remains at the lowest metallization level. In this implementation, the interconnect metal can be reduced due to metal-to-silicon capacitor area, and wider gap spacing (metal-to-metal); but, as the metal via stack penetrates to the input node, the via stack/first level metal-to-silicon capacitor and the line-to-line capacitor is still present.

An RF ESD design practice is as follows:

- Interconnect film capacitance can play a role in the total capacitance load of an RF ESD network.

- Interconnect film capacitance includes the metal-to-silicon capacitance and metal line-to-line capacitance.

- Using via stacks for the metal input, the ESD robustness of the metallurgy and the interconnect capacitance load can be reduced.

- Using the highest metal film for the RF ESD input node and the lowest metal film for ground and V_{DD} connections will provide the lowest interconnect capacitance.

3.4 RF PASSIVES: ESD AND SCHOTTKY BARRIER DIODES

Silicon Schottky Barrier Diodes

Schottky barrier diode (SBD) elements are the key elements in high-speed digital applications and RF communications. In the 1970s and the 1980s, Schottky barrier diodes were used within high performance emitter-coupled logic (ECL) bipolar SRAM cells to prevent transistor saturation. The usage in memory applications lead to significant work in the Schottky barrier diode reliability. In the same time frame, Schottky barrier diodes were used in military applications as mixers and detectors in receiver networks because of the high reliability and RF characteristics. Schottky barrier diodes could achieve low noise figure. Today, interest still exists in high-speed test systems, and as an RF component in modern RF design systems.

Anand first explored the tradeoffs between RF power-to-failure, ESD robustness, and the desired RF characteristics with the development of Schottky barrier diode for 8–12 GHz (e.g., X-band) mixer and detector applications [31–38]. Anand noted that the choice of barrier metal influences two physical parameters. The choice of barrier metals influences the RF characteristics as well as the power-to-failure:

- Schottky barrier height (metal–semiconductor interface).

- Eutectic temperature of the semiconductor–metal interface.

For functional RF performance, providing a low Schottky barrier height improves

- low capacitance;

- improved switching speed;

- low noise figure (NF);

- good RF matching characteristics.

For RF power and ESD robustness, the high eutectic temperature provides a higher breakdown voltage and improved power-to-failure. From the Wunsch–Bell model, the power-to-failure is proportional to the critical temperature. If the eutectic temperature is equal to the critical temperature of failure, then the power-to-failure increases linearly with the increase in the eutectic temperature.

In the work of Anand, Moroney, Morris, Higgins, Hall, and Cook, various metallurgical structures were evaluated that varied both the Schottky barrier height and the eutectic temperature [33]. Schottky barrier diodes were constructed with different refractory metals and composite structure: titanium–gold (Ti–Au), titanium–molybdenum–gold (Ti–Mo–Au), and platinum–titanium–molybdenum–gold (Pt–Ti–Mo–Au). The first two composite structures show a low barrier height and a low breakdown voltage, whereas the latter exhibited a higher barrier height and higher breakdown voltage.

For HBM ESD evaluation, the Schottky barrier diodes were subjected to a single HBM-like RC network (note: a 1000 Ω resistor element and a 100 pF capacitor). Experimental results showed that the Ti–Au and Ti–Mo–Au Schottky diodes failed at

Table 3.1 Schottky barrier diode metallurgy versus HBM failure level

Schottky barrier diode metallurgy	Barrier height (V)	HBM failure level[a](V)
Ti–Mo–Au	0.35	300–400
Ti–Au	0.35	300–400
Pt–Ti–Mo–Au	0.75	1600

[a]$R = 1000\,\Omega$ and $C = 100\,\text{pF}$.

HBM levels of 300–400 V, whereas the Pt–Ti–Mo–Au Schottky barrier diodes failed at 1600 V (Table 3.1).

To evaluate the power-to-failure of the Schottky barrier diodes, the SBD passive elements were evaluated for very short pulse phenomenon (nanoseconds) to long pulse phenomenon (microseconds). For the RF pulse system, pulse widths were varied from 3 to 50 ns (note that the short pulse width is equivalent to present day VF-TLP systems).

Figure 3.10 is a plot of the RF pulse power-to-failure (for 3 ns and 1 μs pulses) as a function of HBM ESD failure level. From the experimental results, as the barrier height and the eutectic temperature increases, the RF pulse power-to-failure and the HBM ESD robustness also increases.

Failure analysis of the circular Schottky barrier diodes for long and short RF pulse phenomenon showed that for the short RF pulse events, the Schottky barrier diode failure occurred at the perimeter of the diode structure, and for long RF pulse width events (1 μs). Schottky barrier diode failure occurred in the center of the diode structure [31–38].

Silicon Germanium Schottky Barrier Diodes

In a Silicon germanium technology, Schottky barrier diodes can be formed using components of the SiGe HBT *npn* bipolar transistor. Using the SiGe HBT extrinsic base, collector

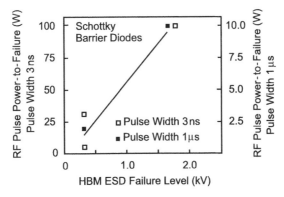

Figure 3.10 RF pulse power-to-failure (for both 3 ns and 1 μs pulse widths) versus the HBM ESD failure level

and subcollector, a Schottky diode can be formed in the emitter window region. First, a subcollector is formed during epitaxial region growth. This is followed by the STI definition and the deposition of the silicon germanium film. A window is formed to prevent the SiGe film to be formed over the device surface. A metal is deposited on the silicon surface to form the Schottky barrier interface between the refractory metal and the silicon surface. Additionally, using the extrinsic base region, p^+ dopant diffuses from the extrinsic base region into the silicon collector region.

Experimental HBM studies of Silicon germanium technology-based Schottky barrier diodes were first reported [40]. Voldman showed that good HBM ESD results are achievable when the structure utilizes a p^+ guard ring region. The p^+ guard ring serves as an anode structure, and the collector/subcollector structure serves as the cathode in the forward bias mode of operation. In the reverse bias mode, Schottky barrier diode structures do not undergo leakage with the presence of the p^+ guard ring structure [40]. Experimental results without the p^+ region lead to poor ESD results and leakage. Comparing four structures in forward bias mode formed out of the SiGe HBT transistor, HBM results were highest in the SiGe-based varactor structure, followed by the SiGe-based Schottky barrier diode, the SiGe HBT (with emitter–base shorted), and lastly, the Schottky barrier diode with no p^+ guard ring. It was concluded that this was proportional to the area of the p^+ base region.

An RF ESD design practice for Schottky barrier diodes can be stated as follows:

- Choice of the barrier metal eutectic temperature can influence the RF power-to-failure and ESD robustness.

- The HBM ESD robustness and RF power-to-failure correlate for short and long pulse width phenomenon.

- Nanosecond short pulse phenomenon demonstrate failure mechanisms near the Schottky barrier diode perimeter.

- Epitaxial-defined silicon germanium technology-based Schottky barrier diodes in forward bias mode of operation achieve high ESD robustness when a p^+ guard ring is formed from the extrinsic base of the SiGe HBT *npn* base region. ESD robustness will be a function of the p^+ guard ring total perimeter and subcollector profile and doping concentration.

3.5 RF PASSIVES: ESD AND INDUCTORS

In RF design, passive elements can undergo ESD degradation, impacting circuit functionality. ESD-induced degradation of passive elements can impact the functionality of a RF circuit or chip [39–41].

Inductors are used in peripheral circuits for resonant circuits, baluns, transformers, and as a.c. current blocks (e.g., also known as a.c. chokes), as well as other circuit applications. High quality factor inductors are important for RF applications [42–50]. On-chip inductors in semiconductor technology are constructed from the interconnect technology. The inductors consist of conductive metal films, metal contacts, metal vias, and interlevel dielectrics. Figure 3.11 shows an example of an inductor coil formed in a semiconductor process.

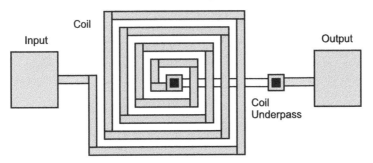

Figure 3.11   RF CMOS passive inductor coil

The quality factor, 'Q' can be defined as associated with the ratio of the imaginary and real part of the self-admittance term,

$$Q = -\frac{\text{Im}\{Y_{11}\}}{\text{Re}\{Y_{11}\}}$$

where at self-resonance, can be expressed as

$$Q = \frac{\omega}{2}\frac{\partial\phi}{\partial\omega}\bigg|_{\omega=\omega_{res}}$$

Assuming the quality factor, 'Q', for an inductor is dependent on the inductor series resistance, and the inductance (ignoring capacitance effects), the Q of the inductor with a series resistance is

$$Q = \frac{\omega L}{R}$$

In the case that there is a permanent shift due to ESD stress in the series resistance, post-ESD stress resistance is $R' = R + \Delta R$, and the quality factor 'Q' can be expressed as [41]

$$Q\prime = \frac{\omega L}{R'} = \frac{\omega L}{R + \Delta R}$$

then the shift in the inductor Q can be expressed as follows:

$$Q' - Q = \frac{\omega L}{R + \Delta R} - \frac{\omega L}{R} = \frac{\omega L}{R}\left[\frac{1}{1 + \dfrac{\Delta R}{R}} - 1\right]$$

Expressing the change in the Q, then

$$\Delta Q = Q' - Q = -\frac{Q}{R + \Delta R}\Delta R$$

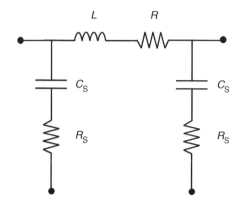

Figure 3.12 RF CMOS passive inductor coil incremental electrical model

where when the $R \gg \Delta R$, then

$$\Delta Q = -\frac{Q}{R}\Delta R$$

For incremental variations, the partial derivative of Q can be taken with respect to resistance, where

$$\frac{\partial}{\partial R}Q = \frac{\partial}{\partial R}\left(\frac{\omega L}{R}\right) = -\frac{Q}{R}$$

Figure 3.12 contains an electrical incremental model for an inductor. Figure 3.13 contains a thermal incremental model of an inductor.

From this analysis, changes in the inductor resistance from high current stress can lead to changes in the quality factor of the inductor. All physical variables associated with the inductor design can influence the ESD robustness of the inductor structure; the ESD robustness of the inductor interconnects are a function of the following layout design and semiconductor process variables [41]:

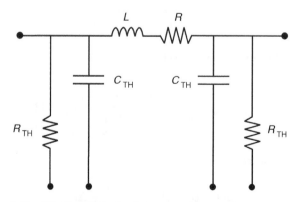

Figure 3.13 RF CMOS passive inductor coil thermal incremental model

- Coil thickness and width.

- Underpass film thickness and width.

- Via resistance.

- Physical distance from the substrate surface or nearest conductive surface.

- Interlevel dielectric materials.

- Interconnect metal fill material (e.g., aluminum, copper, gold).

- Interconnect cladding material (e.g., refractory metal such as titanium, tantalum, tungsten, etc.).

- Interconnect design (e.g., lift-off, damascene, dual damascene structure).

- Ratio of the volume of the interconnect fill material and volume of the cladding material.

- Interconnect design geometry (e.g., square coil, octagonal coil, polygon coil design).

- Interlevel dielectric fill shapes.

During ESD testing, the cross-sectional area through the complete inductor structure influences the current density in the inductor coil and its failure locations. Hence, the materials and physical dimensions of the coil, underpass wire, vias, and contacts all influence the ESD robustness of the inductor structure. In the RF technologies, the metallurgy for the interconnect are aluminum, copper, or gold. In RF CMOS and RF BiCMOS SiGe, the interconnects are typically aluminum or copper; in Gallium Arsenide, the metallurgy is typically gold. In the interconnect films, the films consists of refractory metal cladding materials that are of a different melting temperature and the resistivity. Cladding materials used in RF CMOS, RF BiCMOS Silicon Germanium, and Silicon Germanium Carbon technologies are titanium, tantalum, and tungsten. The interconnect cross-section consists of the composite of the conductor film and the refractory metal cladding material; the composite film cross-section area ratio of fill material versus cladding material influences the ESD robustness [41]. Additionally, the geometric design (e.g., vertical film stack such as Ti/Al/Ti or a damascene trough such as U-shaped trough) also influences the failure location [40,41]. In the via region, whether the vias are a separate film or a dual-damascene structure also influences the via robustness [41].

Additionally, the scaling of the highest film, the last metal (LM) film, relative to the film below the last metal film (denoted LM-1 level) is also critical. In inductor design, there is an input and output to the inductor structure. In inductor design, to input or the output may be in the center of the coil. As a result, an 'underpass wire' must leave the center of the coil underneath the inductor film. In on-chip inductors, the film of the LM-1 film is typically thinner than the LM film. The geometrical scaling of the film thickness, and whether the same cross-section area is maintained between the inductor coil cross-section and the underpass will influence the ESD robustness. Given that the underpass metal lead width is the same as the coil, the scaling of the films of the LM-1 relative to the LM will lead to ESD failure in the underpass structure [41].

Coil geometrical design can also influence the failure level of inductor structures. Inductor coil geometry influences the quality factor, the resistance, and the coil current density within a given cross-section of the coil. Inductors are designed as square coils, octagonal coils, and higher order polygonal coils. In square coils, the corners of the square lead to higher resistance, as well as poor current distribution on the corners. As the order of the polygon increases, the corner resistance and the current distribution improves, as well as the inductor quality factor. A second issue is the coil design influences the current distribution due to the self magnetic field. The current flow in the coil leads to an upward magnetic field. The inductor-induced magnetic field has a radial component (e.g., $1/r$ dependence) which influences the current within the inductor cross-section. As a result, within the coil film itself, the current distribution is nonuniform, leading to an increase in current-crowding. At high current levels, earlier ESD failure is anticipated due to the high thermal gradients.

Additionally, the interlevel dielectrics (ILD) film materials and thicknesses can influence the ESD robustness of the interconnect structure; all the film stack between the inductor and the conductive substrate influences the quality factor, but also the thermal impedance. As the effective thermal impedance between the inductor and the conductive surface, the ability for the inductor to dissipate heat decreases; this leads to degradation of the ESD robustness of the inductor. Hence the interlevel dielectric material thermal conductivity influences the ESD robustness. In RF CMOS and RF BiCMOS SiGe technology, the ILD materials are typically silicon dioxide or low-k materials such as HSQ, FSG, and SiLK^{TM}. Additionally, 'fill shapes' placed in the insulator can lower the 'effective thermal conductivity' of the composite film stack.

With the addition of trench structures, such as deep trench (DT) structures, used in BiCMOS Silicon Germanium, or power technologies, the additional insulator depth leads to an increase in the inductor quality factor. As the effective insulator thickness between the coil and the substrate increases, then the effective thermal impedance increases; this degrades ESD robustness at long pulse width time constants.

In the study of inductor failure, an inductor was tested using an HBM test commercial system by Juliano and Voldman [40,41]. In the structure, the underpass coil width was equal to the coil width. Due to the film thickness scaling of the on-chip film thicknesses, the effective cross-section of the underpass wire was less than the coil itself. The coil film surface was unpassivated. During ESD HBM testing, the coil was observed with a microscope during ESD pulsing. The transitions were observed in the ESD stressing of the coil (Figure 3.14 [41]):

- First indication: Discoloration was observed in the underpass wire.

- Second indication: Interlevel dielectric cracking was observed.

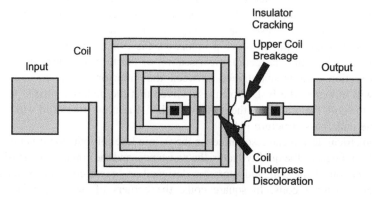

Figure 3.14   RF CMOS passive inductor coil ESD failure

- Third indication: Area of the coil over the underpass failure occurred at the outer most coil.

- Fourth indication: An 'open' occurred in the outermost coil and its output.

- Fifth indication: Continued ESD pulsing lead to an ESD arc between the outermost coil and the output of the coil film. A 'spark gap' was formed between the two points, with the spark being observed through the microscope during the ESD pulse.

3.6 RF PASSIVES: ESD AND CAPACITORS

Dielectrics historically has had significant focus on the reliability due to the need to understand dielectric breakdown [51–60]; MOSFET electric field scaling theory requires the continuous scaling of the thin oxide to achieve MOSFET device performance.

Additionally, dielectric breakdown during electrical overstress (EOS) and electrostatic discharge (ESD) had significant focus due to the concern of MOSFET gate dielectric breakdown and failure during ESD events [61–81]. A primary focus has been on the failure of NMOS and CMOS digital receiver networks where the MOSFET inverter circuit failure occurred in the *n*-channel MOSFET gate dielectric. CMOS digital receiver networks failure concerns occurred as a result of the human body model (HBM), machine model (MM), and charged device model (CDM) [66]. Considerable focus has been given to the CDM and its impacts on MOSFET gates in CMOS receiver networks [62,66–70]. The focus will change in RF applications as a result of the unique placement of the capacitor elements in peripheral circuits, where capacitor passive elements play a new role in circuits with significant topology different from digital CMOS receiver and transmitter circuits.

In RF CMOS and BiCMOS Silicon germanium, on-chip capacitor elements are needed for RF applications. On-chip capacitor passive elements are used in peripheral circuits in RF applications. Capacitors are used in many receiver and transmitters circuits in RF applications from low noise amplifiers (LNA), voltage controlled oscillators (VCO) to RF power amplifiers (PA). Capacitor elements are used in the following applications with RF circuits:

- Direct current (d.c.) blocking capacitors.

- Input and output matching networks.

- Output-to-input feedback subnetworks.

- LC tank circuits.

Capacitor elements are also integrated into semiconductor chips for frequency-triggered ESD networks which respond to the ESD pulse waveform. Capacitor elements are integrated into trigger networks and as impedance isolation elements for RF ESD input circuits and ESD power clamps, for the following:

- capacitor-coupled trigger networks;

- resistor–capacitor (RC) trigger networks;

- LC-tank impedance isolation networks.

The type of elements used for capacitors for RF applications can include:

- MOSFET-based capacitors (MOS-CAP). This is formed with the placement of an *n*-channel MOSFET in an *n*-well region.

- Silicon varactor capacitor. These can be formed using reverse-biased silicon metallurgical junctions (e.g., base–collector regions, *p*-well to triple-well region).

- Hyper-abrupt varactor capacitor. These are formed using reverse-biased silicon metallurgical junction whose junction region is tailored for improved varactor voltage dependence.

- Metal–insulator–metal (MIM) capacitor. These are formed in the interlevel dielectric region using interconnect metallurgy (e.g., refractory metal and conductive metal films) and a thin dielectric film.

- Metal-ILD-Metal capacitor. This is formed with the interconnect metallurgy levels and the interlevel dielectric is utilized as a thick dielectric film.

- Vertical parallel plate (VPP) capacitor. This is formed using two vertically formed plates where the metal capacitor plates are 'stacked vias' and metal films with the interlevel dielectric is utilized as a thick dielectric film.

3.6.1 Metal-Oxide-Semiconductor and Metal–Insulator–Metal Capacitors

Metal-oxide-semiconductor (MOS) and MIM capacitors are used in RF receiver networks as d.c. blocking circuits in CMOS and BiCMOS networks; this is very distinct from CMOS digital receiver networks. In RF receiver circuits, the MIM capacitor is placed in series between an input pad and the RF MOSFET gate (e.g., RF CMOS network), or a bipolar base input (e.g., RF BiCMOS network). Without the presence of an ESD protection network between the RF input pad and the receiver node, ESD failure of the MOS capacitor or MIM capacitor will occur from both HBM and MM type mechanisms. For example, MIM capacitor failure of capacitor elements will occur at 200–400 V HBM protection levels for a positive HBM pulse, and −100 V HBM protection levels for a negative HBM pulse. The two primary reasons for HBM and MM failure are due to the small physical size, and the presence of a dielectric thickness identical or comparable to the MOSFET gate dielectric thickness. In a MIM capacitor, the HBM ESD robustness will increase from 200 to 1000 V HBM levels as a MIM capacitor increase from 0 to 15 pF (Figure 3.15). Hence, an unprotected MIM capacitor element will undergo ESD failure under 2000 V HBM levels without a protection circuit, or protection solution in a RF receiver network.

3.6.2 Varactors and Hyper-Abrupt Junction Varactor Capacitors

Silicon-based varactor and hyper-abrupt varactor structures are constructed as RF passive elements for applications that desire voltage-dependent capacitor elements. Silicon junction varactors can be used in reverse bias mode of operation as a RF passive capacitor as a

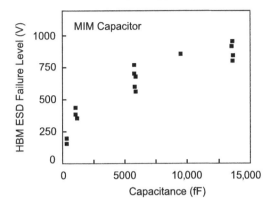

Figure 3.15   MIM capacitor failure level as a function of the capacitor size

varactor, and forward bias mode of operation as a RF diode ESD element to provide ESD protection. Excellent ESD protection can be achieved with varactor structures through proper ESD layout design of the anode and cathode regions, contacts, and metallization. In a RF bipolar or RF BiCMOS technology, these are constructed from the base and emitter regions of the bipolar transistor. In triple-well RF CMOS technology, silicon varactors can be constructed from p-well and n-buried layer regions. These RF passive elements can be cosynthesized to achieve both RF and ESD optimization.

3.6.3 Metal-ILD-Metal Capacitors

The utilization of a MOS capacitor or a MIM capacitor in present day technology will lead to ESD failure without an ESD protection network. An alternate solution that RF circuits designers have undertaken is the avoidance of thin dielectric capacitor elements and utilized the back-end-of-line metal films and ILD films. Figure 3.16 shows an example of the

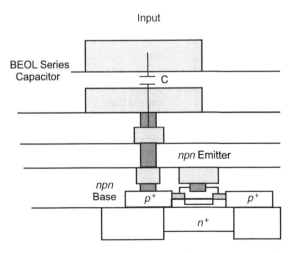

Figure 3.16   RF receiver network using a metal-ILD-metal capacitor structure

utilization of the metal films, and the interlevel dielectric as a RF series input capacitor element. The advantage of the utilization of the M-ILD-M capacitor element is the thick interlevel dielectric film, leading to a high dielectric breakdown voltage. The disadvantage of this technique is a larger capacitor plate structure to form the receiver network and the lower ability to control the interlevel dielectric film (e.g., compared to a MOSFET thin oxide or dual oxide manufacturing control).

3.6.4 Vertical Parallel Plate (VPP) Capacitors

As discussed, the utilization of a MOS capacitor or a MIM capacitor in today's technology will lead to ESD failure without an ESD protection network. An alternate solution that provides a high dielectric breakdown and requires less physical area is a vertical parallel plate (VPP) capacitor. Utilizing the combination of back-end-of-line (BEOL) metal films and via stack a vertical metal plate of a capacitor can be formed. In an RF technology with inductors, the vertical film height is significant to provide a low resistance high Q inductor. Thick metal films, referred to as 'analog metal', and the thick interlevel dielectric add to the vertical height of the BEOL materials, allowing for construction of the VPP capacitor. Using two plates and the interlevel dielectric film between the two capacitor plates, a VPP capacitor can be constructed that does not utilize significant chip area, and has a high breakdown voltage. Figure 3.17 shows an example of the utilization of the metal films, and the interlevel dielectric to form a VPP capacitor as a RF series input capacitor element. The advantage of the utilization of the VPP M-ILD-M capacitor element is the thick interlevel dielectric film, leading to a high dielectric breakdown voltage. Two advantages also exist for the VPP compared to the planar M-ILD-M capacitor. First, the width of the plates can be varied to change or modify the breakdown voltage (note: in the M-ILD-M capacitor the film

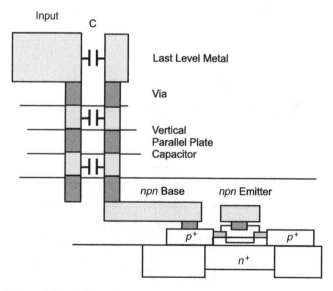

Figure 3.17   RF receiver network using a VPP capacitor structure

is controlled by the ILD thickness). Second, the structure will consume less physical area. VPP breakdown voltages over 100 V have been demonstrated.

An RF ESD design practice for RF capacitor elements can be stated as follows:

- RF MOS and MIM capacitor elements placed in receiver and transmitter networks can be vulnerable to ESD failure from HBM and MM events.

- RF MOS and MIM capacitor elements that are constructed from RF MOSFET gate dielectrics will undergo ESD failure in RF receiver networks when used as d.c. blocking elements in RF inputs and RF output nodes.

- Silicon-based varactor and hyper-abrupt varactor structures can be used in reverse mode of operation for capacitors and forward bias mode of operation to provide ESD protection.

- Choice of the RF passive capacitor element will be a function of the structure size and dielectric breakdown voltage.

- M-ILD-M capacitors have demonstrated high breakdown voltage and are superior to MOS and MIM capacitors for unprotected RF networks.

- VPP capacitors have demonstrated high breakdown voltage and are superior to MOS and MIM capacitors for unprotected RF receiver networks.

3.7 SUMMARY AND CLOSING COMMENTS

In this chapter, ESD protection of RF CMOS technology was addressed. For the ESD design choices, wherein there is no RF-ESD design cosynthesis, the capacitive loading effect of the ESD element constrains the usage of some ESD elements. Additionally, the ESD robustness and effectiveness of these elements for a given capacitive loading effect is addressed. The pros and cons of different ESD elements such as diodes, MOSFETs, and SCRs are highlighted. RF ESD design issues for RF ESD devices are also discussed. In RF ESD design, it is quite popular to utilize both octagonal, circular, and waffle-style layout design methodology; the advantages and disadvantages of these methods and designs are discussed.

This chapter also discussed design choices in Schottky diodes, inductors, and capacitor elements; this is an important issue with RF networks that utilize direct current (d.c.) blocking, or alternating current (a.c.) blocking, matching elements, and ESD inductors and ESD capacitor elements. Today, RF passive element failures contribute to the ESD limitation of RF circuits, and as the increase in focus on cosynthesis increases, this will be a larger issue.

In the next chapter, Chapter 4, we will discuss RF CMOS ESD networks. Chapter 4 will include sections on RF ESD diodes, ESD inductor/diode networks, ESD RF diode strings, ESD RF power clamps, distributed load ESD networks, and distributed coplanar waveguide ESD networks for RF CMOS applications. The chapter will address RF LNA design in RF CMOS. In the discussion of RF LNA, the different choices of ESD elements will be discussed. In addition, RF-ESD cosynthesis in RF CMOS LNA networks will be shown. Additionally, a new area of interest, RF CMOS LDMOS technology for power amplifier (PA) applications will be discussed; ESD protection of LDMOS technology examples will be reviewed.

PROBLEMS

1. Given a shallow trench isolation diode element, derive a model for the capacitance components associated with the geometrical parameters. Exclude resistance and capacitance of metal lines. Derive an RF ESD metric of the ratio of ESD robustness divided by the loading capacitance.

2. Given a grounded gate MOSFET, derive a model for the capacitance components associated with all the physical dimensions of a transistor. Exclude resistance and capacitance of metal lines. Derive an RF ESD metric of the ratio of ESD robustness divided by the loading capacitance.

3. Given a grounded gate MOSFET with a salicide block mask on the drain, derive a model for the capacitance components associated with all the physical dimensions of a transistor. Exclude resistance and capacitance of metal lines. Derive an RF ESD metric of the ratio of ESD robustness divided by the loading capacitance. Compare this to the model without the salicide block mask.

4. Given a grounded gate MOSFET with a salicide block mask on the drain and source region, derive a model for the capacitance components associated with all the physical dimensions of a transistor. Exclude resistance and capacitance of metal lines. Derive an RF ESD metric of the ratio of ESD robustness divided by the loading capacitance. Compare this to the model without the salicide block mask.

5. Given a *pnpn* silicon controlled rectifier, with a STI-bound *p*-diffusion, an *n*-well, a STI-bound n^+ diffusion, derive a model for the capacitance components associated with all the physical dimensions of a high-voltage SCR. Show all dimensional parameters. Exclude resistance and capacitance of metal lines. Derive an RF ESD metric of the ratio of ESD robustness divided by the loading capacitance.

6. Given a medium voltage trigger silicon controlled rectifier, with a STI-bound *p*-diffusion, an *n*-well, an *n*-diffusion extending outside of the *n*-well, and a STI-bound n^+ diffusion, derive a model for the capacitance components associated with all the physical dimensions of a high voltage SCR. Show all dimensional parameters. Derive an RF ESD metric of the ratio of ESD robustness divided by the loading capacitance.

7. Given a low voltage trigger silicon controlled rectifier, with a STI-bound *p*-diffusion, an *n*-well, and an *n*-channel MOSFET whose drain is integrated with the *n*-well, and whose source serves as a cathode, derive a model for the capacitance components associated with all the physical dimensions of a high voltage SCR. Show all dimensional parameters. Derive an RF ESD metric of the ratio of ESD robustness divided by the loading capacitance.

8. Given a three-dimensional shallow trench isolation p^+/n-well diode element, derive a model for the capacitance components associated with the geometrical parameters, including the metal pattern design. Assume the metal is on the first level of metal.

9. Given a three-dimensional shallow trench isolation p^+/n-well diode element, derive a model for the capacitance components associated with the geometrical parameters, including the metal pattern design. Assume the metal is on the first level of metal for the

cathode, and assume the anode input is a stack of via elements to an arbitrary level of metal. Compare this case, with the prior problem (Problem 9) and show a ratio of the total capacitance as the metal levels increase.

10. Derive a capacitance model for a circular design with the p^+ anode as a center circle, an STI space, and an annular n^+ cathode. Derive the capacitance and resistance of this diode element.

11. Compare the model of a rectangular diode to the circular diode. What is the length and width parameters so that the capacitances are equal? What is the width and length of the linear diode so that the resistances are equal?

REFERENCES

1. Voldman S. ESD: Physics and Devices. Chichester: John Wiley & Sons, Ltd., 2004.
2. Voldman S. ESD: Circuits and Devices. Chichester, England: John Wiley & Sons, Ltd., 2005.
3. Voldman S, Gross V, Hargrove M, Never J, Slinkman J, O'Boyle M, *et al.* Shallow trench isolation (STI) double-diode electrostatic discharge (ESD) circuit and interaction with DRAM circuitry. *Proceedings of the Electrical Overstress/ Electrostatic Discharge (EOS/ESD) Symposium,* 1992. p. 277–288.
4. Voldman S, Gross V, Hargrove M, Never J, Slinkman J, O'Boyle M, *et al.* Shallow trench isolation (STI) double-diode electrostatic discharge (ESD) circuit and interaction with DRAM circuitry. *Elsevier Journal of Electrostatics* 1993;**31**(2–3):237–65.
5. Voldman S, Gross V. Scaling, optimization, and design considerations of electrostatic discharge protection circuits in CMOS technology. *Proceedings of the Electrical Overstress/ Electrostatic Discharge (EOS/ESD) Symposium,* 1993. p. 251–60.
6. Voldman S, Gross V. Scaling, optimization, and design considerations of electrostatic discharge protection circuits in CMOS technology. *Journal of Electrostatics* 1994;**33**(3):327–57.
7. Voldman S, Furkay S, Slinkman J. Three dimensional transient electrothermal simulation of electrostatic discharge protection networks. *Proceedings of the Electrical Overstress/Electrostatic Discharge (EOS/ESD) Symposium,* 1994. p. 246–57.
8. Voldman S. Retrograde well implants and ESD. Invited Talk, *SEMATECH Vertical Modulated Well PTAB,* November 1994.
9. Voldman S, Gerosa G. Mixed voltage interface ESD protection circuits for advanced microprocessors in shallow trench isolation and LOCOS isolation CMOS technologies. *International Electron Device Meeting (IEDM) Technical Digest,* 1994. p. 227–30.
10. Voldman S, Gerosa G, Gross VP, Dickson N, Furkay S, Slinkman J. Analysis of a Snubber-clamped diode string mixed voltage interface ESD protection network for advanced microprocessors. *Proceedings of the Electrical Overstress/Electrostatic Discharge (EOS/ESD) Symposium,* 1995. p. 125–38.
11. Never J, Voldman S. ESD failure mechanisms of shallow trench isolated ESD structures. *Proceedings of the Electrical Overstress/Electrostatic Discharge (EOS/ESD) Symposium,* 1995. p. 273–88.
12. Voldman S, Geissler S, Nakos J, Pekarik J. Semiconductor process and structural optimization of shallow trench isolated-defined and polysilicon-bound source/drain diodes for ESD networks. *Proceedings of the Electrical Overstress/Electrostatic Discharge (EOS/ESD) Symposium,* 1999. p. 151–60.
13. Voldman S. The impact of MOSFET technology evolution and scaling on electrostatic discharge protection. *Review Paper, Microelectronics Reliability* 1998;**38**:1649–68.

14. Voldman S, Anderson W, Ashton R, Chaine M, Duvvury C, Maloney T, *et al.* Test structures for benchmarking the electrostatic discharge (ESD) robustness of CMOS technologies. *SEMATECH Technology Transfer Document, SEMATECH TT 98013452A-TR*, May 1998.
15. Voldman S, Anderson W, Ashton R, Chaine M, Duvvury C, Maloney T *et al.* ESD technology benchmarking strategy for evaluation of the ESD robustness of CMOS semiconductor technologies. *Proceedings of the International Reliability Workshop (IRW)*, October 12–16, 1998.
16. Voldman S, Anderson W, Ashton R, Chaine M, Duvvury C, Maloney T, *et al.* A strategy for characterization and evaluation of the ESD robustness of CMOS semiconductor technologies. *Proceedings of the Electrical Overstress/Electrostatic Discharge (EOS/ESD) Symposium*, 1999. p. 212–24.
17. Ashton R, Voldman S, Anderson W, Chaine M, Duvvury C, Maloney T, *et al.* Characterization of ESD robustness of CMOS technology. *Tutorial Notes of the International Conference on Microelectronic Test Structure (ICMTS)*, Sweden, February 1999.
18. Voldman S. Electrostatic discharge protection, scaling, and ion implantation in advanced semiconductor technologies. *Invited Talk, Proceedings of the Ion Implantation Conference (I^2CON)*, Napa, California, 1999.
19. Richier C, Salome P, Mabboux G, Zaza I, Juge A, Mortini P. Investigations on different ESD protection strategies devoted to 3.3 V RF applications (2 GHz) in a 0.18-μm CMOS process. *Proceedings of the Electrical Overstress/Electrostatic Discharge (EOS/ESD) Symposium*, 2000. p. 251–59.
20. Worley E, Bakulin A. Optimization of input protection for high speed applications. *Proceedings of the Electrical Overstress/Electrostatic Discharge (EOS/ESD) Symposium*, 2002. p. 62–73.
21. Worley E, Matloubin M. Diode with variable width metal stripes for improved protection against electrostatic discharge (ESD) current failure. U.S. Patent 6,518,604 (February 11, 2003).
22. Voldman C. The effect of deep trench isolation, trench isolation, and sub-collector doping concentration on the ESD robustness of RF ESD diode structures in a BiCMOS SiGe technology. *Proceedings of the Electrical Overstress/Electrostatic Discharge (EOS/ESD) Symposium*, 2003. p. 214–23.
23. Baker L, Currence R, Law S, Le M, Lin ST, Teene M. A waffle layout technique strengthens the ESD hardness of the NMOS output transistor. *Proceedings of the Electrical Overstress/Electrostatic Discharge (EOS/ESD) Symposium*, 1989. p. 175–81.
24. Beigel DF, Wolfe EL, Krieger WA. Integrated circuit with diode-connected transistor for reducing ESD damage. U.S. Patent 5,637,901 (June 10, 1997).
25. Ker MD, Wu TS, Wang KF. N-sided polygonal cell layout for multiple cell transistor. U.S. Patent 5,852,315 (December 22, 1998).
26. Voldman S. ESD robustness and scaling implications of aluminum and copper interconnects in advanced semiconductor technology. *Proceedings of the Electrical Overstress/Electrostatic Discharge (EOS/ESD) Symposium*, 1997. p. 316–29.
27. Voldman S, Morriseau K and Reinhart D, *et al.* High-current transmission line pulse characterization of aluminum and copper interconnects for advanced CMOS semiconductor technologies. *Proceedings of the International Reliability Physics Symposium (IRPS)*, 1998. p. 293–301.
28. Voldman S. The impact of technology evolution and scaling on electrostatic discharge (ESD) protection on high-pin-count high-performance microprocessors. *Proceedings of the International Solid-State Circuits Conference (ISSCC)*, 1999. p. 366–368.
29. Voldman S, Morriseau K, Hargrove M, McGahay V, and Gross V, *et al.* High current characterization of dual damascene copper/SiO_2 and low-*k* inter-level dielectrics for advanced CMOS semiconductor technologies. *Proceedings of the International Reliability Physics Symposium (IRPS)*, 1999. p. 144–53.
30. Hynoven S, Rosenbaum E. Diode-based tuned ESD protection for 5.25-GHz CMOS LNAs. *Proceedings of the Electrical Overstress/Electrostatic Discharge (EOS/ESD) Symposium*, 2005. p. 9–17.

31. Anand Y, Howell C. The real culprit in diode failure. Microwaves, August 1970; 1–3.

32. Anand Y, Moroney WJ. Microwave mixer and detector diodes. *Proceedings of the IEEE* 1970; **59**:1182–90.

33. Morris GE, Anand Y, Higgins VJ, Cook C, Hall G. RF burnout of mixer diodes as induced under controlled laboratory conditions and correlation to simulated system performance. *Proceedings of the Microwave Technology and Test (MTT) Symposium*, 1975. p. 182–3.

34. Anand Y. RF burnout dependence on variation in barrier capacitance of mixer diodes. *Proceedings of the IEEE (Letters)* 1973;**61**:247–8.

35. Moroney WJ, Anand Y. Reliability of microwave mixer diodes. *Proceedings of the International Reliability Physics Symposium (IRPS)*, Las Vegas, April 1972.

36. Anand Y. High burnout mixer diodes. *Proceedings of the IEEE Electron Device Meeting*, Washington, D.C., December 1974.

37. Anand Y. X-band high-burnout resistance Schottky barrier diodes. *IEEE Transactions on Electron Devices* 1977; **24**:1330–6.

38. Anand Y, Morris G, Higgins V. Electrostatic failure of X-band silicon Schottky barrier diodes. *Proceedings of the Electrical Overstress/Electrostatic Discharge (EOS/ESD) Symposium*, 1979. p. 97–103.

39. Voldman S. The state of the art of electrostatic discharge protection: physics, technology, circuits, designs, simulation and scaling. *Invited Talk, Bipolar/BiCMOS Circuits and Technology Meeting (BCTM) Symposium*, 1998. p. 19–31.

40. Voldman S, Juliano P, Schmidt N, Johnson R, Lanzerotti L, Brennan C, *et al*. ESD robustness of a BiCMOS SiGe technology. *Bipolar/BiCMOS Circuits and Technology Meeting (BCTM) Symposium*, 2000;214–17.

41. Voldman S. ESD Protection and RF Design. *Tutorial J, Tutorial Notes of the Electrical Overstress/Electrostatic Discharge (EOS/ESD) Symposium*, September 10, 2001.

42. Long JR, Copeland MA. Modeling, characterization and design of monolithic inductors for silicon RFIC's. *Proceedings of the Custom Integrated Circuits Conference (CICC)*, 1996. p. 185–8.

43. Long JR, Copeland MA. Modeling, characterization and design of monolithic inductors for silicon RFIC's. *IEEE Journal of Solid State Circuits* 1997;**32**(3):357–69.

44. Burghatz JN, Souyer M, Jenkins KA. Microwave inductors and capacitors in standard multilevel interconnect silicon technology. *IEEE Transactions on Microwave Theory and Technology* 1996; **44**:100–4.

45. Greenhouse HM. Design of planar rectangular microelectronic inductors. *IEEE Transactions on Parts, Hybrids and Packaging* 1974;**PHP-10**:10–9.

46. Yue CP, Wong SS. Physical modeling of spiral inductors on silicon. *IEEE Transactions on Electron Devices* 2000;**47**(3):560–8.

47. Burghhartz JN, Rejaei B. On the design of RF spiral inductors on silicon. *IEEE Transactions on Electron Devices* 2003;**50**(3):718–29.

48. Burghartz JN, Edelstein DC, Soyuer M, Ainspan MA, Jenkins KA. RF circuit design aspects of spiral inductors on silicon. *IEEE Journal of Solid-State Circuits* 1988;**33**:2028–34.

49. Yue CP, Ryu C, Lau J, Lee TH, Wong SS. A physical model for planar spiral inductors on silicon. *International Electron Device Meeting (IEDM) Technical Digest*, 1996. p. 155–8.

50. O KK. Estimation methods for quality factors of inductors fabricated in silicon integrated circuit process technologies. *IEEE Journal of Solid-State Circuits* 1998;**33**:1249–52.

51. Soden JM. The dielectric strength of SiO_2 in a CMOS transistor structure. *Proceedings of the Electrical Overstress/Electrostatic Discharge (EOS/ESD) Symposium*, 1979. p. 176–82.

52. DiStefano TH, Shatzkes M. Dielectric instability and breakdown in SiO_2 thin films. *Journal of Vacuum Science Technology* 1976;**13**:50–4.

53. DiStefano TH, Shatzkes M. Impact ionization model for dielectric instability and breakdown. *Applied Physics Letters* 1974;**25**:685–7.

54. O'Dwyer JJ. Theory of Conduction in Dielectrics., *Journal of Applied Physics* 1969;**39**: 3887–90.

55. DiMaria DJ, Arnold D, Cartier E. Impact ionization and positive charge formation in silicon dioxide films on silicon. *Applied Physics Letters* 1992;**60**:2118–20.

56. Fischetti M, DiMaria DJ, Brorson SD, Theis TN, Kirtley JR. Theory of high field electron transport in silicon dioxide. *Physical Review B* 1985;**31**:8124–42.

57. Schuegraf KF, Hu C. Hole injection oxide breakdown model for very low voltage lifetimeb extrapolation. *Proceedings of the International Reliability Physics Symposium (IRPS)*, 1993. p. 7–12.

58. Klein N, Solomon P. Current runaway in insulators affected by impact ionization and recombination. *Journal of Applied Physics* 1976;**47**:4364–72.

59. Nissan-Cohen Y, Shappir J, Frohman-Bentchkowsky D. High yield and current induced positive charge in thermal SiO_2 layers. *Journal of Applied Physics* 1985;**57**:2830–9.

60. Chen IC, Holland S, Hu C. Electron-trap generation by recombination of electrons and holes in SiO_2. *Journal of Applied Physics* 1987;**61**:4544–8.

61. Amerasekera A, Campbell D. ESD pulse and continuous voltage breakdown in MOS capacitors structures. *Proceedings of the Electrical Overstress/Electrostatic Discharge (EOS/ESD) Symposium*, 1986. p. 208–13.

62. Lin DL, Welsher T. From lightning to charged-device model electrostatic discharges. *Proceedings of the Electrical Overstress/Electrostatic Discharge (EOS/ESD) Symposium*, 1992. p. 68–75.

63. Bridgewood M, Fu Y. A comparison of threshold damage processes in thick field oxide protection devices following square pulse and human body model injection. *Proceedings of the Electrical Overstress/Electrostatic Discharge (EOS/ESD) Symposium*, 1988. p. 126–36.

64. Bridgewood M, Kelly R. Modeling the effects of narrow impulsive overstress on capacitive test structures. *Proceedings of the Electrical Overstress/Electrostatic Discharge (EOS/ESD) Symposium*, 1985. p. 84–91.

65. Tuncliffe MJ, Dwyer VM, Campbell DS. Experimental and theoretical studies of EOS/ESD oxide breakdown in unprotected MOS structures. *Proceedings of the Electrical Overstress/Electrostatic Discharge (EOS/ESD) Symposium*, 1990. p. 162–68.

66. CDM. *ESD Sensitivity Testing: Charged Device Model (CDM)—Component Level ESD Association Standard*, D5.3, 1997.

67. Kitamura Y, Kitamura H, Nakanishi K, Shibuya Y. Breakdown of thin gate-oxide by application of nanosecond pulse as ESD test. *Proceedings of International Test and Failure Analysis (ITSFA)*, 1989. p. 193–9.

68. Maloney TJ. Designing MOS inputs and outputs to avoid oxide failure in the charged device model. *Proceedings of the Electrical Overstress/Electrostatic Discharge (EOS/ESD) Symposium*, 1988. p. 220–27.

69. Renninger RG. Mechanism of charged device electrostatic discharge. *Proceedings of the Electrical Overstress/Electrostatic Discharge (EOS/ESD) Symposium*, 1991. p. 127–43.

70. Renninger RG, Jon M-C, Lin DL, Diep T, Welsher TL. A field-induced charged device model simulator. *Proceedings of the Electrical Overstress/Electrostatic Discharge (EOS/ESD) Symposium*, 1989. p. 59–71.

71. Fong Y-P, Hu C. The effects of high electric field transients on thin gate oxide MOSFETs. *Proceedings of the Electrical Overstress/Electrostatic Discharge (EOS/ESD) Symposium*, 1987. p. 252–7.

72. Lin DL. ESD sensitivity and VLSI technology trends: thermal breakdown and dielectric breakdown. *Proceedings of the Electrical Overstress/Electrostatic Discharge (EOS/ESD) Symposium*, 1993. p. 73–82.

73. Degraeve R, Pangon N, Kaczer B, Nigam T, Groseneken G, Naem A, *et al.* Temperature acceleration of oxide breakdown and its impact on ultra-thin gate oxide reliability. *VLSI Technical Symposium*, 1999. p. 59–60.

74. Nigam T, Martin S, Abusch-Magder D. Temperature dependence and conduction mechanism after analog soft-breakdown. *Proceedings of the International Reliability Physics Symposium (IRPS)*, 2003. p. 417–23.

75. Weir B Silverman P, Manroe D, Krisch K, Alam M, Alers G, Sorsch T, Tim G, Baumann F, *et al.* Ultra-thin dielectrics: they breakdown but do they fail? *International Electron Device Meeting (IEDM) Technical Digest,* 1997. p. 73–6.

76. Stathis J, DiMaria DJ. Reliability projections for ultra-thin oxides at low voltages. *International Electron Device Meeting (IEDM) Technical Digest*, 1998. p. 167–70.

77. Wu J, Juliano P, Rosenbaum E. Breakdown and latent damage of ultra-thin oxides under ESD stress conditions. *Proceedings of the Electrical Overstress/Electrostatic Discharge (EOS/ESD) Symposium*, 2000. p. 287–95.

78. Linder B, Stathis JH, Frank DJ, Lombardo S, Vayshenker A. Growth and scaling of oxide conduction after breakdown. *Proceedings of the International Reliability Physics Symposium (IRPS)*, 2003. p. 402–5.

79. Alam MA, Smith RK. A phenomenological theory of correlated multiple soft-breakdown events in ultra-thin oxides. *Proceedings of the International Reliability Physics Symposium (IRPS)*, 2003. p. 406–12.

80. Cheung KP. Soft breakdown in thin gate oxide—a measurement artifact. *Proceedings of the International Reliability Physics Symposium (IRPS)*, 2003. p. 432–7.

81. Weir B, Leung CC, Silverman PJ, Alam MA. Gate dielectric breakdown: a focus on ESD protection. *Proceedings of the International Reliability Physics Symposium (IRPS)*, 2004. p. 399–404.

4 RF CMOS ESD Networks

ESD networks are necessary for RF CMOS technology to protect the RF CMOS products from entering the high-speed communications, in both the wired and wireless application space. Additionally, RF CMOS will grow significantly as the unity current gain cutoff frequency of MOSFET increases with MOSFET technology scaling. Moreover, with an increase in focus on low-cost applications, the RF CMOS market will enter into the traditional areas dominated by silicon germanium and gallium arsenide technologies.

In this chapter, the RF CMOS ESD input circuits and RF ESD power clamps will be discussed. The RF ESD design practice methods that were discussed in Chapters 2 and 3 will be highlighted and experimental results will be shown. The focus will be more directed toward applications from RF low-noise amplifiers (LNAs) to RF CMOS LDMOS technology.

4.1 RF CMOS INPUT CIRCUITS

4.1.1 RF CMOS ESD Diode Networks

In RF CMOS, a standard ESD protection network is the ESD double-diode (or dual diode) network [1–5]. With the introduction of CMOS technology, the utilization of the ESD double-diode network as an ESD protection network for input circuitry became an industry standard for the digital and analog CMOS applications [2,3]. The ESD double-diode network optimization focus was on the ESD circuit operation, ESD design layout, ESD process optimization, and ESD process sensitivities [2,3]. In CMOS digital design, the ESD design layout and process optimization focus was to provide a low forward on-resistance, uniform current distribution, and high power-to-failure.

In RF CMOS, the ESD double-diode network will have a higher focus on the influence of the specific analog and RF circuit, and the capacitance loading [1]. RF ESD double-diode networks influence the following functional and RF parameters:

- input impedance;
- linearity;

ESD: RF Technology and Circuits Steven H. Voldman
© 2006 John Wiley & Sons, Ltd

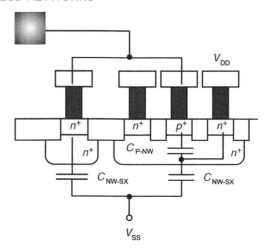

Figure 4.1 RF CMOS double-diode network highlights the metallurgical junction capacitor elements

- leakage;

- noise figure.

Figure 4.1 shows an example of an RF ESD double-diode network in a standard single-well CMOS bulk semiconductor process. In the figure, the dual-diode consists of an n-well diode element and a p^+/n-well diode element. In digital CMOS, significant focus on the $p^+/$ n-well diode has shown that the series resistance of the n-well has a large influence on the ESD robustness of the structure. The ESD robustness has a U-shaped dependency on the n-well sheet resistance [2,3]; at low n-well sheet resistance, the response is dominated by the diode on-resistance, and at higher n-well sheet resistances, the response is influenced by the vertical parasitic pnp element formed between the p^+ anode and the p-substrate. At ultra-low n-well sheet resistance (e.g., utilization of a subcollector), the ESD robustness is influenced by the elimination of lateral ballasting [2,3]. For the n-well to substrate diode element, the operation of the element is a function of the physical dimensions, perimeter-to-area ratio, the substrate doping concentration, and the influence of adjacent structures (e.g., adjacent n-well structures, p-well placement, and deep trench) [2,3].

In Figure 4.1, the capacitance network is shown for the ESD double-diode network. For discussion sake, the interconnect capacitances are ignored (although it can be significant in magnitude). The ESD 'down' diode element forms a capacitance between the input node and the substrate region (e.g., n-well to substrate capacitance). The capacitance of the n-well to substrate diode element is strongly influenced by the p-epitaxial doping concentration, the p-well doping concentration and placement, and the p-substrate resistivity. The ESD 'up' diode element has a first capacitor element between the input pad signal and the second n-well region (e.g., p^+/n-well capacitor). In the 'up' diode element, there is a second n-well to substrate capacitance that is formed between the p^+/n-well capacitor node, and the substrate. In parallel with this capacitor element, a capacitor element exists between V_{DD} and the V_{SS} substrate. In the case of a.c. analysis, assuming V_{SS} and V_{DD} is at ground potential, the two capacitor elements are in a parallel configuration. Note that the nature of the ESD double-diode network is such that the in the d.c. state, one element is in a forward bias state, whereas the second element is in a reverse-bias state [1].

For a p–n junction, the metallurgical junction capacitance can be expressed as

$$C_j(V) = \frac{C_o}{\left[1 - \dfrac{V}{V_o}\right]^{1/2}}$$

where V is the voltage across the metallurgical junction, C_o is the constant associated with the junction capacitance per unit area, and V_o is the junction contact potential. As the d.c. bias condition changes, the voltage dependency of the silicon junction capacitance leads to a variation in the loading capacitance; this again leads to the variation in the linearity of a circuit as a function of the bias voltage.

In the case of a double-diode network, one element is forward biased whereas the other element is reverse biased; this was noted by Richier $et\ al.$ [1] who found that the presence of the forward bias and reverse bias state can compensate for the voltage dependence of a diode element. Figure 4.2 highlights the voltage dependence of a single diode element and the double-diode element.

In the metallurgical junction capacitance relationship, the improvement of the linearity reduction can be achieved by the two variables, the capacitance per unit area (C_o) and the contact potential (V_o). The capacitance per unit area in the relationship can be achieved by the following:

- Type of diode elements (e.g., different elements will have a fundamentally different capacitance per unit area).

- Area ratio of the diode elements (e.g., relative size of the first and second element).

In RF CMOS technology, the substrate resistivity is increased in order to reduce the noise coupling between the digital, analog, and RF circuits. As the substrate resistance increases,

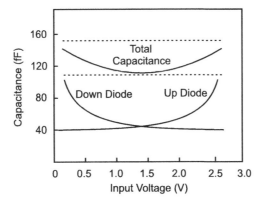

Figure 4.2 RF CMOS voltage dependence of single diode (independently) and double-diode on input signal linearity. Note, this highlights the reduction of the voltage dependence of the ESD double-diode network

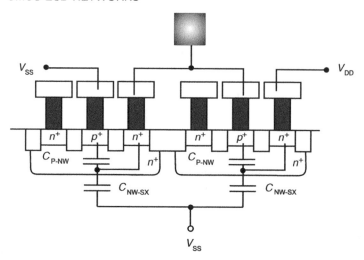

Figure 4.3 RF ESD double-diode element utilizing only the p^+/n-well element. Note that the electrodes are interchanged on the diode to V_{SS} power rail

the effectiveness of the n-well to substrate diode structure decreases. Moreover, in the case of noise injection at the signal pad, the n-well-to-substrate metallurgical junction can be forward biased. Minority carriers can be injected into the substrate region, which can lead to noise coupling and local perturbations of the substrate potential (e.g., 'substrate bounce'). Hence, the element utilized for digital CMOS elements may not be desirable in some RF applications. In RF CMOS and RF BiCMOS applications, the circuit can be modified by using a p^+/n-well diode to both V_{SS} and V_{DD} power rails. Figure 4.3 shows an example of a symmetric double-diode network that utilizes only the p^+/n-well diode element.

Utilizing the same physical element in the 'up' and 'down' diode has the following RF ESD design advantages:

- diode series on-resistance will be dependent on the n-well and less dependent on the substrate resistivity and wafer;

- reduced injection into the substrate wafer;

- single voltage dependence;

- better matching and tracking of the two diode elements (e.g., photolithographic, etch, mask design, and process variations);

- single d.c. and RF diode model.

In the ESD testing of the circuit, the limitation of the network will be a function of the n-well sheet resistance. Hence, the circuit response will be less dependent on the substrate wafer. Besides, with the secondary ground electrode, there will be less injection to the chip substrate wafer. This reduces the likelihood of noise injection and CMOS latchup. The usage of a single element also has advantages from a semiconductor process and model development. For an improved linearity from the input node, the two p^+/n-well capacitors are in parallel (note: the V_{DD} and V_{SS} are both at ground potential during a.c. analysis), where one element is in a forward bias d.c. voltage and the other element is in reverse bias.

In the RF ESD design of ESD input networks, the following RF ESD design practices can be applied:

- Linearity improvement by utilization of a reverse-bias and forward-bias element in a common circuit to compensate the voltage dependence of a single element.

- Linearity improvement by utilization of different sizes of reverse and forward bias elements (e.g., using the same element).

- Linearity improvement by utilization of different types of reverse and forward bias elements.

- Utilization of the same element for the 'up' and 'down' diode to provide independence from the substrate wafer.

- Utilization of the same element for the 'up' and 'down' diode for an improved power-to-failure symmetry.

- Utilization of the same element for the 'up' and 'down' diode for an improved tracking and matching.

4.1.2 RF CMOS Diode String ESD Network

In digital CMOS technology, CMOS ESD diode string networks were used for peripheral circuits between two chips of different power supply voltages [2,3]. Whereas, in the CMOS digital applications, ESD diode strings were utilized on receiver networks, off-chip driver (OCD) networks, and bi-directional peripheral circuits. With the introduction of mixed voltage interface OCD networks, the utilization of the ESD diode string (also referred to as series diode or stacked diode ESD networks) became an industry standard for DRAMs, SRAMs, and microprocessor applications [2,3]. The ESD diode string network optimization focus was on the ESD circuit operation, ESD design layout, ESD process optimization, ESD process sensitivities, and leakage amplification [2,3].

ESD diode strings were also utilized for ESD power clamps between power and ground (e.g., V_{DD} to V_{SS}), for ESD networks between power rails (e.g., V_{CC} to V_{DD}), and for ESD rail-to-rail networks between electrical ground rails (e.g., V_{SS} to V_{SS}). For digital circuits, the number of series diode elements is determined by the application voltage conditions (e.g., the differential voltage required between power supplies).

In RF CMOS, the utilization of multiple diodes in a series configuration can play multiple roles. Firstly, the number of diodes in series influences the loading capacitance and impedance of the network. Secondly, the number of diodes in series influences the circuit linearity (e.g., capacitance of the network as a function of the voltage). Thirdly, it can influence the capacitance coupling between power rails (e.g., digital V_{SS} to RF V_{SS}). And fourthly, it can influence the circuit stability.

- input impedance;
- linearity;

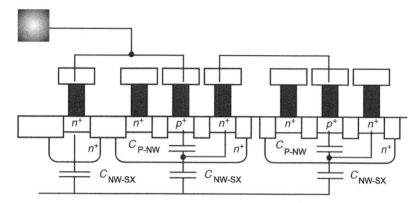

Figure 4.4 RF CMOS diode string network highlighting the metallurgical junction capacitor elements

- noise;

- stability.

Figure 4.4 shows an example of an RF ESD diode string network in a standard single-well RF CMOS bulk semiconductor process. In the figure, the circuit consists of an n-well diode element and a plurality of series p^+/n-well diode element. In digital CMOS, significant focus on the p^+/n-well diode has shown that the n-well has an influence on the ESD robustness, and leakage amplification [2,3]. In the figure, the capacitance network is shown for the ESD diode string network. The ESD 'down' diode element forms a capacitance between the input node and the substrate region (e.g., n-well to substrate capacitance). The ESD 'up' diode string elements have a first capacitor element between the input pad signal and the second n-well region (e.g., p^+/n-well capacitor). In the 'up' diode element, there is a second n-well to substrate capacitance that is formed between the p^+/n-well capacitor node, and the substrate. The cathode of the first p^+/n-well diode is electrically connected in series with the anode of the successive element.

In the case of a diode string, as in the RF ESD double-diode network, firstly, one element is forward biased whereas the other element is reverse biased; the presence of the forward bias and reverse bias state can compensate for the voltage dependence of a diode element leading to improved linearity [1]. Secondly, as the number of diodes are placed in series, the ESD diode string capacitance decreases; as the number of diodes are placed in series the total capacitance of the diode string is reduced. Simplistically, without the n-well to substrate capacitance terms, and the n-well to substrate 'down' diode,

$$\frac{1}{C_{\text{T}}} = \sum_{i=1}^{i=N} \frac{1}{C_i} = \sum_{i=1}^{i=N} \frac{\left[1 - \frac{V_i}{V_{\text{o}}}\right]^{1/2}}{\{C_{\text{o}}\}_i}$$

Thirdly, as the diode are added in series, the voltage dependence is also modified. Hence, the dependence of the capacitance on voltage is decreased, leading to improved linearity. Richier

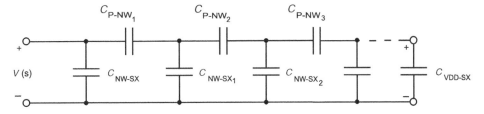

Figure 4.5   RF CMOS diode string network capacitance distributed network

et al. [1] demonstrated that by adding multiple diode in series, the capacitance in a ESD diode string network improved.

Addressing the presence of the *n*-well to substrate capacitance, the circuit can be represented as a transmission line of capacitor elements. Figure 4.5 is a representation of the ESD diode string. Note that the first segment contains the ESD 'down diode' *n*-well to substrate shunt capacitance term. Each successive element of the transmission line contains a series p^+/*n*-well capacitance term and a shunt *n*-well capacitor term. The network contains *N* stages, where in the *N*th stage the last series p^+/*n*-well capacitor is electrically connected to a termination (e.g., V_{DD}). Hence, ignoring the resistance terms and the interconnect capacitance terms, the ESD diode string representation can be shown as a capacitive distributed network with each successive diode element serving as a series element, and the *n*-well to substrate term serving as a shunt capacitance. Another perspective is that each successive stage can be represented as a capacitor divider term. Note that this network can include the diode resistance and interconnect terms by providing more complex RC distributed network representation.

From the discussion on double-diode networks and the ESD diode string networks, it should be apparent that the type, size, and number of 'up' diodes and 'down' diodes can be modified to achieve capacitive decoupling and improved linearity. Hence in general, the network can be adjusted to achieve improved impedance, linearity, stability, noise rejection, and noise injection through modification of these elements.

In the RF ESD design of ESD diode string networks, the following RF ESD design practices can be applied:

- A minimum number of diode elements are needed to address the d.c. voltage differential conditions.

- The capacitance loading can be reduced by the introduction of series diode elements.

- The linearity improvement can be modified by the placement of series diode elements.

- Capacitance, linearity, and loading improvements can be obtained by the utilization of different types of reverse and forward bias elements, the number of elements, and the physical size of the various elements in the network.

4.2 RF CMOS: DIODE–INDUCTOR ESD NETWORKS

As discussed, using resonance conditions of the LC resonator circuits, the capacitance load or the inductor load of a circuit can be tuned out so as to reduce the load on the RF input or

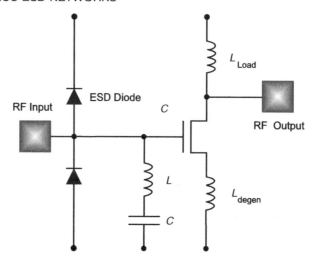

Figure 4.6   ESD double-diode network and a cancellation ESD shunt inductor element and capacitor element

output circuit. Hence, the ESD elements can be the following [5–9]:

- ESD element can be an ESD capacitive load, tuning out an inductance of the circuit whose capacitance value is chosen for a desired resonance frequency.

- ESD element can be an ESD inductive load, tuning out a capacitance of the circuit whose ESD inductor value is chosen for a desired resonance frequency.

- ESD elements can be both the inductive and capacitance load whose values are chosen on the basis of a desired resonant frequency.

Figure 4.6 shows an example of an RF ESD circuit with an ESD double-diode network. ESD double-diode elements are designed as low resistance structure. The ESD double-diode network can be represented as a capacitor element. In the a.c. analysis, the two ESD diode elements are in parallel configuration between the input pad and the a.c. ground reference. This capacitive element is also in parallel with the input pad capacitance. In the case of ESD double-diode networks, both the capacitor elements are voltage dependent (i.e., capacitance variation is a function of the metallurgical junction). Hence, the d.c. bias state of the ESD double-diode circuit alters the capacitance load condition.

Hynoven *et al.* [6–8] noted that a cancellation method can use an ESD inductor element to cancel the loading effect of the ESD double-diode element. The inductor element value is chosen such that during the combined loading effect of the ESD double-diode and inductor is minimized at the functional application frequency. It was also noted that a direct electrical connection of the ESD inductor to a ground reference is not desirable. Hence, placement of a capacitor element in series with the inductor is desirable (as shown in the Figure 4.6).

Figure 4.7 shows an RF ESD double-diode network, an ESD shunt capacitor and series capacitor element, and a d.c. blocking capacitor in series with the input of the RF circuit. Hynoven *et al.* [6–8] noted that in this case, the addition of a d.c. blocking capacitor in series

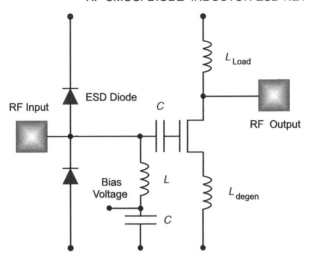

Figure 4.7 ESD double-diode network, cancellation of ESD shunt inductor element with series capacitor, and an additional RF input series d.c. blocking capacitor element. Note the addition of the d.c. bias electrical connection

with the input node of the RF circuit allows biasing of the ESD double-diode/inductor network to a d.c. potential to tune out the capacitance of the RF input pad and the ESD diode elements.

4.2.1 RF Inductor–Diode ESD Networks

As an alternative ESD network, the inductor can be substituted for one of the ESD diode elements (Figure 4.8). In the first case, the ESD inductor element is placed between the input node and the V_{DD} power supply node. The advantages of the diode/inductor implementation are as follows:

- Elimination of the ESD diode (e.g., diode to V_{DD}) reduces the total capacitance load and design area.
- Elimination of ESD diode capacitance requires larger shunt impedance with reduced impact on RF performance.
- The d.c. bias condition of the single diode element reduces the capacitance load of the diode element.

  Disadvantages of this network are as follows:

- inductor area;
- ESD robustness limited to the inductor coil ESD robustness;
- input voltage transient $L(di/dt)$ response to HBM and MM pulse;

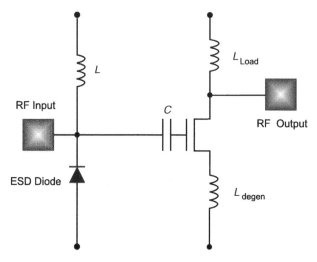

Figure 4.8 ESD diode/inductor network with series d.c. blocking capacitor element for RF inputs (inductor to V_{DD} power supply)

- CDM transient response (pre-charged V_{DD} power rail) leads to voltage overshoot and resonance oscillation of cancellation network.

4.2.2 RF Diode–Inductor ESD Networks

As a second alternative ESD network, the inductor can be substituted for the ESD diode to the ground reference (Figure 4.9). In the first case, the ESD inductor element is placed

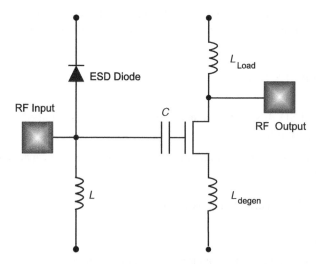

Figure 4.9 ESD diode/inductor network with series d.c. blocking capacitor element for RF inputs (inductor to ground reference)

between the input node and the V_{DD} power supply node. The advantages of the diode/inductor implementation are as follows:

- Elimination of the ESD diode (e.g., diode to V_{SS}) reduces the total capacitance load and design area.
- Elimination of ESD diode capacitance requires larger shunt impedance with reduced impact on RF performance.

Disadvantages of this network are as follows:

- inductor area;
- ESD robustness limited to the inductor coil ESD robustness;
- input voltage transient $L(di/dt)$ response to HBM and MM pulse;
- CDM transient response (pre-charged V_{SS} power rail) initiates voltage undershoot and resonance oscillation.

An RF ESD design practice known as the cancellation method is as follows:

- In an RF application, an ESD inductor element can be chosen such that the combined capacitance load of an input pad and the ESD diodes forms an LC resonance cancellation circuit.
- The ESD inductor element value is chosen such that the load is minimized during RF functional applications.
- An alternative ESD implementation includes the usage of a single inductor and a single diode element to form the LC resonance cancellation circuit.
- RF ESD optimization factors between the ESD double-diode versus ESD inductor–diode networks include capacitance load, design area, ESD robustness of diode and inductor elements, and response to ESD events.

4.3 RF CMOS IMPEDANCE ISOLATION LC RESONATOR ESD NETWORKS

A method to reduce the effective loading in a circuit can utilize resonant conditions of LC resonator circuits. With the placement of a LC resonator circuit in series with an ESD network, the impedance of the network will approach infinity at the LC resonator resonant frequency [10–12]. Hence, by the placement of an LC resonator network in series with ESD elements, the loading effect of the complete network approaches infinity (e.g., open circuit) at the resonant state. By choosing the resonance condition of the LC network to be equal to the application frequency, the ESD loading does not impact the circuit performance. This RF ESD technique of 'impedance isolation' was demonstrated by Ker and Lee [10–12]. Ker and Lee demonstrated two types of impedance-isolated RF ESD networks.

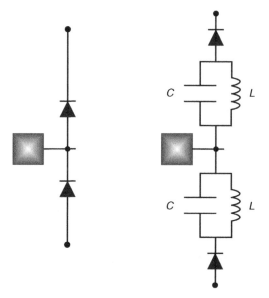

Figure 4.10 Standard digital ESD double-diode network and the RF LC resonator–diode ESD network

An RF ESD network can be constructed as follows:

- an ESD element connected to a signal pad in series with a LC network connected to a power rail (whose LC resonant frequency equals the application frequency);

- an LC network (whose resonant frequency equals the application frequency) connected to a signal pad in series with an ESD element connected to a power rail.

In this manner, four types of ESD networks are possible for a two-power supply system (e.g., V_{DD} and V_{SS}). Ker and Lee demonstrated two such types of networks: the first network where the LC resonator is connected to the RF input pad, followed by a series p–n diode element (e.g., LC-D); the second network of a diode element connected to the RF input pad, followed by a series LC resonator (e.g., D-LC).

4.3.1 RF CMOS LC–Diode ESD Networks

Figure 4.10 shows an example of a standard double-diode ESD network and an impedance isolation ESD network with the LC resonator connected to the RF input pad (e.g., LC-D).

4.3.2 RF CMOS Diode–LC ESD Networks

Figure 4.11 shows again the standard digital CMOS double-diode ESD network, and an impedance isolation ESD network with the diode connected to the RF input pad followed by the LC resonator network connected to the power rails (e.g., D-LC).

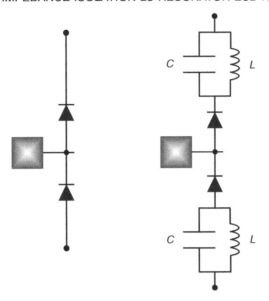

Figure 4.11 Standard digital ESD double-diode network and the RF diode–LC resonator ESD network

4.3.3 Experimental Results of the RF CMOS LC–Diode Networks

Ker and Lee [10–12] experimentally demonstrated the HBM and MM sensitivity of the RF LC resonator–diode ESD network by varying the inductor element (e.g., number of coil turns), capacitor element (e.g., capacitor area) and ESD diode element (number of fingers). In the first experimental study, the size of the ESD diode element was kept constant at a load capacitance of 600 fF, while the inductor element was increased in magnitude, as the capacitor was reduced for a fixed LC resonant frequency.

Figure 4.12 shows the HBM results as a function of the capacitor C and inductor L (for a fixed diode capacitance of 600 fF). Experimental results for HBM and MM ESD stress

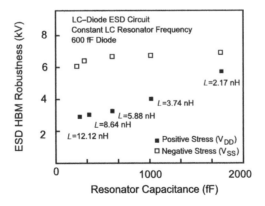

Figure 4.12 HBM ESD results of the RF LC–diode network as a function of the LC parameters (e.g., capacitance and inductance) for a fixed ESD diode size ($C_{\text{Diodes}} = 600$ fF)

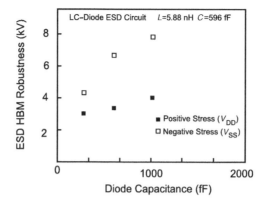

Figure 4.13   HBM ESD results of the RF LC–diode network as a function of the diode size for a fixed LC network parameters (e.g., $L = 5.88$ nH, $C = 596$ fF)

indicated that as the size of the capacitor decreased (and the inductor increased), the HBM and the MM results decreased by approximately 2×.

Figure 4.13 shows the HBM ESD stress results as a function of the RF ESD diode element capacitance for a fixed capacitor C and inductor L (e.g., $L = 5.88$ nH, and $C = 596$ fF). The ESD diode capacitance was reduced from 1200 to 200 fF. Experimental results for HBM and MM ESD stress indicated that as the size of the ESD diode element decreased, the HBM and the MM results also decreased [10–12].

4.4 RF CMOS LNA ESD DESIGN

In recent times, there has been a large interest in providing ESD protection for RF CMOS LNA designs [13–40]. In the optimization of RF CMOS LNA design, the integration of ESD elements and RF matching networks must address both the ESD robustness and the RF response [13–22]. Hence it is preferable to have ESD elements with a high quality factor and RF matching components that are ESD robust.

In LNA design, LNA requirements consist of the following RF design characteristics:

- good matching characteristics;
- low noise figure;
- high voltage gain;
- high power gain;
- good intermodulation free dynamic range (IMFDR);
- good reverse isolation;
- stability.

In digital CMOS receiver design, it is preferable to have the input pad signal transfer to the receiver gates of the n-channel and p-channel transistor. Historically, it was not preferable to have resistor elements placed in series with the CMOS receiver MOSFET gate structures. Resistor elements were added to minimize CMOS receiver ESD concerns caused by charged device model (CDM) failures of the MOSFET gate structure. In digital CMOS, these receivers became more complex to provide good digital signal transfer.

In LNA receiver design, to achieve good matching and power transfer, it is not desirable to have only the MOSFET gate as the receiving signal that is purely capacitive. Hence, it is preferable to have a physical element that serves an equivalency of a resistor shunt during an a.c. response. Moreover, a load resistor is added between the output and the V_{DD} power supply to provide voltage amplification. This unfortunately leads to a high Miller effect as well as resistor noise generation. Instead of using resistor elements in LNA designs, the utilization of MOSFETs in series cascode and the utilization of load inductors is preferred. As a result, a preferred network that can use inductive load devices on the output is evolved. On the output stage, it is considered better to place a degeneration inductor element between the MOSFET source and the ground rail. This element is 'visible' to the input network as it is in series configuration with the MOSFET gate-to-source capacitance.

Besides, as stated above, it is not desirable to have a purely capacitive element, hence an input network that is equivalent to a pure resistive load is preferred. Hence, a design practice is to provide a resonant state at the operation frequency where the load and degeneration inductors are codesigned with the MOSFET gate-to-source capacitance. In this case, the nonideal nature of the capacitor and inductors can influence the quality factor of the network. As a result, the inductor resistance plays an important role in the output gain, as well as in the input matching characteristics.

With the optimization of the LNA characteristics, matching can be achieved using L-match network and Π-match networks. In the case of the L-match network, the presence of a series gate inductor and a capacitive shunt element are relatively preferred. The capacitive shunt element can be an ESD diode element as discussed. The series inductor element resistance and the ESD shunt diode element resistance will play a role in the input characteristics (e.g., influencing noise figure and gain properties). When the size of the ESD element increases, the noise figure will increase and the gain will decrease.

Using a Π-match network, a network can be synthesized in which the series element is an inductor and the two shunt elements are diodes (represented as capacitor elements). With the introduction of an inductor shunt in parallel with the capacitor element in the network, cancellation techniques can be applied using the resonance condition of the shunt capacitor and the shunt inductor. Hence, a good circuit topology for RF ESD LNA cosynthesis is a topology that utilizes a series inductor element and an inductor–capacitor shunt pair. In this configuration, the resistance of the inductor is critical for a good ESD design.

4.4.1 RF LNA ESD Design: Low Resistance ESD Inductor and ESD Diode Clamping Elements in Π-Configuration

RF LNA design solutions include cosynthesis of the ESD and RF design requirements. In these requirements, the integration of the ESD elements, the input and output matching networks and the LNA network are critical to achieve optimized solutions. With the need for RF ESD cosynthesis, the active and the passive elements will be preferred to be ESD robust.

Utilization of ESD inductor elements that are considered suitable for RF characteristics and ESD robustness are as follows:

- high quality factor (Q) at the RF application frequency and current magnitude;

- low series resistance;

- large cross sectional area of coil wire and coil underpass;

- high via number.

The high 'Q' is important to provide good RF response. This high 'Q' is commonly achieved using an inductor that is far from the silicon substrate, utilizing an inductor shield element, and placing insulators in the chip substrate; these factors reduce the substrate eddy currents, as well as the parasitic capacitance. Besides, the inductor resistance is also important to prevent Q degradation; hence, coil cross sectional area, via density, and the coil underpass design are the keys that provide a low resistance design of an inductor. Moreover, low resistance shunt inductors also provide a low electrical resistance shunt for an ESD discharge. To achieve ESD robust inductors, the large cross sectional area of the coil and the coil underpass should allow a low current density in the interconnect structure. In addition, a high via number in the coil inductor can also prevent high current density and low resistance in the inductor structure.

Figure 4.14 shows an example of an RF series cascode LNA design that utilizes a low resistance inductor structure and ESD diode clamping elements. Thijs *et al.* [23] utilized the following elements in the RF LNA design:

- ESD shunt inductor element;

- metal–insulator–metal (MIM) series capacitor;

- shunt inductor element;

- shunt diode elements in both reverse bias and forward bias configuration between the input signal and ground rail;

- diode elements between the first and series cascode LNA output stage (e.g., across the RF MOSFET drain-to-gate nodes of the lower RF MOSFET).

In the implementation of the concept by Thijs *et al.* [23], the ESD inductor element has an inductance of approximately 3 nH and a peak Q of 42 at an application frequency of 3 GHz. The MIM capacitor element is used for d.c. biasing of the RF LNA input whose value was 700 fF. ESD octagonal RF ESD diodes are used whose physical diameter was approximately 25 µm.

With the addition of the ESD diode networks, a Π-type network is formed. In this network, the RF input is dominated by an inductor–capacitor–capacitor type configuration, where the first and the last element in the Π-network are ESD elements (e.g., shunt ESD inductor and the back-to-back diode ESD diodes), and the center element of Π-network

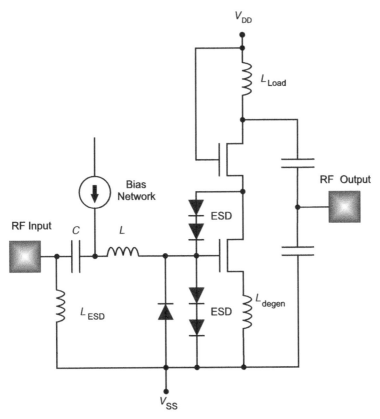

Figure 4.14 RF CMOS-based low-noise amplifier (LNA) with low resistance high Q shunt ESD inductor and diode ESD clamping elements

serves as a d.c. bias capacitor (e.g., the MIM capacitor element). Thus, the Π-network utilizes both an inductive shunt and a capacitor shunt.

Without the inductor element or the ESD diode elements, the RF LNA input failed at an HBM ESD level of less than 10 V. From this result, it is clear that with the lack of inductive shunt element, the series d.c. bias MIM capacitor element and RF LNA input MOSFET gate dielectric are vulnerable to ESD events with no ESD structures between the RF input pad and the series MIM capacitor element. Moreover, the MOSFET gate dielectric is also vulnerable during ESD stress. Thijs *et al.* [23] stated that the failure mechanism without the ESD diode elements (and without the inductor) is the RF LNA MOSFET gate dielectric.

Without ESD diode elements, using a low Q ESD shunt inductor with high series resistance (e.g., 5.5 Ω series resistance), the ESD results increased to a 2500 V HBM and a 225 V MM failure level. Using a high-Q ESD shunt inductor with a low series resistance (e.g., 0.6 Ω series resistance), the ESD results increased to 6000 V HBM and 550 MM failure levels. Hence, the reduction in the series resistance of the ESD shunt inductor lowered the applied voltage observed on the MIM capacitor element, allowing the current discharge to the ground plane (Figure 4.15).

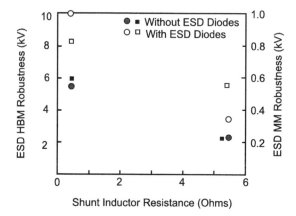

Figure 4.15 RF CMOS-based low-noise amplifier (LNA) ESD results with and without ESD inductors and clamping ESD diode elements

With the addition of the ESD clamping diode elements, with the low Q ESD shunt inductor with high series resistance (e.g., 5.5 Ω series resistance), the ESD results increased from a 2500 to 5500 V HBM level, and from a 225 to a 350 V MM failure level (e.g., in the order of a 2× improvement in the HBM and MM level events). Using a high Q ESD shunt inductor with a low series resistance (e.g., 0.6 Ω series resistance), the ESD results increased from 6000 to over 8000 V HBM failure levels. For the machine model, the ESD results increased from 550 to 1000 V MM failure levels. Hence in both cases, the ESD results increased approximately by 2× with the additional diode clamping elements [23].

Simplistically, ignoring the series inductor, we can assume that a voltage drop across the inductor must be less than the failure levels of the capacitor elements, V_{MIM}, and V_G. Hence, simply,

$$L\frac{dI_{ESD}}{dt} + I_{ESD}R_L < V^* = V_{MIM} + V_G$$

From this expression, we can solve for the magnitude of the inductor resistor,

$$R_L \le \frac{V_{MIM} + V_G}{I_{ESD}} - L\left\{\frac{1}{I_{ESD}}\frac{dI_{ESD}}{dt}\right\}$$

Hence, a resistance requirement can be obtained on the magnitude of the allowable series resistance of the inductor element for a given ESD current magnitude. Additionally, in the case of no MIM capacitor the voltage is directly across the LNA input MOSFET gate structure (with no ESD elements).

It was noted by Thijs *et al.* that when there was no inductive element, the ESD results improved with the use of ESD diode elements across the MOSFET LNA gate input; in that case, the MOSFET LNA gate was not the limiting failure mechanism, but the MIM capacitor element that failed. Note that when the diode elements are placed across the RF LNA input structure, the node is 'pinned' to the ground potential by the ESD diode elements. As and when observed in RF digital network half-pass networks, the 'pinning' of the far end of the receiver network can lead to the failure of the first element [3]. In that case, the voltage across the MIM capacitor element (and the two diode elements) and the resistance must be chosen to avoid the failure of the MIM capacitor element. The ESD failure can then be anticipated to be associated with the voltage across the ESD elements, V_{ESD},

$$R_L \leq \frac{V_{MIM} + V_{ESD}}{I_{ESD}} - L \left\{ \frac{1}{I_{ESD}} \frac{dI_{ESD}}{dt} \right\}$$

where for two diode elements, V_{ESD} is the two forward bias diode voltages V_{BE},

$$R_L \leq \frac{V_{MIM} + 2V_{BE}}{I_{ESD}} - L \left\{ \frac{1}{I_{ESD}} \frac{dI_{ESD}}{dt} \right\}$$

The key RF ESD design practices from this example are as follows:

• An ESD shunt inductor series resistance requirement is critical in avoiding electrical overstress of the RF LNA MOSFET gate.

• An ESD shunt inductor series resistance requirement is critical in avoiding electrical overstress of the series a.c. blocking MIM capacitor elements (e.g., when no ESD elements are placed before the MIM capacitor element).

4.5 RF CMOS T-COIL INDUCTOR ESD INPUT NETWORK

In RF ESD design, inductors and capacitor elements can be used to improve the broadband response of circuit networks. Inductors at low frequency can be utilized to provide a low impedance path when the voltage across the inductor (e.g., $V_L = L(di/dt)$) is low. Capacitor elements at high frequency can serve as low impedance paths when the current flow through the capacitor (e.g., $I_C = C(dv/dt)$) is high.

Galal and Razavi [27,28] showed the utilization of a T-coil network with a bridge capacitor C (Figure 4.16). During ESD events, the first inductor L_1 is chosen such that the voltage drop across the inductor is low, allowing the current to flow to the ESD network that is placed after the inductor elements. Given the ESD robustness of the inductor coil as sufficient, the ESD current will flow from the RF input pad to the ESD network. Given an ESD element with high impedance (e.g., reverse biased diode, MOSFET, or other elements), at low frequency, the inductor coil impedance will be such that the input impedance is determined by the series resistor impedance and the resistor load, R. When the series impedance is zero, and all elements have relatively higher impedance, the load resistor determines the effective input impedance. At high frequency, to prevent the inductor coil from causing detrimental impact to functionality, the bridging capacitor serves as a capacitive short across the inductor L_1 and L_2; this preserves a broadband response of the network. Also note, simplifying the network without the bridge capacitor, the inductor L_1 and ESD element form an 'LC' L-match network. But with the capacitor element, this does not limit the broadband response of the network. The optimization of the network will be a function of the ESD capacitance term, inductor coil terms, the coupling factor 'k', the capacitor bridging element, and the resistor load [27,28].

Note in this implementation, the ESD element capacitance and the T-coil implementation must be cosynthesized in order to achieve functional and ESD optimization of the design variables (e.g., load resistor, series resistance, inductors, coupling factor, and capacitor).

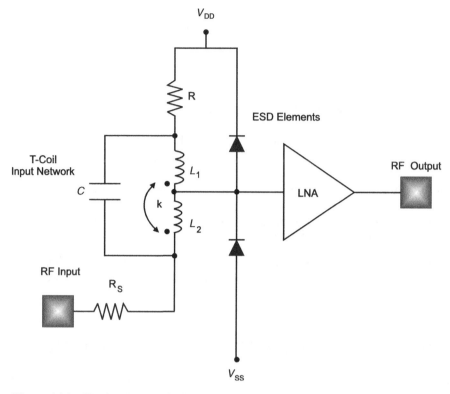

Figure 4.16 Circuit schematic highlighting the T-coil network on the RF input circuitry

From an ESD perspective, to avoid the failure of the T-coil and bridge capacitor element the following precautions are suggested:

- Inductor wire cross sectional area is adequate to prevent inductor coil failure.

- Load resistor impedance is chosen to provide a high series impedance with the bridge capacitor and the inductor coil. Note that during ESD stress, relative to the V_{DD} power supply, voltage stress will be applied across the parallel combination of the inductors and capacitor. The series impedance of the load resistor should be larger than the ESD network series impedance relative to the power supply.

- The turn-on voltage of the ESD network should be below the dielectric breakdown voltage of the bridge capacitor element.

An RF ESD practice for LDMOS must address the following:

- With the utilization of a T-coil element, where one side serves as a series inductor element, low frequency signals can transmit to the RF input node, as well as to the ESD element (placed after the T-coil). This will provide low frequency response as well as ESD protection.

- With the utilization of a T-coil element, with a bridge capacitor element, high frequency signals can transmit to the RF input node by shorting out the T-coil inductors.

- Choosing an adequate size load resistor is valuable to prevent over-voltage of the bridge capacitor and inductor elements during ESD stress.

- ESD network 'turn-on' voltage must be below the dielectric breakdown voltage of the capacitor.

- Inductor coil L_1 critical current-to-failure should be above the ESD element critical current-to-failure to avoid inductor degradation prior to ESD element failure.

4.6 RF CMOS DISTRIBUTED ESD NETWORKS

An interesting synergy of design practices can be established for RF ESD design. In standard ESD design, there are ESD design practices that utilize multiple element networks to reduce the d.c. voltage through the network prior to the sensitive circuit. In standard digital or analog ESD design practices, it is common to introduce the following design practices in a given ESD implementation [3]:

- multistage design with nonidentical elements;
- resistor ballasting;
- segmentation.

Multistage design practices are used to lower the voltage state in successive stages through the ESD network and to reduce the observed voltage at the sensitive circuit node. Multistage ESD networks take advantage of the high voltage elements near the signal pad (which are less sensitive to the ESD event itself), followed by the lower voltage elements. Moreover, using resistor elements in conjunction with active circuit elements, voltage division is established in the network to lower the voltage condition through the physical design. For example, in NMOS technology, 'thick oxide' MOSFET elements (with high MOSFET snapback voltage) are used at the signal pad, followed by a resistor and thin-oxide MOSFET element (with a lower MOSFET snapback voltage) [3]. Hence, a π-type network is established that consists of MOSFET/resistor/MOSFET combination. In the digital electronics, the objective was to lower the breakdown voltage successively as one approaches the sensitive MOSFET receiver gate structure. Additionally, in the digital and analog CMOS and bipolar ESD design, resistor elements and segmentation are introduced to provide an improved current distribution in the ESD network [3]. Hence this is achieved using multiple elements in a parallel configuration. Therefore, in digital and analog ESD design, multistage elements are introduced for the utilization of the allowable d.c. voltage condition, desired trigger conditions, and voltage division networks to lower the voltage observed on the sensitive nodes.

4.6.1 RF CMOS Distributed RF ESD Networks

In RF amplifier design, the introduction of multiple stages also has a value of an RF functional advantage [29–33]. RF multistage ESD networks can be constructed that can achieve the following [24–26,34–40]:

- improved matching conditions;

- improved power transfer;

- lower effective loading effect;

- improved RF broadband response.

Hence, there is a natural design synergy of the ESD multistage network and the RF distributed network to achieve both RF and ESD objectives. The early ESD networks provide the advantage of the different turn-on conditions (e.g., trigger voltage states and time response) and the voltage tolerance associated with the transient current and voltage states; the RF distributed network, in the frequency domain, provides an advantage of the functional response of a lumped versus distributed load.

Difference between the RF ESD distributed design network and the digital multistage design is as follows:

- MOSFET ESD multistage designs utilize series resistor elements for buffering (current limiting elements), establishing of voltage division (e.g., resistive divider conditions), and providing series impedance for CDM events.

- RF distributed networks utilize series inductor elements for providing lossless transmission and impedance optimization.

For an ideal lossless circuit topology for RF ESD networks, the RF ESD networks must have an LC transmission line characteristic; for example, this can be achieved with a plurality of series inductor elements and a plurality of shunt capacitor elements. In these implementation, the capacitor element can also serve as a current shunt to the power supply rails (e.g., V_{DD} or V_{SS} power rails). Examples of such RF networks can have the following stage topology:

- successive stages of series inductor and shunt grounded gate MOSFET (e.g., MOSFET drain and gate at V_{SS});

- successive stages of series inductor and double-diode network;

- successive stages of series inductor and shunt silicon controlled rectifier.

In the above implementation, the inductor and shunt element can be of identical and nonidentical value.

4.6.2 RF CMOS Distributed RF ESD Networks Using Series Inductor and Dual-Diode Shunt

RF distributed ESD networks for matching can be formed using series inductors and ESD diode networks. With the integration of a 'two-element stage' of a series inductor and shunt parallel capacitor element, a distributed network for RF matching, signal transition, and ESD can be synthesized. In the synthesis of the RF distributed ESD network, the inductor element

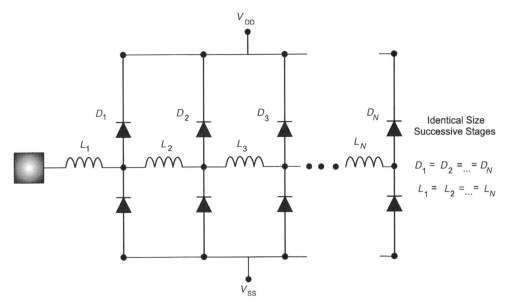

Figure 4.17 RF distributed inductor–double-diode ESD network forming a distributed LC transmission line (comprising series inductors and parallel shunt capacitors). This network has identical size stages

can be a transmission line or a coil inductor element. Low resistance diode elements can be assumed to be a lumped capacitor element. In the a.c. response, an ESD double-diode (also known as a dual diode ESD network), the V_{DD}, and the V_{SS} are at an a.c. ground potential. From an a.c. response, two diode elements are in parallel configuration as a capacitor element.

Figure 4.17 shows an example of a RF distributed ESD network consisting of lumped series inductor and a double-diode ESD network. In this implementation, the ESD double-diode element serves as an a.c. shunt capacitor; this forms a lossless two-element LC transmission line.

An interesting synergy of design practices can be established for this RF ESD design. In digital CMOS networks, a commonly used ESD network for receivers is a two-stage network consisting of a double-diode, a series resistor element, followed by a second double-diode network. In this implementation, the first double-diode element serves the role of providing a discharge current path to the power rails for positive and negative ESD events that occur on the pad. The resistor element serves as a series resistance element to provide impedance in series with the receiver network for both HBM and CDM events. The second double-diode element serves as a CDM solution and is physically placed adjacent to a CMOS receiver network. The placement of a resistor element in an RF signal path impacts power transfer and load-to-source matching, and introduces noise.

In the distributed RF ESD network, the front-end of the RF network has a similar topology, but with a key difference is the resistor element is substituted with lossless inductor elements. Hence, with the RF-distributed two-element ESD network, given the resistance in the ESD network as minimal, the network forms a lossless network with improved matching and power transfer. A second advantage of this topology is that the multiple stages can assist in CDM events. The response and current distribution to CDM events of this RF distributed circuit will be a function of the choice of the inductance magnitude of the coil inductor.

At operational frequencies, the dual-diode will have one diode element in forward bias whereas the second element will be in reverse; this configuration will improve linearity (e.g., the diode capacitance voltage dependence is cancelled because of the forward and reverse nature of the two elements). During the ESD events, current flows through the first inductor element, followed by the forward bias of the ESD diode elements. In addition, the transient current response leads to a voltage drop across the series element (e.g., $V_L = L(di/dt)$) that will be in series with each successive stage. In this RF ESD distributed design, there is an RF ESD design tradeoff in a given two-element segment of the size of the inductor (which determines voltage drop during transient pulse event) and the ESD diode perimeter (which determines on-resistance and total current discharge capability of the given stage). Figure 4.18 shows an example of a RF distributed ESD network consisting of lumped series inductor and a double-diode ESD network with nonidentical stages.

The RF ESD design practices can be as follows:

- An RF ESD design structure comprises a multistage two-element LC network using a series inductor element and a double-diode ESD network for improved power transfer and ESD protection.

- Optimization of the network is a function of the inductor magnitude, diode perimeter, and the successive stages of the network.

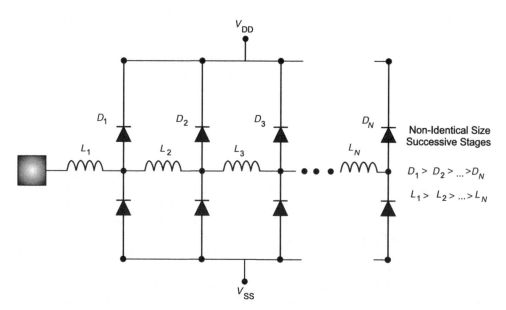

Figure 4.18 RF distributed inductor–double-diode ESD network forming a distributed LC transmission line (comprising series inductors and parallel shunt capacitors). This network has nonidentical size stages

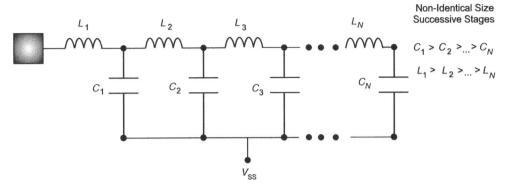

Figure 4.19   Distributed LC transmission line comprising series inductors and parallel shunt capacitors

4.6.3 RF CMOS Distributed RF ESD Networks Using Series Inductor and MOSFET Parallel Shunt

Distributed ESD networks have the advantages of providing better broadband response and better power transfer characteristics and lowering the effect load of the ESD network. RF ESD design techniques can establish a synergy of design techniques; this is achieved by using RF methods for RF functional response and ESD structural elements.

Figure 4.19 shows an example of a lossless LC transmission line consisting of lumped series inductor elements and shunt capacitor elements. In an ideal lossless LC transmission line, power transfer and matching can be optimized. In this implementation, letting the shunt capacitor elements be substituted into an ESD element, ESD events at the input node can discharge current to the a.c. ground potential.

An interesting synergy of design practices can be established for an RF ESD design. By a replacement of the capacitor element with a MOSFET structure, a distributed RF ESD network can be constructed. Figure 4.20 shows an example of a distributed two-element RF

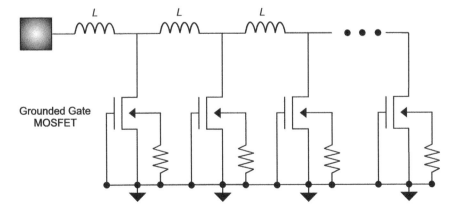

Figure 4.20   Distributed lumped LC transmission line comprising series inductors and grounded gate MOSFET (GG MOSFET) shunt element

ESD network comprising a series inductor element and a MOSFET. The MOSFET has the gate structure at a ground potential and the substrate that is also at the ground forming a grounded-gate MOSFET network. A capacitance load consists of the MOSFET drain-to-gate and MOSFET drain-to-substrate capacitance. During an RF normal operation condition, the grounded-gate MOSFET structure is in 'off-state'. The distributed network represents an LC transmission line of lumped elements at the RF functional frequency. During the event, current flows through the first inductor element. When the voltage at the MOSFET drain element reaches the MOSFET drain avalanche breakdown voltage, MOSFET snapback occurs, leading current discharge to the ground rail. Moreover, the transient current response leads to a voltage drop across the series element (e.g., $V_L = L(di/dt)$). As the size of the inductor element increases, the voltage drop increases. In the case of large inductor voltage drops, each successive grounded gate MOSFET stage will turn-on sequentially as its MOSFET drain-to-source voltage reaches the MOSFET snapback voltage. Hence, an optimum RF ESD design is such that the early stages must not reach MOSFET second breakdown prior to the MOSFET snapback of the latter stages. To avoid MOSFET failure of the earlier stages prior to the initiation of the latter stages, there is a RF ESD design tradeoff in a given two-element segment of the size of the inductor, MOSFET channel length (which determines MOSFET snapback voltage), and MOSFET channel width (which determines total capacitance of the MOSFET gate structure, and the total current discharge capability).

In ESD design practices, multistage structures introduce resistor ballasting and segmentation using either separate physical element or through physical design layout techniques [3]. A lossy transmission line structure can be formed using a three-element transmission line segment with a series inductor element and a parallel shunt element comprising of resistor and capacitor elements (Figure 4.21). Again, a series inductor can be utilized for the series element. For the transmission line shunt element, a resistor-ballasted MOSFET structure is introduced; this element can have a MOSFET drain and source resistor element. In this implementation, with the utilization of MOSFET ballast resistor elements, ESD current distribution through the successive transmission line stages will be improved. In this LRC transmission line, the optimization of this RF ESD network will be a function of the inductor magnitude, MOSFET channel length and width, as well as the resistor magnitude of the ballast resistor elements.

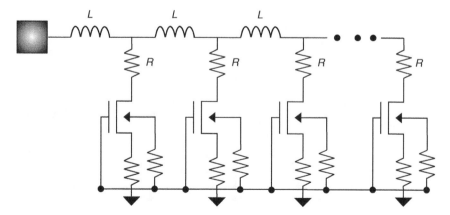

Figure 4.21   Distributed lumped LC transmission line comprising series inductors and grounded gate MOSFET (GG MOSFET) shunt element

The RF ESD design practices can be as follows:

- An RF ESD design structure comprises a multistage two-element LC network using a series inductor element, and a shunt MOSFET element for an improved power transfer and ESD protection.

- A RF ESD design structure comprises a multistage three-element LCR network using a series inductor element, and a shunt resistor ballasted MOSFET element for an improved ESD network current distribution.

- Optimization of the network is a function of the inductor magnitude, MOSFET channel length, MOSFET channel width, and resistor.

4.7 RF CMOS DISTRIBUTED ESD NETWORKS: TRANSMISSION LINES AND COPLANAR WAVEGUIDES

RF CMOS distributed ESD networks will provide improved broadband response [24–26, 34–40]. In designing protection circuits for RF applications, the loading capacitance will be a concern as the reactance continues to decrease. Given a scaling condition of a constant reactance, where reactance is

$$X_C = 1/(2\pi f C_{ESD})$$

when the application frequency increases, the ESD loading capacitance would have to be reduced. In the case of a narrow-band application, an ESD resonant cancellation technique can be utilized to eliminate the loading effect (e.g., adding a parallel inductor element). Hence, for a single narrow band application, any inductive element can be utilized to circumvent the impact of the capacitor element (e.g., at a given d.c. voltage state). This resonance cancellation technique is not acceptable for a broadband applications.

Hence, an RF ESD protection device is needed that has the following characteristics:

- low impedance during ESD events;
- high impedance during RF functional application;
- low capacitance load during RF functional application;
- low impact on bandwidth.

A common RF design technique for broadband applications has been to use distributed circuit networks for amplifier and oscillator networks [29–33]. Kleveland proposed using the same concept for RF ESD networks for broadband applications [34–37]. Using a distributed RF ESD network, a segmented transmission line is placed between successive stages of an ESD network [34–40]. The RF ESD elements are distributed along the transmission line in such a manner so as to provide matching with the external signal line. The transmission line can consist of the following elements:

- microstrip transmission line;
- coplanar waveguide (CPW);

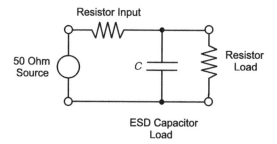

Figure 4.22   Example of a 50 Ω source, a series resistor, a ESD equivalent capacitance load, and a 50 Ω circuit load

- coplanar strip-line;

- bond wire;

- inductor element.

As an explanation, Figure 4.22 shows an example of a 50 Ω source, a series resistor, an ESD capacitance load, and a circuit load [34–37]. The series resistance represents the interconnect wiring, pad and ESD element resistance. The ESD capacitance load is the total capacitance coupling to the ground or power supply rails of the ESD element and other parasitic capacitance elements. In this case, the series resistor element and the ESD element are chosen to provide the optimum ESD protection scheme. Also, the power loss occurs because of the mismatch between the input source and the effective load condition. An additional issue is that the voltage variation of the capacitor element can lead to poor linearity. ESD implementations that utilize metallurgical junctions can initiate poor linearity in RF circuits; this is true for RF GG MOSFET, single diode, dual-diode, and bipolar ESD networks.

Figure 4.23 shows an example of a coplanar waveguide segment replacing the series resistance element [34–37]. In this manner, the effective ESD network comprises an LC component with a capacitor shunt element. The LC transmission line can provide matching conditions for the 50 Ω input source. Additionally, the linearity of the circuit can improve as the LC transmission line is not a voltage dependent capacitor element.

Figure 4.24 shows an example of the subdivision of the same ESD capacitor element into multiple segments. In this manner, the coplanar waveguide splits the load of the ESD

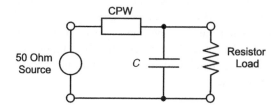

Figure 4.23   Example of a 50 Ω source, a coplanar waveguide (CPW), a ESD equivalent capacitance load, and a 50 Ω circuit load

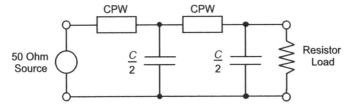

Figure 4.24 Example of a 50 Ω source, a dual segment coplanar waveguide (CPW), and an ESD equivalent capacitance load

element into a first segment, followed by another segment. Figure 4.25 shows the subdivision of the ESD element into four segments. As more segments are added, and the size of the network load is more distributed, the network becomes a more ideal transmission line, and the broadband matching of the 50 Ω input source and the 50 Ω output load improves [34–40].

The RF ESD design practices can be as follows:

- With a distributed RF ESD network, the deleterious loading effects of an ESD network can be reduced (e.g., impedance mismatch, reflection, and power transfer).

- With a distributed RF ESD network, s-parameter degradation of reflection parameter S_{11} and transmission parameter S_{21} is minimized (compared with single lumped RF ESD network).

- With transmission lines in a distributed RF ESD network, circuit linearity can be improved because of the nonvoltage dependent nature of LC transmission lines.

4.8 RF CMOS: ESD AND RF LDMOS POWER TECHNOLOGY

Bipolar transistors have traditionally used power applications because of their ability to provide high breakdown voltages and large currents. From the Johnson Limit relationships (e.g., the power–cutoff frequency relationship and the breakdown voltage–cutoff frequency relationships), bipolar transistors demonstrate the ability to provide RF performance and breakdown product. For RF power applications, the ability to achieve both good power characteristics and high frequencies is preferred. Hence, it is preferable to have a circuit

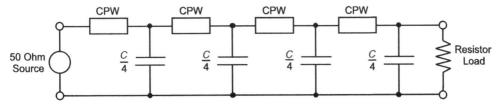

Figure 4.25 Example of a 50 Ω source, a four-segment coplanar waveguide (CPW), and an ESD equivalent capacitance load

element that can achieve power transfer, power gain, linearity, and high switching speed under high voltage conditions.

With CMOS scaling of the MOSFET gate thickness, standard CMOS transistors are not preferred for RF power application. But, with the MOSFET channel length scaling and the introduction of laterally diffused metal oxide semiconductor (LDMOS) MOSFET transistors, the usage of MOSFETs as discrete or integrated elements can provide the preferred breakdown voltage and the switching speed needed for power applications. LDMOS transistors have had an increase of interest for linear high power amplifier (PA) applications for cellular base stations [41–44]. In power environments, LDMOS transistors are being utilized in 25 and 40 V applications. An advantage of LDMOS technology is that it can be integrated with the standard CMOS foundry technology. This will provide advantages of low cost and high level integration.

Disadvantages of LDMOS are linearity and stability issues associated with hot electron injection. LDMOS threshold shifts due to hot electron injection can occur over the LDMOS power device lifetime; these LDMOS threshold shifts influence the gain, output power, and linearity [41–44]. Moreover, time-independent degradation effects associated with LDMOS second breakdown have been observed; these are apparent from the LDMOS transistor I–V transfer characteristics. LDMOS second breakdown occurs because of the parasitic bipolar transistor in the LDMOS transistor element. LDMOS second breakdown is more likely to occur from slow events, which can lead to LDMOS parasitic bipolar turn-on. LDMOS second breakdown can lead to functional degradation mechanisms, causing nonlinear harmonics and impacting intermodulation distortion (IMD). In practice, ballast resistors are used to lessen the likeliness of LDMOS second breakdown [41–45].

Owing to the sensitivity issues of LDMOS technology, ESD protection of LDMOS is important to avoid LDMOS MOSFET gate failure and electrical overstress that can lead to hot electron injection. Avoidance of LDMOS second breakdown is also important to prevent nonlinear harmonic distortions. Hence, ESD protection must be provided to LDMOS devices that prevents ESD-induced second breakdown degradation mechanism (preserving gain and stability characteristics) and permanent damage. At the same time, the ESD protection devices must satisfy the application voltage conditions and RF characteristics (e.g., not impact matching, the gain [Gp], and power added efficiency [PAE]).

An RF ESD practice for LDMOS must address the following conditions:

- LDMOS RF ESD devices breakdown voltages must be above the LDMOS application condition (e.g., power supply and the signal swing).

- LDMOS RF ESD devices must prevent LDMOS gate hot electron injection and MOSFET second breakdown (e.g., prevent turn-on of the LDMOS parasitic bipolar transistor to avoid nonlinear harmonic distortion, IMD, etc.).

- LDMOS RF ESD devices must not have significant impact on the RF input matching characteristics.

- LDMOS RF ESD devices must not impact the Gp and PAE.

Smedes *et al.* [45] evaluated potential ESD protection devices for RF LDMOS transistors and LDMOS technology. In the study, Smedes *et al.* [45] addressed both the ESD device appropriateness of ESD protection of the LDMOS gate input and the LDMOS output; the parameters of interest included the ESD breakdown voltages, ESD robustness of the ESD structure, RF matching characteristics, and the impact on the Gp and PAE RF performance characteristics. Three types of potential ESD networks were evaluated for the LDMOS input network:

- reversed-biased avalanche breakdown diodes of various breakdown voltages;

- single grounded-gate *n*-channel MOSFET (GGNMOS);

- an integrated series cascode *n*-channel MOSFET.

For an RF input circuit, the reverse biased shunt diode element can serve as an RF matching capacitor and prevent breakdown of the gate input. But, in the study by Smedes *et al.* [45], for their given technology, it was found that the diode current-to-failure, breakdown voltages, and the required structure size were inappropriate to provide LDMOS gate protection.

Smedes *et al.* showed that the single grounded-gate *n*-channel MOSFET (GGNMOS) impacted the Gp and the PAE characteristics. At low power levels for a single tone RF signal, this was not a concern; but as the output power increased, Gp and PAE did not increase beyond a given power level. Smedes *et al.* [45] showed with a series cascode GGNMOS device, there was insignificant distortion of the Gp and PAE characteristics over the desired output power. Figure 4.26 shows an example of a RF ESD protection device for the RF LDMOS input node. Figure 4.27 shows a series cascode MOSFET structure constructed in a RF LDMOS technology.

Figure 4.27 shows the RF ESD circuit for a series cascode GGNMOS network. The RF ESD circuit contains a series drain resistance and the series GGNMOS network. In this implementation, an RF T-type matching network was created using two low inductance bond wires and a shunt capacitor element.

In the implementation, it was found that the T-match shunt capacitor became charged during ESD-simulated testing; this T-match capacitor is discharged to the RF LDMOS gate

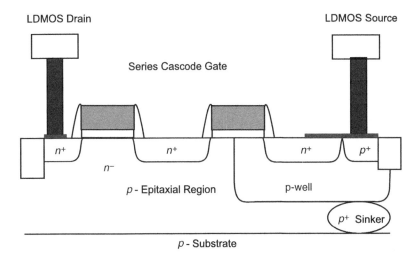

Figure 4.26   Cross section of a series cascode grounded gate NMOS (GGNMOS) structure used for ESD protection of an RF LDMOS power amplifier

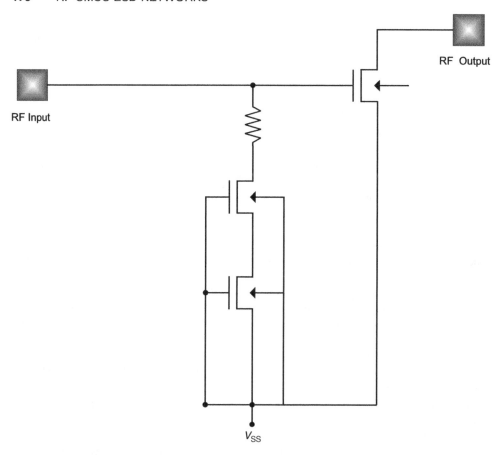

Figure 4.27   Circuit schematic of a series cascode grounded gate NMOS (GGNMOS) structure used for ESD protection of an RF LDMOS power amplifier

structure. Through the optimization of the LC T-match network and adding some series resistance to the cascode GGNMOS ESD structure, optimum matching and ESD results were achieved. In this case, Smedes *et al.* [45] noted that the RF matching network influenced the ESD response; during ESD test simulation, the precharged source capacitor charged the RF matching shunt capacitor. The shunt capacitance element used for this application ranged from 1 to 50 pF. Note that a 50 pF matching capacitor is comparable to the ESD source capacitor (e.g., HBM capacitance is 100 pF). This charging of the shunt capacitor leads to a second discharge process of the RF matching network into the RF circuit. Through optimization of the inductor, capacitor, and a resistor element, the peak current was evaluated to produce the desired ESD response.

4.9 RF CMOS ESD POWER CLAMPS

In CMOS technology, MOSFET-based ESD power clamps have become a standard ESD design practice in chip design [46–58]. In early implementations, grounded-gate NMOS

(GGNMOS) ESD network were utilized as ESD power clamps. When a MOSFET undergoes MOSFET second breakdown, nonuniform conduction occurs; this leads to inadequate predictability as well as lack of ESD protection scaling with the MOSFET width. RC-triggered MOSFET ESD power clamps introduced two concepts:

- gate-coupling;

- frequency triggering.

There are a number of advantages that have led to the widespread introduction of RC-triggered MOSFET ESD power clamps. These advantages are as follows:

- frequency triggered;

- compatibility with CMOS technology;

- design integration with digital circuits;

- use of supported MOSFET devices (e.g., nonuse of parasitic devices);

- circuit simulation;

- scalable;

- utilization of single-gate and dual-gate oxides;

- latchup immune.

Frequency-triggered networks have the advantage that it is not dependent on the turn-on voltage. Voltage-triggered ESD power clamps have a delay of operation until a certain d.c. voltage level is achieved. Frequency-triggered ESD power clamps are a.c. responsive instead of a d.c. voltage level.

RC-triggered MOSFET ESD power clamps are compatible with CMOS technology. The basic elements are scaled with every technology generation, and are compatible with digital CMOS circuitry. The compatibility with CMOS digital logic and memory prevents any additional integration issues (e.g., $1/f$ noise, voltage level incompatibility, scaling).

RC-triggered MOSFET ESD power clamps do not use parasitic elements. ESD solutions that utilize parasitic devices (e.g., parasitic *pnp*, *npn*, diodes, etc) are typically not supported with characterization, device models, and circuit simulation models, and are not scalable and well controlled in a manufacturing environment.

By using MOSFETs in a low voltage regime, MOSFET device current models exist. As a result, the RC-triggered MOSFET ESD power clamps can be simulated using standard circuit simulation; in these ESD power clamps, MOSFET device simulation (e.g., ambient and/or electrothermal simulation) is not required to demonstrate the operability of the ESD network. Electrical characterization, device models, and circuit models in the low voltage regime (e.g., below avalanche breakdown) can be used to demonstrate operability of the MOSFET power clamp. In this manner, the ESD design synthesis including the ESD network, the power bussing, and the ESD networks can be completed using circuit simulation. High voltage, electrothermal device simulation, and electrothermal circuit simulation can be avoided.

4.9.1 RC-Triggered MOSFET ESD Power Clamp

Using RC-triggered MOSFET ESD power clamps, these structures scale with the technology generation. Many ESD power clamp solutions do not contain scalable trigger conditions (e.g., Zener diode, SCR, and MLSCR ESD power clamps). With a scalable ESD design solution, design migration through technology generations or design 'shrinks' is simplified.

With the MOSFET scaling and mixed voltage interface environments, the gate dielectric scaling can be addressed by producing series-cascode RC-triggered ESD power clamps, or utilizing both single gate and dual-gate oxides.

CMOS latchup is a concern in CMOS technology. RC-triggered MOSFET ESD power clamps are typically CMOS latchup immune and do not introduce CMOS latchup concerns in the digital segments of a semiconductor chip design.

RC-triggered MOSFET ESD power clamps consist of the following elements:

- RC-frequency discrimination circuit;

- inverter drive circuit;

- MOSFET output clamp element.

In the RC-frequency discriminator circuit, resistor and capacitor elements are needed to form the RC discriminator network. The 'resistor' elements can be supported by resistor elements (e.g., p-type, n-type, polysilicon) or MOSFETs in an 'on' state. Capacitor elements can be MOSFET (MOS) capacitors, MIM capacitor, varactors, hyper-abrupt varactors, or other capacitor structures.

The inverter drive circuit is formed using a series of inverter elements between the RC-discriminator circuit and the MOSFET output clamp element. The inverter drive circuit has two roles; first, it isolates the frequency discrimination circuit from the MOSFET output clamp, and secondly, it drives the MOSFET output clamp MOSFET gate voltage level. In the first issue, without the inverter drive stage, the output capacitance of the MOSFET clamp would be in parallel with the capacitor element; this would change the frequency response of the network as the MOSFET output clamp MOSFET width is varied. In an ideal implementation, the RC-trigger discriminator is isolated from the output. Secondly, to drive the MOSFET output clamp, the size of the inverter stage increases to provide adequate drive strength for the MOSFET output clamp gate capacitance.

In this circuit, using additional MOSFET device elements, the operability of the RC-triggered MOSFET clamp can be further improved. Additional design objectives include the following:

- desensitize the inverter stage during ESD events because of power bus voltage drops, noise triggering, power-up, and oscillations;

- improve the triggering and logic control;

- improve the stand-by power consumption and leakage.

In this network, it is preferable to prevent false triggering, noise triggering, and power supply reduction by using resistive elements in the inverter stage. These elements serve as a

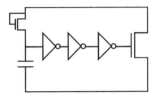

Figure 4.28   RC-triggered MOSFET ESD power clamp

means to desensitize noise, and provide a level-shift element (e.g., shifts the inverter switching condition).

To improve the triggering and logic control during ESD events, feedback elements can be used in the inverter drive stage. One solution is to use resistor feedback elements between the input and output of alternating inverter stages. Another solution is to use feedback 'keeper' network (e.g., half-latch network) in the inverter stages to improve inverter stage response. In addition, MOSFETs can be placed between the driven signal node and the power rails that hold the logic state during chip operation.

As the RC-triggered MOSFET power clamp is large, it can contribute to a semiconductor 'off' state power consumption. This can be addressed using isolated well structures, triple well, or dynamic threshold control of the MOSFET body.

Figure 4.28 shows an RC-triggered MOSFET power clamp. A MOSFET and a capacitor element are used as the RC-discriminator circuit. The RC discriminator is activated by the ESD pulse. This signal is transmitted to the inverter stages that drive the MOSFET to discharge the ESD current. In this design, the size of the inverters are designed so as to avoid false triggering of the RC-triggered network.

Table 4.1 shows the HBM and MM ESD results as a function of MOSFET width. When the correct RC time is established, the circuit ESD response improves with the structure size. In this structure, the design was increased in MOSFET total width by the addition of parallel MOSFET fingers. As the MOSFET width increased, the HBM and MM ESD failure levels increased [3].

In the optimization of the RC-triggered ESD power clamp, a correct resistor–capacitor value is required to observe the increase in the ESD results with the MOSFET width. Figure 4.29 shows an example where the capacitor element size was varied. In both cases, the ESD power clamp results increased with the MOSFET width. The RC-triggered MOSFETs were optimized to find the best point of operation. From these results, the MOSFET clamp structure achieves less than 2.5 V/μm [3].

Table 4.1   RC-triggered MOSFET ESD results as a function of MOSFET width

Type	Size (μm)	HBM (kV)	MM (kV)
RC MOSFET	1000	2.3	0.3
	2000	3.8	0.6
	4000	10.00	1.15
	8000	>10	1.75

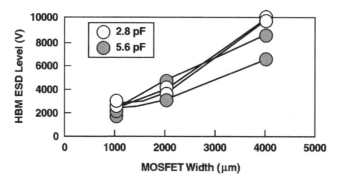

Figure 4.29 RC-triggered MOSFET ESD clamp with different RC values

4.9.2 High Voltage RC-Triggered MOSFET ESD Power Clamp

In a mixed voltage interface environment, the peripheral power supply rail can exceed the native power supply voltage [3]. In a case where two or multiple power supply voltages exist on a common semiconductor chips, the ESD power clamps must be able to withstand the voltage conditions without degradation. For CMOS, or BiCMOS applications that utilize MOSFET-based ESD power clamps, MOSFET device dielectric breakdown, or MOSFET hot electron degradation must not occur. In an RC-triggered power clamp, all elements are potentially subject to electrical overstress. For example, the resistor, the capacitor, the inverter stages, and the output clamp must all avoid electrical overstress from the functional voltage conditions, reliability accelerated voltage stress, and ESD conditions. To provide RC-triggered MOSFET ESD power clamps in a mixed voltage environment, two ESD design strategies can be incorporated as follows:

- Additional ESD device elements are to be placed between the peripheral power supply voltage conditions, and the native power supply voltage V_{DD} to 'level shift' the voltage state on the RC-triggered MOSFET network rail clamp.

- 'ESD Dummy Power Rails' are to be used within the ESD current path to avoid electrical connection to the higher power supply voltages.

- Dual-gate MOSFET (e.g., thick oxide MOSFETs) and triple-gate MOSFET devices are to be used in the RC-clamp and inverter structures.

In the first case, the use of additional elements in series with the RC-triggered MOSFET power clamp allows for the utilization of the native power supply RC-triggered network in the core networks. The use of the additional elements serve the purpose as a voltage level shifting element during normal functional conditions. The type of elements that can be utilized can be series diode elements or MOSFET elements.

Figure 4.30 shows an example of a mixed voltage interface RC-triggered MOSFET ESD power clamp using a MOSFET in an 'on' state to lower the voltage stress. In this implementation, the MOSFET gate is connected to the MOSFET drain. The MOSFET drain is connected to the higher power supply voltage and the MOSFET source is connected to the RC-triggered MOSFET power clamp power rail. In this configuration, the MOSFET

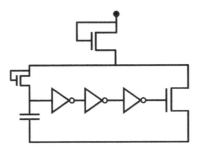

Figure 4.30 Mixed voltage interface RC-triggered MOSFET ESD power clamp using a MOSFET level-shifting element

level-shifting element serves as a means to provide a voltage drop between the higher voltage power supply node and the power rail of the RC-triggered MOSFET. The level-shifting MOSFET is an 'on-state' and provides a voltage drop to lower the voltage on the RC-triggered MOSFET power rail. In this implementation, the level-shifting element reduces the voltage stress on the RC-trigger discriminator network, the inverter drive network, and the output clamp device. Another perspective is that the element serves as a transmission gate that is tied to the input voltage (e.g., whose transmission is the $V_{DD} - V_T$). Without the 'level-shifting' element (or transmission gate), the power rail of the RC-triggered clamp can be integrated with the core V_{DD} power supply. In this manner, the power rail is separated and serves as a pseudo-V_{DD} dummy rail for the ESD power clamp network. Another perspective of this network is that the two MOSFETs output clamp device are in a series-cascode configuration, where the first MOSFETs gate is set to an 'on' state and the second element is RC-triggered (e.g., akin to a mixed voltage n-channel MOSFET pull-down segment of a OCD).

In order to provide good ESD protection for this ESD network, the ESD level-shifting element must not limit the ESD robustness of the network when the size of the MOSFET output clamp width increases. Experimental results shows that if the MOSFET output clamp width is increased, while the MOSFET level-shifting element remains fixed in width, the ESD results do not increase with the increasing MOSFET output clamp width. But, if the MOSFET level-shifting element width increases with the MOSFET output clamp width, the ESD results are not limited by the MOSFET level-shifting element and the ESD results are as good as the RC-triggered clamp without a MOSFET level-shifting element. Hence, an ESD design practice for this network is that the MOSFET level-shifting element must scale with the MOSFET output clamp element in order to achieve improved results with MOSFET output clamp width.

4.9.3 Voltage-Triggered MOSFET ESD Power Clamps

ESD power clamps can be triggered by either frequency or the voltage or current conditions. In some applications, frequency-triggered networks are not preferred. Applications where voltage triggering is preferred are as follows [3]:

- RF applications that do not prefer an RC frequency pole in the circuit network.

- Applications with inductive loads on the power grid structure.

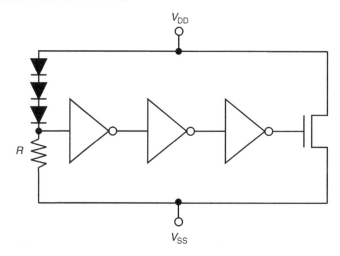

Figure 4.31 Voltage-triggered MOSFET ESD power clamp

In the first example, RF designers may not want additional frequency poles in the signal response. RC networks can modify the stability of an RF circuit. With the additional RC response, the frequency response of the power grid, and the semiconductor chip may be influenced in small circuit number and small chip applications. The RC network can influence the stability criteria in a circuit network.

In the second case, inductive loads in peripheral circuits can interact with RC trigger network leading to chip and peripheral circuit LRC oscillations that impact functionality, or LC oscillations that affect the operability of the ESD power clamp.

A voltage-triggered MOSFET ESD triggered power clamp does not have the concerns of the RC-triggered ESD power clamp. In these implementations, the voltage triggering can be established by different trigger networks:

- resistor divider trigger network;

- diode string – resistor trigger network.

Figure 4.31 shows an example of a voltage-triggered MOSFET ESD power clamp. In the second case, a diode string is placed in series with a resistor element are placed between the V_{DD} and V_{SS} power rails. The center node of the trigger network is electrically connected to a series of inverters; the output of the inverters is connected to the MOSFET gate electrode that initiates the ESD discharge between V_{DD} and V_{SS}. When an overvoltage condition is established, the diode string trigger element is initiated leading to the increase in the voltage at the first inverter input; this initiates the inverter stage, leading to gate-coupling of the MOSFET power clamp element.

4.10 SUMMARY AND CLOSING COMMENTS

In this chapter we have addressed the state-of-the-art on ESD circuit design in RF CMOS technology. RF ESD diodes, ESD inductor/diode networks, RF ESD diode strings, RF

ESD-power clamps, distributed load ESD networks, and distributed coplanar waveguide ESD networks were shown demonstrating the RF ESD design techniques in examples and in application to circuit functions such as RF CMOS LNA design. Also, we have highlighted the ESD solutions and RF ESD design techniques from cancellation techniques, distributed loads to integration with the RF matching networks. A new growing area of interest is the use of LDMOS transistors for RF applications as the transistor speed continues to increase. In these RF LDMOS applications, the RF CMOS ESD-RF cosynthesis techniques are being utilized in recent times.

The next chapter, will address RF bipolar transistors. We will discuss bipolar transistor physics and bipolar transistor device parameters. The chapter will also address the bipolar transistor under high voltage and high current conditions; this will be followed by future chapter discussions of RF bipolar ESD networks in silicon germanium, silicon germanium carbon, gallium arsenide, indium gallium arsenide, and other RF technologies.

PROBLEMS

1. Derive the general relationship for a distributed two-port ESD network representing the network as the Stieljes continued fraction. Representing the impedance term as an impedance and admittance term as follows:

$$z_{11}(s) = Z_1(s) + \cfrac{1}{Y_2(s) + \cfrac{1}{Z_3(s) + \cfrac{1}{Y_4(s) + \cfrac{1}{Z_5(s) + Z_d(s)}}}}$$

and repeated to the last admittance element $Y_n(s)$,

$$z_{11}(s) = Z_1(s) + \cfrac{1}{Y_2(s) + \cfrac{1}{Z_3(s) + \cfrac{1}{Y_4(s) + \cfrac{1}{\ \bullet\ }}}}$$
$$\bullet$$
$$\bullet$$
$$\cfrac{1}{Y_n(s)}$$

Draw the transmission line of lumped elements and derive the above relationship.

2. In the above relationship, let the admittance terms, Y_2, Y_4, ... , Y_n be ESD shunt elements and the series impedance terms, Z_1, Z_3, Z_{N-1} be inductor elements. Substitute into the Stieljes continued fraction form.

3. As in the above section, let the RF CMOS ESD network consist of MOSFETs shunt elements and inductor series elements. Derive the impedance in the Stieljes continued fraction form.

4. As in the above section, let the shunt element be transmission lines. Derive the Steiljes continued fraction form.

5. Derive the current and voltage relationship of a RGLC incremental model for a transmission line with a series resistor, shunt conductance, series inductor, and shunt capacitor. Solve for the current and voltage differential equations. Let R be an ESD series resistor, G an ESD shunt conductance element, L a series inductance, and C a capacitive ESD shunt.

6. Derive the current and voltage relationship of a RLC incremental model for a transmission line with a series resistor, series inductor, and shunt capacitor. Solve for the current and voltage differential equations. Let R be an ESD series resistor, G an ESD shunt conductance element, L a series inductance, and C a capacitive ESD shunt. Show the current and voltage distribution through the network.

7. Derive the current and voltage relationship of an LC incremental model for a transmission line with a series inductor and shunt capacitor. Solve for the current and voltage differential equations. Let R be an ESD series resistor, G an ESD shunt conductance element, L a series inductance, and C a capacitive ESD shunt. Let the capacitance shunt be a grounded gate MOSFET and in the second case, a dual-diode element. Show the voltage and current distribution.

8. Show on a Smith chart, the change in the impedance for a circuit that is formed from a plurality of identical stages where each stage is formed from a dual diode network and inductor. Treat the dual diode network as a capacitor only.

9. Show on a Smith chart, the change in the impedance for a circuit that is formed from a plurality of nonidentical stages where each stage is formed from a dual diode network and an inductor, and in each case, the stages of the diode elements are smaller by a parameter α. Treat the dual diode network as a capacitor only.

10. In the design of ESD protection of an RF LDMOS input network, a series cascode GGNMOS is used. Calculate the necessary condition for Vt1, and Vt2 for the GGNMOS device to avoid dielectric breakdown of the LDMOS input.

REFERENCES

1. Richier C, Salome P, Mabboux G, Zaza I, Juge A, Mortini P. Investigations on different ESD protection strategies devoted to 3.3 V RF applications (2 GHz) in a 0.18-μm CMOS process. *Proceedings of the Electrical Overstress/Electrostatic Discharge (EOS/ESD) Symposium*, 2000. p. 251–9.
2. Voldman S. *ESD: circuits and devices*. Chichester, England: John Wiley and Sons, Ltd.; 2005.
3. Voldman S. *ESD: physics and devices*. Chichester, England: John Wiley and Sons, Ltd.; 2004.
4. Worley ER, Bakulin A. Optimization of input protection for high speed applications. *Proceedings of the Electrical Overstress/Electrostatic Discharge (EOS/ESD) Symposium*, 2002. p. 62–72.
5. Voldman S. ESD protection and RF design, *Tutorial J, Tutorial Notes of the Electrical Overstress/Electrostatic Discharge (EOS/ESD) Symposium*, 2001.
6. Hynoven S, Rosenbaum E. Diode-based tuned ESD protection for 5.25 GHz CMOS LNAs. *Proceedings of the Electrical Overstress/Electrostatic Discharge (EOS/ESD) Symposium*, 2003. p. 188–94.

7. Hynoven S, Joshi S, Rosenbaum E. Comprehensive ESD protection for RF inputs. *Proceedings of the Electrical Overstress/Electrostatic Discharge (EOS/ESD) Symposium*, 2003. p. 188–94.

8. Hynoven S, Joshi S, Rosenbaum E. Cancellation technique to provide ESD protection for multi-GHz RF inputs. *IEEE Transactions of Electron Device Letters* 2003;**39**(3):284–6.

9. Rosenbaum E. ESD protection for multi-GHz I/Os. *Proceedings of the Taiwan Electrostatic Discharge Conference (T-ESDC)*, 2004. p. 2–7.

10. Ker MD, Lee CM. ESD protection design for GHz RF CMOS LNA with novel impedance isolation technique. *Proceedings of the Electrical Overstress/Electrostatic Discharge (EOS/ESD) Symposium*, 2003. p. 204–13.

11. Ker MD, Lo WY, Lee CM, Chen CP, Kao HS. ESD protection design for 900 MHz RF receiver with 8 kV HBM ESD robustness. *Proceedings of the IEEE Radio Frequency Integrated Circuit (RFIC) Symposium*, 2002. p. 427–30.

12. Lee CM, Ker MD. Investigation of RF performance of diodes for ESD protection in GHz RF circuits. *Proceedings of the Taiwan Electrostatic Discharge Conference (T-ESDC)*, 2002. p. 45–50.

13. Leroux P, Steyart M. High performance 5.25 GHz LNA with on-chip inductor to provide ESD protection. *IEEE Transactions of Electron Device Letters* 2001;**37**(7):467–9.

14. Janssens J. Deep submicron CMOS cellular receiver front-ends. PhD thesis, K.U., Leuvens, Belgium, July 2001.

15. Leroux P, Janssens J, Steyaert M. A 0.8 dB NF ESD-protected 9 mW CMOS LNA. *Proceedings of the IEEE International Solid State Circuits Conference (ISSCC)*, 2001. p. 410–1.

16. Leroux P, Steyaert M. A high performance 5.25 GHz LNA with an on-chip inductor to provide ESD protection. *IEEE Electron Device Letters* 2001;**37**(5):467–9.

17. Leroux P, Janssens J, Steyaert M. A new ESD protection topology for high frequency CMOS Low noise amplifiers. *Proceedings of the IEEE International Symposium on Electromagnetic Compatibility*, September 2002. p. 129–33.

18. Leroux P, Vassilev V, Steyaert M, Maes H. A 6 mW 1.5 dB NF CMOS LNA for GPS with 3 kV HBM protection. *Proceedings of the Electrical Overstress/Electrostatic Discharge (EOS/ESD) Symposium*, 2002. p. 18–25.

19. Leroux P, Steyaert M. RF-ESD co-design for high performance CMOS LNAs. *Proceedings of the Workshop on Advances in Analog Circuit Design*, Graz, Austria, 2003.

20. Leroux P, Vassilev V, Steyaert M, Maes H. High performance, low power CMOS LNA for GPS applications. *Journal of Electrostatics* 2003;**59**(3–4):179–92.

21. Vassilev V, Thijs S, Segura PL, Leroux P, Wambacq P, Groseneken G, et al. Co-design methodology to provide high ESD protection levels in the advanced RF circuits. *Proceedings of the Electrical Overstress/Electrostatic Discharge (EOS/ESD) Symposium*, 2003. p. 195–203.

22. Leroux P, Steyaert M. *LNA-ESD co-design for fully integrated CMOS wireless receivers.* Dordecht, The Netherlands: Springer; 2005.

23. Thijs S, Linten D, Natarajan MI, Jeamsaksiri W, Mercha A, Ramos J, et al. Class 3 HBM and class 4 ESD protected 5.5 GHz LNA in 90 nm RF CMOS using above-IC inductor. *Proceedings of the Electrical Overstress/Electrostatic Discharge (EOS/ESD) Symposium*, 2005. p. 25–32.

24. Ker MD, Kuo BJ. New distributed ESD protection circuit for broadband RF ICs. *Proceedings of the Taiwan Electrostatic Discharge Conference (T-ESDC)*, 2003. p. 163–8.

25. Ker MD, Kuo BJ. Optimization of broadband RF performance and ESD robustness by π-model distributed ESD protection scheme. *Proceedings of the Electrical Overstress/Electrostatic Discharge (EOS/ESD) Symposium*, 2004. p. 32–9.

26. Hsiao YW, Kuo BJ, Ker MD. ESD protection design for a 1–10 GHz wideband distributed amplifier in CMOS technology. *Proceedings of the Taiwan Electrostatic Discharge Conference (T-ESDC)*, 2004. p. 90–4.

27. Galal S, Razavi B. Broadband ESD protection circuits for CMOS technology. *IEEE Journal of Solid State Circuits* 2003;**38**(12):2334–40.

28. Galal S, Razavi B. 40 Gb/s amplifier and ESD protection circuit in 0.18-mm CMOS technology. *IEEE Journal of Solid-State Circuits* 2004;**39**(12):2389–96.

29. Ginzton E, Hewlett WR, Jasberg JH, and Noe JD, *et al*. Distributed amplification. *Proceedings of the IRE*, 1948. p. 956–69.

30. Sarma DG. On distributed amplification. *Proceedings of the Institute of Electrical Engineering*, 1954. p. 689–97.

31. Majidi-Ahy R, Nishimoto CK, Riaziat M, Glenn M, Silverman S, Weng S-L, Pao Y-C, Zdasiuk GA, Bandy SG, Tan ZCH, *et al*. A 5–100 GHz InP coplanar waveguide MMIC distributed amplifier. *IEEE Transactions on Microwave Theory and Techniques* 1990;**38**(12):1986–94.

32. Sullivan PJ, Xavier BA, Ku WH. An integrated CMOS distributed amplifier utilizing packaging inductance. *IEEE Transactions on Microwave Theory and Techniques* 1997;**45**(10):1969–77.

33. Lee TH. *The design of CMOS radio frequency integrated circuits*. Cambridge University Press; 1998.

34. Kleveland B, Diaz CH, Vook D, Madden L, Lee TH, Wong SS. Monolithic CMOS distributed amplifier and oscillator. *Proceedings of the International Solid State Circuits Conference (ISSCC)*, 1999. p. 70–1.

35. Kleveland B, Lee TH. Distributed ESD protection device for high speed integrated circuits. U.S. Patent No. 5,969,929 (October 19, 1999).

36. Kleveland B, Maloney TJ, Morgan I, Madden L, Lee TH, Wong SS. Distributed ESD protection for high speed integrated circuits. *IEEE Transactions Electron Device Letters* 2000;**21**(8): 390–2.

37. Kleveland B, Diaz CH, Vook D, Madden L, Lee TH, Wong SS. Exploiting CMOS reverse interconnect scaling in multigigahertz amplifier and oscillator design. *IEEE Journal of Solid State Circuits* 2001;**36**:1480–8.

38. Ito C, Banerjee K, Dutton RW. Analysis and design of ESD protection circuits for high frequency/ RF applications. *IEEE International Symposium on Quality and Electronic Design (ISQED)*, 2001. p. 117–22.

39. Ito C, Banerjee K, Dutton R. Analysis and optimization of distributed ESD protection circuits for high-speed mixed-signal and RF applications. *Proceedings of the Electrical Overstress/Electrostatic Discharge (EOS/ESD) Symposium*, 2001. p. 355–63.

40. Ito C, Banerjee K, Dutton R. Analysis and design of distributed ESD protection circuits for high-speed mixed-signal and RF applications. *IEEE Transactions on Electron Devices* 2002;**49**:1444–54.

41. Rice J. LDMOS linearity and reliability. *Microwave Journal* 2002;**45**(6):64.

42. Rabany A, Nguyen L, Rice D. Memory effect reduction for LDMOS bias circuits. *Microwave Journal* 2003;**46**(2):124.

43. Rice J. Gaining LDMOS device linearity and stability. ED Online ID#5899. *Microwaves and RF: Trusted Resource for Working RF Engineer*, September 2003.

44. Hammes PCA, Jos HFF, Rijs F van, Theeuwen SJCH, and Vennema K, *et al*. High efficiency high power WCDMA LDMOS transistors for base stations. *Microwave Journal* 2004;**47**(4):94–7.

45. Smedes T, de Boet J, Rodle T. Selecting an appropriate ESD protection for discrete RF power LDMOSTs. *Proceedings of the Electrical Overstress/Electrostatic Discharge (EOS/ESD) Symposium*, 2005. p. 1–9.

46. Mack W, Meyer R. New ESD protection schemes for BiCMOS Processes with application to cellular radio designs. *Proceedings of the IEEE International Symposium on Circuits and Systems*, 1992. Volume 6, p. 2699–2702, May 10–12.

47. Merrill R, Issaq E. ESD design methodology. *Proceeding of the Electrical Overstress/Electrostatic Discharge (EOS/ESD) Symposium*, 1993. p. 233–8.

48. Dabral S, Aslett R, Maloney T. Designing on-chip power supply coupling diodes for ESD protection and noise immunity. *Proceeding of the Electrical Overstress/Electrostatic Discharge (EOS/ESD) Symposium*, 1994. p. 239–49.

49. Maloney T. Novel clamp circuits for IC power supply protection. *Proceedings of the Electrical Overstress/Electrostatic Discharge (EOS/ESD) Symposium*, 1995. p. 1–12.

50. Dabral S, Maloney TJ. *Basic ESD and I/O design.* New York: John Wiley and Sons, Ltd.; 1998.

51. Dabral S, Aslett R, Maloney T. Core clamps for low voltage technologies. *Proceeding of the Electrical Overstress/Electrostatic Discharge (EOS/ESD) Symposium*, 1994. p. 141–9.

52. Maloney T. Stacked PMOS clamps for high voltage power supply protection. *Proceedings of the Electrical Overstress/Electrostatic Discharge (EOS/ESD) Symposium*, 1999. p. 70–7.

53. Smith J, Boselli G. A MOSFET power clamp with feedback enhanced triggering for ESD protection in advanced CMOS technologies. *Proceedings of the Electrical Overstress/Electrostatic Discharge (EOS/ESD) Symposium*, 2003. p. 8–16.

54. Torres C, Miller JW, Stockinger M, Akers MD, Khazhinsky MG, Weldon JC. Modular, portable, and easily simulated ESD protection networks for advanced CMOS technologies. *Proceedings of the Electrical Overstress/Electrostatic Discharge (EOS/ESD) Symposium*, 2001. p. 82–95.

55. Stockinger M, Miller J, Khazhinsky M, Torres C, Weldon J, Preble B, *et al.* Boosted and distributed rail clamp networks for ESD protection in advanced CMOS technologies. *Proceedings of the Electrical Overstress/Electrostatic Discharge (EOS/ESD) Symposium*, 2003. p. 17–26.

56. Stockinger M, Miller J. Advanced ESD rail clamp network design for high voltage CMOS applications. *Proceedings of the Electrical Overstress/Electrostatic Discharge (EOS/ESD) Symposium*, 2003. p. 280–8.

57. Juliano PA, Anderson WR. ESD protection design challenges for a high pin-count alpha microprocessor in a 0.13-μm CMOS SOI technology. *Proceedings of the Electrical Overstress/ Electrostatic Discharge (EOS/ESD) Symposium*, 2003. p. 59–69.

58. Poon S, Maloney T. New considerations for MOSFET power clamps. *Proceedings of the Electrical Overstress/Electrostatic Discharge (EOS/ESD) Symposium*, 2002. p. 1–5.

5 Bipolar Physics

In recent times, the majority of RF applications are based on bipolar transistor technology as opposed to MOSFET technology. Bipolar technology had significant advantage in performance compared to CMOS technology prior to 1985. With the scaling of CMOS, the advantage of MOSFET technology displaced bipolar transistors in a number of applications from static random access memory (SRAMs), dynamic memory (DRAM), and logic circuitry. Bipolar transistors remained important for analog, high voltage, power, and microwave applications. In recent times, the bipolar transistor as both a homo-junction and heterojunction structure exists in silicon, silicon germanium, silicon germanium carbon, gallium arsenide, indium phosphide, and other exotic material structures.

In this chapter, the focus will be on fundamentals of bipolar device physics. This will serve as background for the understanding of high current, high voltage, and power characteristics. These characteristics are important for ESD as well.

5.1 BIPOLAR DEVICE PHYSICS

In this chapter, bipolar device physics equations, definitions, and terms will be discussed that are needed for the understanding of RF devices and ESD. This chapter will also discuss the high-current phenomenon, power and speed limitations, device design ordering, and thermal properties of ESD phenomenon in bipolar transistors.

5.1.1 Bipolar Transistor Current Equations

Evaluation of bipolar transistor current density from the emitter to the collector for an *npn* transistor can be represented as [1]

$$J = \frac{qn_i^2\left[\exp\left(\dfrac{qV_{\mathrm{BC}}}{kT}\right) - \exp\left(\dfrac{\alpha V_{\mathrm{BE}}}{kT}\right)\right]}{\displaystyle\int_0^{x_B} \frac{p(x)}{D_n(x)}\,\mathrm{d}x}$$

where we can express the current in the form,

$$I = I_S \left[\exp\left(\frac{qV_{BC}}{kT} \right) - \exp\left(\frac{qV_{BE}}{kT} \right) \right]$$

$$I_S = \frac{q^2 A^2 n_i^2 \langle Dn \rangle}{Q_B}$$

$$\langle D_n \rangle = \frac{qA \displaystyle\int_0^{x_B} p(x)dx}{\displaystyle\int_0^{x_B} \frac{p(x)}{D_n} dx}$$

and

$$Q_B = qA \int_0^{x_B} p(x)dx$$

The base current can be expressed in a similar form where current is a function of the excess charge in the base divided by the recombination time,

$$Q'_B = qA \int_0^{x_B} [n(x) - n_0] \, dx$$

$$I_{rB} = \frac{Q'_B}{\tau_n}$$

5.1.2 Bipolar Current Gain and Collector-to-Emitter Transport

Important parameters of bipolar junction transistors are the metrics associated with the efficiency of transporting the current from the collector to the emitter [1–5]. The effectiveness of the emitter of an *npn* bipolar junction transistor is associated with the emitter efficiency, γ. The emitter efficiency is associated with the ratio of the electrons injected into the base region over the total emitter current (where the total emitter current is the electrons and the holes), expressed as follows:

$$\gamma = \frac{I_{nE}}{I_E} = \frac{I_{nE}}{I_{nE} + I_{pE}} = \frac{1}{1 + \left(\dfrac{I_{pE}}{I_{nE}} \right)}$$

As the emitter electron current increases well above the emitter hole current, the emitter efficiency approaches unity.

A second key parameter is the efficiency at which electrons are transmitted from the collector to the emitter; this is referred to as the base transport factor, or the collector-to-emitter transport factor, α. For an *npn* bipolar transistor, this is the ratio of electron current that reach the collector to the emitter injection current;

$$\alpha = \frac{I_C}{I_E}$$

This can be expressed as a function of the number of electrons that recombine in the base region [1],

$$\alpha = \frac{I_{nE} - I_{rB}}{I_{nE}}$$

A third parameter is the bipolar current gain, β. The bipolar current gain is the ratio of the collector current to the base current,

$$\beta = \frac{I_C}{I_B}$$

The bipolar current gain is the key parameter for current amplification, and is related to the unity current gain cutoff frequency, f_T. The bipolar current gain is a function of the magnitude of the collector current and has both low-current and high-current limitations. Equivalently, the unity current gain cutoff frequency also has the same limitations. As a result, the bipolar current gain undergoes a peak bipolar current gain, β_{MAX}.

5.1.3 Unity Current Gain Cutoff Frequency

A key FOM for a SiGe HBT device is the unity current gain cutoff frequency, f_T. From a frequency response perspective, there is a time constant associated with the ability to transport an electron from the emitter region to the collector region. This is the sum of the times through all physical regions of the transistor. The unity current gain cutoff frequency f_T is inversely related to the emitter-to-collector transit time, τ_{EC} [6].

$$\frac{1}{2\pi f_T} = \tau_{EC} = \tau_E + \tau_B + \tau_{CSL} + \tau_C$$

where

$$\tau_E = \frac{C_{eb} + C_{bc}}{g_m}$$

$$\tau_B = \frac{W_B^2}{KD_B}$$

$$\tau_{CSL} = r_C(C_{C\text{-}SX} + C_{bc})$$

$$\tau_C = \frac{x_C}{v_{sat}L}$$

The emitter transit time is a function of the emitter–base capacitance (C_{eb}), base–collector capacitance (C_{bc}), and transconductance (g_m). The base transit time is a function of the base width (W_B), diffusion coefficient (D_B), and base grading factor (K). The collector terms consist of the collector capacitance ($C_{\text{C-SX}}$, C_{bc}), collector resistance (r_C), and velocity saturation term (v_{sat}).

5.1.4 Unity Power Gain Cutoff Frequency

A key metric for power transistors is the unity power gain cutoff frequency, f_{MAX}. A common form is as follows [6],

$$f_{\text{MAX}} = \sqrt{\frac{f_T}{8\pi R_b C_{\text{bc}}}}$$

The base resistance is defined as an intrinsic base resistance and an extrinsic base resistance. Minimizing the collector–base capacitance, C_{cb}, collector-substrate, C_{csx}, and base resistance, R_b improves f_{MAX}.

5.2 TRANSISTOR BREAKDOWN

Breakdown voltages in bipolar transistors are important for both RF functional issues as well as ESD protection [7–27]. Avalanche breakdown initiates electrical instability in a bipolar transistor structure by inducing negative resistance regimes. Given current magnitude is significant, self-heating can occur, leading to thermal instability. When the device is in the thermal safe-operational region, the bipolar transistor can still be used without damage and failure, but with excessive current, thermal failure and permanent damage can be observed in transistor devices [12–24]. Hence, breakdown can lead to destruction of the component from ESD events.

Electrical breakdown can also be utilized as a 'trigger' for ESD networks. Hence a common ESD design practice is to utilize the breakdown voltage states to initiate 'turn-on' of ESD elements or higher level circuits. This is successful given that the 'trigger element' remains in the thermal safe operating area (SOA) prior to initiation of the discharging element.

Breakdown voltages are also related to the fundamental limits of transistor operation [29–37]. As a result, relationship between the breakdown voltage and RF performance is related in bipolar transistors.

5.2.1 Avalanche Multiplication and Breakdown

Impact ionization leads to the generation of carriers, leading to a cascade effect in a region with an electric field applied [7–11]. Avalanche multiplication occurs for both the negative and positive species when the energy level exceeds the impact ionization threshold for both species.

Let there be an incident current I_{p_0} and total current I which is the sum of the positive and negative species (e.g., electron and hole in semiconductors). At the position $x = W$, the hole current is $M_p I_p$. The differential generation in an increment is associated with the product of the impact ionization coefficient and current at the position in space at that location in space.

$$dI_p(x) = (\alpha_p I_p + \alpha_n I_n)dx$$

This can be put in the as a function of the total current

$$\frac{dI_p(x)}{dx} = \alpha_n I + (\alpha_p - \alpha_n)I_p$$

or equivalently

$$\frac{dI_p(x)}{dx} - (\alpha_p - \alpha_n)I_p(x) = \alpha_n I$$

In this form, the equation is a first-order ordinary differential equation with a variable coefficient. The integration factor for the integral can be expressed as

$$\mu(x) = \exp\left[-\int_0^x (\alpha_p - \alpha_n)dt\right]$$

with a general solution of

$$I_p(x) = \frac{1}{\mu(x)}\left[\int_0^x \mu(s)(\alpha_n I)\,ds + C\right]$$

where C is a constant. Then we obtain,

$$I_p(x) = \exp\left[\int_0^x (\alpha_p - \alpha_n)dt\right]\left[\int_0^x \exp\left[-\int_0^s (\alpha_p - \alpha_n)dt\right](\alpha_n I)ds + C\right]$$

This can be expressed as a function of the total current as

$$I_p(x) = I\exp\left[\int_0^x (\alpha_p - \alpha_n)dt\right]\left[\int_0^x \alpha_n \exp\left[-\int_0^s (\alpha_p - \alpha_n)dt\right]ds + K\right]$$

Letting the hole current at $x = 0$ be I_{p_0}, we can solve for the constant where $I_{p_0} = M_p I = K$ which is the hole multiplication times the total current at the electrode at $x = 0$. Then $K = 1/M_p$. The general solution is then,

$$I_p(x) = I\exp\left[\int_0^x (\alpha_p - \alpha_n)dt\right]\left[\int_0^x \alpha_n \exp\left[-\int_0^s (\alpha_p - \alpha_n)dt\right]ds + \frac{1}{M_p}\right]$$

From this form, we can solve the avalanche multiplication term M_p from the boundary condition at the position $x = W$, where $I_p(W) = M_p I_{p_0}$.

$$\frac{1}{M_p} = \exp\left[-\int_0^{x=W} (\alpha_p - \alpha_n)dx\right] - \left[\int_0^{x=W} \alpha_n \exp\left[-\int_0^{x'=x} (\alpha_p - \alpha_n)dx'\right]dx\right]$$

To integrate the integral expression, change the form to

$$
\left[\int_0^{x=W} (\alpha_n - \alpha_p) \exp\left[-\int_0^{x'=x} (\alpha_p - \alpha_n)dx' \right] dx \right] = \exp\left[-\int_0^W (\alpha_p - \alpha_n)dx \right] - 1
$$

Hence by adding the impact ionization term for holes, we obtain an expression equal to the first term of the RHS. Adding this to the equation and then subtracting out the common integral terms, the multiplication term can be placed in the form as follows

$$
1 - \frac{1}{M_p} = \left[\int_0^{x=W} \alpha_p \exp\left[-\int_0^{x'=x} (\alpha_p - \alpha_n)dx' \right] dx \right]
$$

Let us establish a relationship between the hole and electron impact ionization terms,

$$
\alpha_p = \gamma \alpha_n
$$

$$
1 - \frac{1}{M_p} = \frac{\gamma}{1 - \gamma} \left\{ \exp\left[-\left(1 - \frac{1}{\gamma}\right) \int_0^W (\alpha_p)dx \right] - 1 \right\}
$$

and

$$
1 - \frac{1}{M_n} = \frac{1}{\gamma - 1} \left\{ \exp\left[\int_0^W (\gamma - 1)(\alpha_n)dx \right] - 1 \right\}
$$

From this formulation, we obtain a modified Townsend criterion of the form

$$
\int_0^W \alpha_n dx = \frac{\ln \gamma}{\gamma - 1}
$$

$$
\int_0^W \alpha_p dx = \frac{\gamma \ln \gamma}{\gamma - 1}
$$

5.2.2 Bipolar Transistor Breakdown

In bipolar transistors, there are different breakdown criteria that are established because of the multiple number of metallurgical junctions and the potential state of the additional electrode. Additionally, analytical relationships exist between the different breakdown states [1–6].

Emitter–Base Breakdown Voltage with Open Collector (BV_{EBO}). The emitter–base breakdown voltage is defined as the breakdown voltage when the collector is unbiased or floating; this is referred to as BV_{EBO}. During ESD stress of the base–emitter junction, the collector junction may be floating. For example, in the ESD stress of a bipolar receiver network, the emitter node may be the reference ground electrode and the ESD pulse is incident on the base electrode.

Collector–Base Breakdown Voltage with Open Emitter (BV$_{CBO}$). The collector-to-base breakdown voltage is defined as the breakdown voltage when the base is unbiased or floating; this is referred to as BV$_{CBO}$. During ESD stress of collector-to-base, the emitter may be floating; this can occur when a bipolar transistor is used as an ESD diode element. Another case of interest is a pin-to-pin RF input-to-RF output stress of a bipolar receiver network; in this case, the emitter may be in a floating state.

A method to obtain BV$_{CBO}$ is to extract the value from a plot of $M-1$ versus V_{CB} (collector–base voltage) [6]. Using the Miller equation,

$$M = \frac{1}{1 - \left(\dfrac{V_{CB}}{BV_{CBO}}\right)^n}$$

the value of BV$_{CBO}$ is extracted.

Collector–Emitter Breakdown Voltage with Open Base (BV$_{CEO}$). The collector-to-emitter breakdown voltage is defined as the breakdown voltage when the base is unbiased or floating; this is referred to as BV$_{CEO}$. During ESD stress of collector-to-emitter, the base may be floating. For example, in the ESD stress of a bipolar receiver network, the emitter node may be the reference ground electrode and the ESD pulse is incident on the collector electrode. The breakdown value of BV$_{CEO}$ can be extracted experimentally or from the extraction of the BV$_{CBO}$. The breakdown voltage BV$_{CEO}$ can be obtained from BV$_{CBO}$ from

$$BV_{CEO} = BV_{CBO}(1 - \alpha)^{1/n}$$

Collector–Emitter Breakdown Voltage with Base Resistor (BV$_{CER}$). The collector-to-emitter breakdown voltage is defined as the breakdown voltage when the base is unbiased or floating; this is referred to as BV$_{CEO}$. During ESD stress of collector-to-emitter, the base may be floating. For example, in the ESD stress of a bipolar receiver network, the emitter node may be the reference ground electrode and the ESD pulse is incident on the collector electrode. Given a base resistor, R_b, the BV$_{CER}$ value can be derived from the BV$_{CBO}$ value as

$$BV_{CER} = BV_{CBO} \sqrt[n]{1 - \frac{\alpha}{1 + \dfrac{kT}{q}\dfrac{1}{R_b}\dfrac{1}{I_E}}}$$

Breakdown Voltage Collector-to-Substrate (BV$_{C-SX}$). The collector-to-substrate isolation breakdown voltage is defined as the breakdown voltage between the subcollector and the p^- substrate region. During ESD stress between the npn sub-collector region and the substrate, the collector can be pulsed positive as the p^- substrate is reference ground; the p^- base and n^--emitter are left floating or shorted to the subcollector.

Breakdown Voltage to the Isolation (BV$_{ISO}$). The isolation breakdown voltage is defined as the breakdown voltage relative to the isolation regions. The isolation breakdown voltage,

BV_{ISO}, can be defined as the *npn* base region to the p^- substrate region. During ESD stress between the *npn* p^- base region and the substrate, the collector can be left floating; the p^- base can be pulsed as the p^- substrate is at a reference ground potential.

5.3 KIRK EFFECT

During an ESD event, high current will flow in a bipolar transistor in bipolar transistor peripheral circuits or ESD power clamps. At these high currents, the frequency response of the transistor is impacted by the current flowing in the base region. As the current density increases, the space-charge region in the base–collector junction is pushed outward. This effect is known as the Kirk effect [28]. In analysis of the electrostatics in a junction it is assumed that the carriers are swept out of the depletion region and do not modulate the electric field. At high currents, the ESD current flowing through the junction modifies the space-charge region leading to a space-charge region modulation.

$$\frac{dE}{dx} = \frac{1}{\varepsilon}\left[qN(x) - \frac{J_c}{v(x)}\right]$$

where J_c is the current density from the ESD current flowing through the transistor element. Assuming that the collector–base voltage is constant, then the collector–base voltage and the built-in potential will equal the voltage drop across the space-charge layer. The voltage drop across the depletion region is the integral of the electric field over the junction of the collector to base region. This can be expressed as

$$V_{CB} = \frac{1}{\varepsilon}\int_{x_b}^{x_c} x\left[qN(x) - \frac{Jc}{v(x)}\right]dx - \phi_i$$

This can be simplified by assuming that the doping concentration is constant in the epitaxial region,

$$V_{CB} = \frac{1}{2\varepsilon}\left[qN_{epi} + \frac{J_c}{v_{sat}}\right](x_c - x_b)^2 - \phi_i$$

Defining a saturation current,

$$J_{sat} = qN_{epi}v_{sat}$$

Then

$$V_{CB} = \frac{qN_{epi}}{2\varepsilon}\left[1 + \frac{J_c}{J_{sat}}\right](x_c - x_b)^2 - \phi_i$$

From this expression, the variation of the base width as a function of the ESD current can be solved as

$$W_{CB}(J_{ESD}) = \frac{W_{CBO}}{\left(1 + \dfrac{J_{ESD}}{J_{sat}}\right)^{1/2}}$$

where

$$W_{CBO} = \left[\frac{2\varepsilon(V_{CB} + \phi_i)}{qN_{epi}}\right]^{1/2}$$

The ESD collector-to-emitter current will flow through the collector region. As the ESD current increases, the collector–base depletion width will decrease and the base width will increase, leading to a lower response to the ESD event, if the ESD event is slower than the charge relaxation time. When the ESD event achieves the peak current, the transistor response will be least responsive. Hence we can express a time dependent depletion width relationship during an ESD event as

$$W_{CB}(t) = \frac{W_{CBO}}{\left(1 + \dfrac{J_{ESD}(t)}{J_{sat}}\right)^{1/2}}$$

where the depletion width is a function of the ESD pulse width time response.

5.4 JOHNSON LIMIT: PHYSICAL LIMITATIONS OF THE TRANSISTOR

The question of the physical limitation of the bipolar transistor in frequency response and power has been a fundamental issue from the onset of development of the transistor. Early bipolar transistor researchists, such as Pritchard, Giacoletto, Early, Goldey, Ryder, and Johnson [29–37], addressed the question of the maximum speed and maximum power and the relationship to the material properties.

From this work, Johnson [37] established equations that addressed the interrelationship of the performance properties of the transistor to its material properties; this is known as the Johnson Limit. Johnson's objective was to develop a relationship that was fundamental in nature between the performance of the transistor and the material properties. As a result, the physical development intentionally avoided device design and structural issues. But, at the same time, the analytical development must be useful as a design metric for transistor device design and transistor scaling.

5.4.1 Voltage–Frequency Relationship

To determine the relationship between the maximum allowed voltage and the frequency response of the transistor there are fundamental limits of transport of a minority carrier in a semiconductor material [37]. The maximum velocity to transport a carrier across a gap L, in time τ, is the saturation velocity, v_s. The saturation velocity is achieved when the electric field across the gap L is maximum. The maximum electric field across a medium can be referred to as E_m.

Hence for a length scale, L, and transit time, t, the shortest time to transport an electron across L is

$$\frac{L}{\tau} = v_s$$

Hence the fastest frequency to transport an electron across the length scale L is

$$f_T = \frac{1}{2\pi}\frac{1}{\tau} = \frac{v_s}{2\pi L}$$

From this expression, the maximum voltage across the medium can be expressed as the maximum voltage or voltage breakdown across the length scale L, or

$$V_m f_T = \frac{V_m}{L}\frac{v_s}{2\pi}$$

When the breakdown voltage is achieved, this is associated with the maximum electric field that can be formed across the structure. This can also be expressed in terms of maximum voltage and maximum electric field condition,

$$V_m f_T = \frac{E_m v_s}{2\pi}$$

The formulation states that the product of the maximum velocity with which an electron can traverse a medium and the maximum electric field across that region is a constant.

From this relationship, it can be seen that there is an inverse relationship between the maximum voltage and the frequency response (on the left hand side of the equation), and that the right hand side of the equation is a function of the material properties of the system. Restructuring this equation, the relationship states that there is a maximum voltage condition across a radio frequency (RF) transistor element that satisfies the following

$$V_m = \frac{\left\{\dfrac{E_m v_s}{2\pi}\right\}}{f_T}$$

From an RF ESD design perspective, there exists a breakdown voltage in the transistor which is proportional to the ratio of the maximum electric field to velocity saturation product over the unity current cutoff frequency. The importance of this condition for RF ESD design is as follows:

- For a given unity current gain cutoff frequency, different RF components will have different breakdown voltage states based on the value of the maximum electric field–velocity saturation product.

- As the unity current gain cutoff frequency increases with technology scaling, the breakdown voltages within the transistor will decrease.

The first key point is that different materials (e.g., silicon, germanium, silicon germanium, and gallium arsenide) will have different breakdown voltages for the same cutoff frequency. The second key point is that as technologies are scaled, the breakdown voltage will decrease.

On the second issue, the scaling of the breakdown voltages will lead to the decreased voltage margin to protect RF circuits prior to electrical overstress. But, with technology scaling there will be a transistor element whose breakdown voltage can be utilized as an ESD network, or a trigger circuit within an ESD network for other elements.

Additionally, inverse relationship of breakdown voltage and unity current cutoff frequency exists within a given technology with multiple transistor elements. Hence, within a given technology with a multiple number of transistors, this relationship can be utilized for RF ESD protection networks.

5.4.2 Johnson Limit Current–Frequency Formulation

From a charge control perspective, Johnson [37] related the current through the transistor to the base charge storage,

$$I = \frac{Q}{\tau}$$

It was noted that for a constant time constant, the expression can be put in the form,

$$\frac{1}{Q} = \frac{1}{\tau} = \text{const.}$$

Because of the Kirk effect, there is a maximum current condition prior to the saturation of the transistor response. Hence, there is a peak unity current cutoff frequency associated with the maximum current condition (e.g., there is a relationship between the maximum current and the peak unity current cutoff frequency). Then the maximum current, the charge (e.g., the maximum voltage V_m and the capacitance of the collector–base capacitance C_o)

$$\frac{I_m}{C_o} = \frac{V_m}{\tau} = \left(\frac{E_m v_s}{2\pi f_T}\right)\frac{1}{\tau}$$

Johnson noted that the time constant for base charge storage is related to the transit time (e.g., ignoring recombination time terms and charge loss), thus the expression simplifies to

$$\frac{I_m}{C_o} = \frac{V_m}{\tau} = E_m v_s$$

This expression has value in the relationship between the maximum current and capacitance ratio is equal to the maximum electric field and saturation velocity product. In defining a reactive impedance for the capacitance term, the Johnson Limit can be expressed as a function of maximum current and reactive impedance, as

$$(I_m X_c)f_T = \frac{E_m v_s}{2\pi}$$

where

$$X_c = \frac{1}{2\pi f_T C_o}$$

5.4.3 Johnson Limit Power Formulation

The Johnson Limit can be expressed as a function of maximum power from the maximum current–frequency limitation relationship. From the relationship of power to current [37],

$$P_m = I_m^2 X_c = V_m I_m$$

Substituting in for the maximum current–frequency limitation form of the Johnson Limit can be expressed as,

$$(P_m X_c)^{1/2} f_T = \frac{E_m v_s}{2\pi}$$

where the reactive impedance X_c is defined as

$$X_c = \frac{1}{2\pi f_T C_0}$$

where P_m is the maximum power, f_T is the unity current gain cutoff frequency, E_m is the maximum electric field, and v_s is the electron saturation velocity. In this form, it states that there is an inverse relationship between the maximum power and frequency response.

In this formulation of the power relationship, the limitations of the maximum current and maximum power are associated with the onset of base-width broadening (e.g., also known as the Kirk effect). Hence, beyond the onset of base-width broadening, a transistor can still provide a higher current carrying capability. In the Johnson Limit analysis, thermal effects are intentionally not included. Johnson noted that there is an inverse relationship between frequency and power parameters that is not dependent on thermal properties but the fundamental time constants of the transistor. As a result, there is a distinction between the limitations of current and power based on the functionality and under self-heating conditions.

Electrostatic discharge (ESD) events can exceed the maximum current and maximum power limit of the transistor. In electrothermal ESD models, the analysis is the transient solution to the partial differential equation, thermal diffusion equation, with an initial condition of a thermal pulse instead of an electrical pulse. Analogous to the Johnson Limit relationship, the physical limitations of the power parameters of a device are related to the material properties; in electrothermal analysis the power-to-failure is based on the thermal material properties.

In the derivations of Tasca, Wunsch, and Bell, the maximum power-to-failure is derived by relating the thermal pulse width to the thermal electrical parameters (e.g., heat capacity, thermal conductivity, and melting temperature of the material). Analogous to the Johnson Limit relationship, the relationship is not dependent on the device design details and parameters. As a result, the Wunsch–Bell model has also served as an 'yard-stick' for the understanding of the power-to-failure of semiconductor components. Analogous to the Johnson relationships, the equation can be formed in such a manner that there is an inverse relationship between the power-to-failure and the pulse width of the applied pulse which is equal to a constant, where the constant is a function of material properties and area.

For the ESD RF design, there are two fundamental metrics (in current or power formulations) that exist:

- Johnson Limit formulation (e.g., current–frequency and power–frequency relations).

- Wunsch–Bell formulation (e.g., current-to-failure pulse width and power-to-failure pulse width relations).

The two independent relationships—one addressing the power–speed functional inverse relationship and the other is the power-to-failure pulse width inverse relationship form a fundamental set of relations for understanding of both the power performance tradeoffs and power-to-failure versus pulse width tradeoffs. As a result, these serve as key metrics for the understanding of RF ESD design.

5.5 RF INSTABILITY: EMITTER COLLAPSE

Thermal instability can occur in heterojunction bipolar transistors due to the thermal response associated with increase in temperature [38–41]. In a silicon bipolar transistor, as a transistor collector-to-emitter voltage increases and as self-heating occurs, avalanche multiplication increases and the forward bias emitter voltage decreases; this leads to increase in the collector current. But, given the thermal response of a device, such as a heterojunction bipolar transistor, is different, the collector current can decrease with the increase in the collector-to-emitter voltage associated with a phenomenon called 'emitter collapse.' In the case of a multiple emitter finger structure, given nonuniformities in the temperature field, one emitter finger may 'current rob' from the other emitter fingers, which has the response of decreased.

Given a multifinger implementation, the collector current through any emitter (e.g., jth emitter) can be expressed as a function of the thermal electrical feedback coefficient, ϕ [41],

$$(I_{\mathrm{C}})_j = (I_{\mathrm{CO}})_j \exp\left\{\frac{q}{nkT}\left[(V_{\mathrm{BE}})_j - \phi(T_j - T_A)\right]\right\}$$

From this expression, the forward bias state of jth emitter can be shown as

$$(V_{\mathrm{BE}})_j = \frac{nkT_A}{q}\ln\left[\frac{(I_{\mathrm{C}})_j}{I_{\mathrm{CO}}}\right] - \phi(T_j - T_A)$$

The local temperature can be solved from the relationship of temperature to power. The temperature in the emitter can be expressed in tensor form as

$$T_j - T_A = \theta_{jk}P_k$$

where

$$P_k = (I_{\mathrm{C}}V_{\mathrm{CE}})_k$$

In a multiemitter structure, assuming the collector-to-emitter voltage is the same, hence

$$P_k = (I_C)_k V_{CE}$$

Hence, the general expression for temperature in a given emitter structure in tensor representation can be presented as follows,

$$T_j - T_A = \theta_{jk}(I_C)_k V_{CE}$$

In the analysis, Liu and Khatibzadeh [41] represented the transistor as having a local thermal resistor element associated with each emitter, R_{th}, and substrate thermal resistor element, $(R_{th})_{sx}$. In this case, the substrate thermal resistance is common to all emitters in the multiple emitter structure. As a result, the thermal coupling is contained in the substrate thermal resistance element, $(R_{th})_{sx}$. Given two emitters, this can then be represented as the sum of the power in the self-emitter thermal resistance and power in the substrate thermal resistance,

$$T_1 - T_A = (R_{th})_1 I_{C1} V_{CE} + R_{thsx}(I_{C1} V_{CE} + I_{C2} V_{CE})$$

$$T_2 - T_A = (R_{th})_2 I_{C2} V_{CE} + R_{thsx}(I_{C1} V_{CE} + I_{C2} V_{CE})$$

or,

$$T_1 - T_A = (R_{th1} + R_{thsx}) I_{C1} V_{CE} + (R_{thsx}) I_{C2} V_{CE}$$

$$T_2 - T_A = (R_{thsx}) I_{C1} V_{CE} + (R_{th2} + R_{thsx}) I_{C2} V_{CE}$$

or in matrix form,

$$\begin{bmatrix} T_1 - T_A \\ T_2 - T_A \end{bmatrix} = \begin{bmatrix} R_{th1} + R_{thsx} & R_{thsx} \\ R_{thsx} & R_{th2} + R_{thsx} \end{bmatrix} \begin{bmatrix} I_{C1} V_{CE} \\ I_{C2} V_{CE} \end{bmatrix}$$

From this representation, the thermal coupling is evident through the substrate thermal resistance term.

The interaction of the stability is associated with the emitter–base voltage of the emitters in the structure. To establish the electrical relationship between the emitter electrical response, the voltage drops across the emitter–base junction and the emitter resistors are obtained from the voltage loop formed by the emitter–base junctions and the emitter resistance terms,

$$V_{BE1} + I_{C1} R_{E1} - V_{BE2} - I_{C2} R_{E2} = 0$$

or expressed as the difference between the emitter–base voltages that is equal to the difference in the IR drops across the emitter resistors,

$$V_{BE1} - V_{BE2} = I_{C2} R_{E2} - I_{C1} R_{E1}$$

Hence, three equations exist relating the emitter–base voltages, the current, and temperatures in the emitters. Letting the emitter resistor and the thermal resistors to be equal in all emitters, then we can express the current in the collector current as a function of the current in the two elements. Substituting in the relationship of forward bias voltage to the collector current and temperature,

$$(V_{BE})_1 - (V_{BE})_2 = \frac{nkT_A}{q}\left\{\ln\left[\frac{(I_C)_1}{I_{CO}}\right] - \ln\left[\frac{(I_C)_2}{I_{CO}}\right]\right\} - \phi\{(T_1 - T_A) - (T_2 - T_A)\}$$

From the Kirchoff's loop equation,

$$I_{C2}R_{E2} - I_{C1}R_{E1} = \frac{nkT_A}{q}\left\{\ln\left[\frac{(I_C)_1}{I_{CO}}\right] - \ln\left[\frac{(I_C)_2}{I_{CO}}\right]\right\} - \phi\{(T_1 - T_A) - (T_2 - T_A)\}$$

Hence, from the matrix equation pair, the last term in the expression can be expressed as

$$T_1 - T_A = (R_{th1} + R_{thsx})I_{C1}V_{CE} + (R_{thsx})I_{C2}V_{CE}$$

$$T_2 - T_A = (R_{thsx})I_{C1}V_{CE} + (R_{th2} + R_{thsx})I_{C2}V_{CE}$$

Hence, subtracting the two equations, the expression is the solution of the Kirchoff's thermal loop,

$$\{(T_1 - T_A) - (T_2 - T_A)\} = R_{th1}I_{C1}V_{CE} - R_{th2}I_{C2}V_{CE}$$

Substituting into the above terms,

$$I_{C2}R_{E2} - I_{C1}R_{E1} = \frac{nkT_A}{q}\left\{\ln\left[\frac{(I_C)_1}{I_{CO}}\right] - \ln\left[\frac{(I_C)_2}{I_{CO}}\right]\right\} - \phi\{R_{th1}I_{C1}V_{CE} - R_{th2}I_{C2}V_{CE}\}$$

Factoring,

$$I_{C2}(R_{E2} - \phi R_{th2}V_{CE}) - I_{C1}(R_{E1} - \phi R_{th1}V_{CE}) = \frac{nkT_A}{q}\left\{\ln\left[\frac{(I_C)_1}{(I_C)_2}\right]\right\}$$

Hence, the collector current of the second emitter can be expressed as a function of the first emitter, as follows,

$$I_{C2}(R_{E2} - \phi R_{th2}V_{CE}) = \frac{nkT_A}{q}\left\{\ln\left[\frac{(I_C)_1}{(I_C)_2}\right]\right\} - I_{C1}(R_{E1} - \phi R_{th1}V_{CE})$$

or,

$$I_{C2} = \frac{\frac{nkT_A}{q}\left\{\ln\left[\frac{(I_C)_1}{(I_C)_2}\right]\right\} - I_{C1}(R_{E1} - \phi R_{th1}V_{CE})}{(R_{E2} - \phi R_{th2}V_{CE})}$$

Given the special case when emitter electrical and thermal resistors are the same, the above expression can factor out the thermal feedback term,

$$I_{C2} = (R_E - \phi R_{th} V_{CE}) - I_{C1}(R_E - \phi R_{th} V_{CE}) \frac{nkT_A}{q} \left\{ \ln \left[\frac{(I_C)_1}{(I_C)_2} \right] \right\}$$

Factoring out the collector current, then

$$I_{C2} \left(1 - \frac{I_{C1}}{I_{C2}}\right)(R_E - \phi R_{th} V_{CE}) = \frac{nkT_A}{q} \left\{ \ln \left[\frac{(I_C)_1}{(I_C)_2} \right] \right\}$$

Hence,

$$I_{C2} = \frac{\dfrac{nkT_A}{q} \left\{ \ln \left[\dfrac{(I_C)_1}{(I_C)_2} \right] \right\}}{\left(1 - \dfrac{I_{C1}}{I_{C2}}\right)(R_E - \phi R_{th} V_{CE})}$$

Liu and Khatibzadeh [41] noted that this equation is to address the onset of emitter collapse, hence the domain boundary between stability and instability. Letting the case where the first and second emitter currents are equal will be a location of stability. Hence, defining the variable, χ

$$\chi = \frac{I_{C1}}{I_{C2}}$$

$$\lim_{\chi \to 1} I_{C2} = \lim_{\chi \to 1} \frac{\dfrac{nkT_A}{q} \left\{ \ln \left[\dfrac{(I_C)_1}{(I_C)_2} \right] \right\}}{\left(1 - \dfrac{I_{C1}}{I_{C2}}\right)(R_E - \phi R_{th} V_{CE})} = \frac{nkT_A}{q} \frac{1}{(R_E - \phi R_{th} V_{CE})} \lim_{\chi \to 1} \frac{\left\{ \ln \left[\dfrac{(I_C)_1}{(I_C)_2} \right] \right\}}{\left(1 - \dfrac{I_{C1}}{I_{C2}}\right)}$$

$$\lim_{\chi \to 1} I_{C2} = \lim_{\chi \to 1} \frac{\dfrac{nkT_A}{q} \left\{ \ln \left[\dfrac{(I_C)_1}{(I_C)_2} \right] \right\}}{\left(1 - \dfrac{I_{C1}}{I_{C2}}\right)(R_E - \phi R_{th} V_{CE})} = \frac{nkT_A}{q} \frac{1}{(R_E - \phi R_{th} V_{CE})} \lim_{\chi \to 1} \frac{\ln[\chi]}{(1 - \chi)}$$

From L'Hopital's rule, in the limit the two emitters are equal,

$$\lim_{\chi \to 1} \frac{\ln[\chi]}{1 - \chi} = \frac{\dfrac{d \ln \chi}{d\chi}}{\dfrac{d(1 - \chi)}{d\chi}} = \lim_{\chi \to 1} \frac{\dfrac{1}{\chi}}{-1} = -1$$

Hence,

$$\lim_{\chi \to 1} I_C = \frac{\dfrac{nkT_A}{q}}{(\phi R_{th} V_{CE} - R_E)}$$

From this expression, the condition for emitter thermal stability is a function of the thermal feedback and emitter resistance. This can be expressed in normalized form, as

$$\lim_{\chi \to 1} I_C = \frac{\left\{ \dfrac{\dfrac{nkT_A}{q}}{\phi R_{th} V_{CE}} \right\}}{\left(1 - \dfrac{R_E}{\phi R_{th} V_{CE}} \right)}$$

From this two finger case, Liu and Khatibzadeh generalized the expression to the N finger case. Note that in the above analysis, all the emitters conduct current until the instability state occurs. Hence, we can consider that all emitters have the same collector current. But, as one emitter increases above the others, redefining as the higher emitter as the second emitter, and the second 'emitter' can be $N{-}1$ emitters. Liu and Khatibzadeh developed instability criteria for emitter collapse. The critical current where thermal collapse occurs is equal to the following [41],

$$I_{critical} = N \frac{nkT_A}{q} \left[\frac{1}{\phi_{fb} R_{TH} V_{CE} - R_E} \right]$$

where N is the number of emitter fingers, n is the ideality factor, ϕ_{fb} is the base–emitter junction voltage, R_{TH} is the thermal resistance, V_{CE} is the collector-to-emitter voltage, and R_E is the emitter resistance. Modifying the form of this equation we can express it as follows,

$$I_{critical} = N \left\{ \frac{\dfrac{nkT_A}{q}}{\phi_{fb} R_{TH} V_{CE}} \right\} \left[\frac{1}{1 - \dfrac{R_E}{\phi_{fb} R_{TH} V_{CE}}} \right]$$

In this form, the critical current for emitter collapse is the thermal potential divided by the effective 'resistance' of the device times the number of emitters. In the bracket, the stability phenomenon is clear. The emitter collapse condition is related to the ratio of the emitter resistance and the product of the emitter–base potential, thermal resistance, and the collector-to-emitter voltage. Redefining the condition a second time, it can be expressed in this form,

$$I_{critical} = N \left\{ \frac{\dfrac{nkT_A}{q}}{\phi_{fb} R_{TH} V_{CE}} \right\} \left[\frac{1}{1 - \dfrac{\left(\dfrac{R_E}{\phi_{fb}} \right)}{R_{TH} V_{CE}}} \right]$$

In this form, the stability term is expressed as the emitter resistance over its forward bias voltage over the thermal resistance and collector-to-emitter voltage. Hence, the stability is associated with the current through the emitter structure.

In another form, this can be expressed in terms of the emitters that are in the collapse state and the standard state; I_{C2} is the collector current not in collapse, I_{C1} is emitter current in the collapsed collector, and I_{CT} is the total collector current [41],

$$I_{C1} = \frac{\frac{nkT_A}{q}\left\{\ln\left[\frac{(I_C)_2}{(I_C)_1}\right]\right\}}{\left(1 - \frac{I_{C2}}{I_{C1}}\right)(R_E - \phi R_{th}V_{CE})}$$

and the current in the other emitters is

$$I_{C2} = \frac{1}{N-1}\{I_{CT} - I_{C1}\}$$

Hower established a thermal instability metric called the S-factor to quantify the thermal instability condition. Thermal instability criterion, the S-factor, proposed by Hower [42], is

$$S = R_{TH}V_{CE}\frac{\partial I_C}{\partial T}\bigg|_{V_{BE}=const}$$

The S-factor is the product of the thermal resistance, the collector-to-emitter voltage, and the partial derivative of the current as a function of temperature for a constant emitter–base voltage. Hence, a set of forward bias contours can be created in an S-factor versus temperature, T, for a constant forward bias condition. When S is less than 1, the system is thermally stable; when S is greater than unity it is thermally unstable.

For an S-factor loci, solving for the forward bias voltage from the current equation, the forward bias voltage can be expressed as a function of the voltage drop across the emitter, the bandgap equilibrium bandgap, and an additional term to address the temperature variation in the bandgap.

$$V_{BE} = \frac{n_f kT}{q}\ln\frac{I_C}{I_S} + R_E I_C + \frac{E_{GO}}{q} - \frac{\beta^* T}{q}$$

Taking the partial derivative of the forward bias voltage,

$$\frac{\partial}{\partial T}V_{BE} = \frac{\partial}{\partial T}\left(\frac{n_f kT}{q}\ln\frac{I_C}{I_S} + R_e I_C + \frac{E_{GO}}{q} - \frac{\beta^* T}{q}\right)$$

On a constant forward bias contour,

$$0 = \frac{\partial}{\partial T}\left(\frac{n_f kT}{q}\ln\frac{I_C}{I_S} + R_e I_C + \frac{E_{GO}}{q} - \frac{\beta^* T}{q}\right)$$

Expanding the partial derivative terms,

$$0 = \left(\frac{n_f kT}{q}\right)\frac{\partial}{\partial T}\ln\frac{I_C}{I_S} + \left(\frac{n_f kT}{q}\right)\ln\frac{I_C}{I_S}R_e\frac{\partial I_C}{\partial T} - \frac{\beta^*}{q}$$

Solving for the partial derivative of current with temperature,

$$-\frac{\partial I_C}{\partial T}\left(\frac{n_f k}{q}\frac{1}{I_C}+R_e\right)=\left(\frac{n_f k}{q}\right)\ln\frac{I_C}{I_S}-\frac{\beta^*}{q}$$

$$\frac{\partial I_C}{\partial T}=-\frac{\left(\dfrac{n_f k}{q}\right)\ln\dfrac{I_C}{I_S}-\dfrac{\beta^*}{q}}{\dfrac{n_f k}{q}\dfrac{1}{I_C}+R_e}$$

From this, we can define the S-factor term for a constant forward bias voltage contour,

$$S=-R_{TH}V_{CE}\frac{\left(\dfrac{n_f k}{q}\right)\ln\dfrac{I_C}{I_S}-\dfrac{\beta^*}{q}}{\dfrac{n_f k}{q}\dfrac{1}{I_C}+R_e}$$

In the above development, it is assumed that the thermal resistance is not a function of temperature. In reality, in semiconductors, the thermal resistance is a function of temperature, and hence a more general development is needed to address the stability criteria. Addressing the temperature dependence of the thermal resistance, the S-factor term can be generalized as follows,

$$S=\frac{d}{dT}(R_{TH}V_{CE}I_C)\big|_{V_{BE}=\text{const}}$$

Expanding the partial derivative, it is clear that a new term is present in the S-factor expression,

$$S=R_{TH}V_{CE}\left(\frac{\partial I_C}{\partial T}+\frac{I_C}{R_{TH}}\frac{\partial R_{TH}}{\partial T}\right)\bigg|_{V_{BE}=\text{const}}$$

The critical current expression can be expressed as follows [41],

$$I_{\text{critical}}=N\frac{nkT_A}{q}\left[\frac{1}{\phi_{\text{fb}}R_{TH}V_{CE}\left\{1-\left(\dfrac{T-T_A}{R_{TH}}\right)\left(\dfrac{\partial R_{TH}}{\partial T}\right)\right\}^{-1}-R_e}\right]$$

Hence, a general criterion can be established at the condition of thermal instability in a multiemitter structure. The critical current of thermal instability is associated with the emitter electrical resistance, the thermal resistance, the temperature sensitivity of the thermal resistance with temperature, the collector-to-emitter voltage condition, thermal feedback coefficient, and the number of emitter stripes.

5.6 ESD RF DESIGN LAYOUT: EMITTER, BASE, AND COLLECTOR CONFIGURATIONS

In single- and multiemitter structures, the configuration of the emitter, base, and collector can influence the ESD robustness of the homo- or heterojunction bipolar transistor.

There are many combinations and permutations of the ordering of the emitter, base, and collector regions, where there are a multiple number of emitters, or base regions, or collectors. The ordering of the regions of the transistor with a single emitter (E), at least one base (B), and one collector (C) can be designated include the following:

- EBC;

- ECB;

- BEC;

- BEBC;

- BECB;

- CBEBC;

- BCECB;

- CBCBEBCBC.

For the ordering with transistors with multiple emitters, for example, the ordering combinations and permutations can include the following:

- BECEB;

- CEBEB;

- CEBEC;

- CEBEBC;

- CBEBEBC;

- CBEBEBC.

In these single and multiple-emitter configurations, there are ESD advantages and disadvantages associated with the ordering of the transistor regions. The number of physical structures and the ordering can influence the following ESD design layout issues:

- Asymmetric or symmetric current flow.

- Nonuniform current distribution.

- Peak current density and self-heating.

- Thermal stability and emitter collapse loci.

- Parasitic parallel current paths.

For a single emitter transistor, with a single-side base and single-side collector (e.g., EBC, ECB, and BEC), the current-flow from the emitter-to-base or the emitter-to-collector is asymmetric. The current-flow through the emitter to either element does not distribute symmetrically; this leads to a higher series resistance, base or collector internal voltage drops, current crowding, and nonuniform current flow; this can lead to higher peak current density. With higher peak current density, the peak self-heating will be more significant.

For a single emitter transistor in the case of a two-sided base regions (e.g., 'wrap-around base' BEBC, CBEBC, BCECB, or CBCBEBCBC), or two-sided collector (e.g., 'wrap-around collector' CBEBC, BCECB, or CBCBEBCBC) can lead to symmetric current distribution, lower series resistance, and higher ESD robustness.

For a single emitter transistor structure, Minear and Dodson showed that parasitic parallel current paths can exist in bipolar *npn* transistors because of internal voltage drops [43,44]. Minear and Dodson showed that a transistor in a CBE configuration has low HBM ESD robustness when the emitter is pulsed with a positive HBM pulse relative to the base region; this is consistent with the literature. A key discovery of Minear and Dodson is that when the collector is pulsed positively relative to the base, the HBM ESD results were comparable, even though the emitter was in a floating state. Additionally, when HBM stressing with a positive polarity of the emitter referenced relative to the combination of the base and collector, the HBM ESD results were 7× higher and also when the HBM testing was completed from the emitter to only the collector (e.g., base floating). Minear and Dodson concluded that although the ESD current path was intended to remain within the extrinsic part of the *npn* transistor through the base and collector regions, ESD current flowed from the extrinsic collector contact, through the collector region, the intrinsic region of the *npn*, and then back to the extrinsic base contact; it was determined that even though the emitter floated, there was an alternative current path through the intrinsic section of the *npn*. From the Gummel plots, after ESD stress, the base current versus the forward bias emitter base voltage increased with an enhanced slope at voltage below 0.7 V. The primary reason why this occurred was associated with a base region voltage drop leading to the forward biasing of the emitter–base junction. Hence voltage drops internal to the structure associated with the ordering of the elements can influence the ESD robustness.

For multiple emitter structures, the ordering of the emitter, base, and collector can influence the temperature field, leading to thermal instability and 'emitter collapse' in heterojunction bipolar transistors. Emitter collapse can be reduced in heterojunction bipolar transistors by the design symmetry and the spatial separation of the emitter regions. Hence, ordering of the emitter, base, and collector that are asymmetric in nature associated with the emitter design symmetry or allowing emitter spacing adjacency are more prone to these issues.

Wu, Chang, and Ker [45] showed in a four emitter CBEBEBEBEBC Silicon Germanium transistor that the ESD robustness is a function of the number of base regions grounded during collector-to-emitter HBM positive stress. When a positive HBM pulse was applied to the collector and all the emitter regions were ground references, the HBM ESD robustness is a function of the base regions that are grounded. When all the base regions were grounded or 'odd' base regions (e.g., B1, B3, B5), the HBM ESD results were 100 V HBM. This is indicative of an emitter–base failure level. For the case when only the first and the third base regions are grounded, the HBM ESD results increased to 1800 V (18× increase). As in Minear and Dodson's study [43,44], the response of the transistor varied between an emitter–base HBM failure level and a collector-to-emitter HBM failure level. Hence, the layout as well as the voltage state of the various regions can lead to different ESD results.

ESD design layout practices that can be learned from these results are:

• Different permutations and combinations of ordering the emitter, base, and collector exist in homo- and heterojunction bipolar transistor design.

- Ordering of the emitter, base, and collector can influence the ESD robustness.

- Symmetrical ordering can lead to lower current density and better thermal distribution.

- Symmetrical ordering can lead to improved thermal instability and thermal safe operation area (SOA) to avoid emitter collapse.

- The voltage state (e.g., biased, floating) adds an additional variable to the permutations and combinations (e.g., each order has additional conditions) that can influence the ESD robustness.

5.7 ESD RF DESIGN LAYOUT: UTILIZATION OF A SECOND EMITTER (PHANTOM EMITTER)

In single- and multiemitter structures, the configuration of the emitter, base, and collector can influence the ESD robustness of the homo- or heterojunction bipolar transistor [43,44]. For a single emitter transistor, with a single-side base and single-side collector (e.g., EBC, ECB, and BEC), the current-flow from the emitter-to-base or the emitter-to-collector is asymmetric. The current-flow through the emitter to either element does not distribute symmetrically; this leads to a higher series resistance, base or collector internal voltage drops, current crowding, and nonuniform current flow; this can lead to higher peak current density.

Figure 5.1 shows an example of a silicon bipolar transistor. In this figure, a deep trench structure borders the subcollector and collector region. An electrical connection is established between the subcollector and surface implants using a reach-through implant in the extrinsic part of the transistor. In the intrinsic transistor region, the base and collector are under the emitter window forming a one-dimensional vertical profile. Outside this region, the extrinsic transistor region extends from the intrinsic region to the base and collector

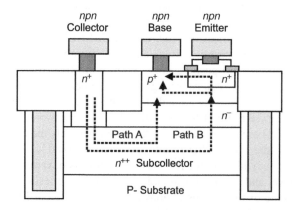

Figure 5.1 It shows a planar bipolar transistor highlighting current paths through the subcollector, collector, base, and emitter regions. Path (A) extends between the extrinsic collector and base region. Path (B) extends between the extrinsic collector region and traverses through and under the emitter region

contacts. In the extrinsic region, voltage drops in the collector or base region can influence the current path during ESD stress.

During ESD stress, only two of the three terminals are electrically biased; one of the three bipolar terminals is stressed, while the second terminal is at a reference ground potential. During ESD stress, for a single emitter transistor structure, Minear and Dodson showed that parasitic parallel current paths can exist in bipolar *npn* transistors due to internal voltage drops [43,44]. Minear and Dodson showed that during ESD stress of the collector-to-base junction (for a CBE configuration), there are two current paths. The first path is through the extrinsic transistor region from the collector contact to the base contact (Path A). The second path is through both the extrinsic and intrinsic regions; these paths are from the collector contact, through the low resistance subcollector region, through the intrinsic region of the vertical *npn*, and then back to the extrinsic base contact (Path B).

During ESD collector-to-base stress, current flows from the collector to base region when the voltage is the collector-to-base avalanche breakdown voltage, BV_{CBO}. After the transistor undergoes avalanche breakdown at the base region, the current flows through the complete base region, both the intrinsic and extrinsic regions; lateral current flow can exists under the emitter structure to the extrinsic base contact. When the voltage drop in the base exceeds the emitter–base voltage for becoming forward active, the emitter can become forward active. In this state, a current path is established through the emitter region. Current will flow laterally through the emitter and towards the base contact; it was noted that the emitter edge nearest the base contact will be forward bias.

The voltage states for the two paths are as follows:

- Extrinsic Only Path: avalanche breakdown voltage of collector-to-emitter BV_{CBO}.

- Extrinsic and Intrinsic Path: sum of avalanche breakdown voltages of collector-to-emitter BV_{CEO} and emitter–base BV_{EBO}.

Given that the collector-to-base breakdown voltage exceeds the sum of the collector-to-emitter and emitter base breakdown voltages (e.g., $BV_{CBO} > BV_{CEO} + BV_{EBO}$), the ESD current path through the emitter structure will be preferred.

Minear and Dodson showed that the integration of a butted junction structure (referred to as the 'phantom' emitter) near the base contact can prevent the alternative current path through the intrinsic part of the device. Figure 5.2 shows the bipolar transistor with the additional n^+ implant in the base region. The authors added an additional emitter region in the transistor structure local to the base contact. Forming a 'local' emitter region near the base contact in a CBE configuration, the structure is converted to a CBE'E or CE'BE structure, where E is the emitter associated with the intrinsic transistor and E' is a second emitter or n^+ diffusion (e.g., 'phantom' emitter). The emitter E' is electrically shorted to the base contact either through electrical interconnects or by abutting the p^+ base contact and the n^+ emitter contact. With this additional emitter, a vertical *npn* transistor is formed in the extrinsic part of the transistor structure whose breakdown characteristics is identical to the intrinsic vertical *npn* transistor.

There are two cases of interest of ESD stress conditions: first, the emitter–base stress and second, the collector–base stress. Figure 5.3 shows the case of the structure during emitter–base stress. During emitter–base ESD stress, the collector node is floating. With the addition of a second emitter structure, when the emitter is biased positive and the base is referenced

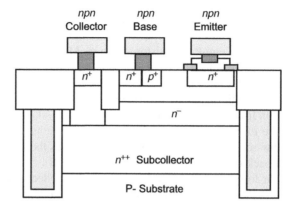

Figure 5.2 A CBE bipolar transistor structure with a 'phantom emitter' structure contained in the extrinsic base region (e.g., at the extrinsic base contact)

as ground potential, two paths exist: the first path between the actual emitter and the base contact through the base region that is initiated at the emitter–base breakdown voltage; the second path through the intrinsic transistor from emitter-to-collector, the subcollector and the vertical *npn* formed by the 'phantom emitter,' base, and collector. Given that both current paths are established during ESD stress, the current in the emitter structure will be less preferred towards the edge near the extrinsic base contact and will distribute vertically through the intrinsic transistor.

In the second stress case, the collector-to-base reverse bias, there are two current paths without the 'phantom' emitter structure (Figure 5.4). By introducing a vertical *npn* phantom transistor within the extrinsic region, with the lower series resistance, the preferred current path through the intrinsic functional transistor during collector-to-base stress will not occur and reduce to a single preferred current path.

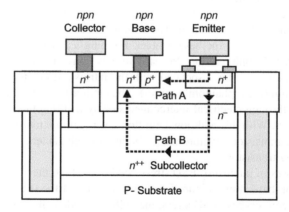

Figure 5.3 Emitter–base stress of a CBE bipolar transistor (with a phantom emitter structure)

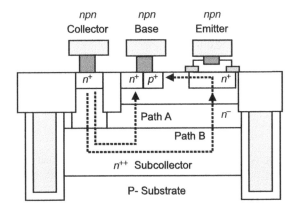

Figure 5.4 Collector–base stress of a CBE bipolar transistor (with a phantom emitter structure)

Table 5.1 shows a comparison of the HBM ESD results with and without the additional extrinsic region vertical transistor. In the ESD testing, Minear and Dodson tested the structures as a function of different terminal states (Figure 5.5). In the table, the data represent the results of different test modes. The key items to note from the experimental data for the additional emitter structure are as follows:

- The emitter–base HBM failure problem is solved, with the HBM failure levels increased to the magnitude of the emitter–base/collector HBM failure levels.

- The emitter/collector–base HBM failure problem is solved, with the HBM failure levels even exceeding the emitter/collector HBM failure levels.

- The collector/base HBM failure problem is solved, with the HBM failure levels exceeding the emitter/collector HBM failure levels.

This indicates the addition of the 'local' vertical *npn* transistor near the base contact and eliminates the ESD failure concern typically observed in the homo- and hetero-junction bipolar transistor structures.

Table 5.1 HBM ESD results of a CBE bipolar transistor with and without additional extrinsic region 'phantom emitter' (as a function of terminal conditions)

Transistor terminal stress	Transistor terminal ground reference	Standard transistor HBM results (V)	Phantom transistor HBM results (V)
Emitter	Base	600	2500
Emitter, collector	Base	600	>3000
Collector	Base	700	>3000
Emitter	Base, collector	2400	2500
Emitter	Collector	2600	2600

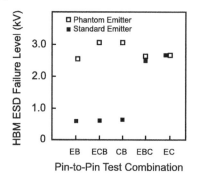

Figure 5.5   HBM ESD stress of a CBE bipolar transistor with and without phantom emitter structure

ESD design layout practices that can be learned from these results are as follows:

- Ordering of the collector, base, and emitter regions can influence the current paths within a bipolar transistor and influence the ESD robustness (e.g., ordering of bipolar transistors are CBE, CEB, CEBC, CBEBC, etc.).

- Placement of the extrinsic collector and base contacts can lead to parasitic current paths and influence the ESD robustness of the bipolar transistor.

- Addition of a second emitter structure (e.g., in a single emitter structure) that is nearest to the extrinsic base contact can lead to an avoidance of the E–B ESD failure mechanism.

- Addition of a phantom emitter is a 'local' emitter to the extrinsic base contact region.

- Additional phantom emitter structure can be a butted-contact structure (n^+ emitter and p^+ base implant).

- Electrically connecting the additional emitter (e.g., dummy emitter or 'phantom' emitter) to the base region (e.g., biased or shorted) can influence the ESD robustness.

- Addition of 'dummy emitter' regions between the actual emitters (e.g., in a multiple emitter structure).

5.8 ESD RF DESIGN LAYOUT: EMITTER BALLASTING

For an ESD RF design, emitter ballasting is used in heterojunction bipolar transistor (HBT) design for improvement in the thermal stability [46–53]. The thermal stability can be improved by providing better power distribution and avoiding 'emitter collapse.'

Emitter ballasting resistors can be integrated into RF circuits, ESD input node circuitry, or ESD power clamp networks. RF ESD design practices can introduce emitter resistance for ballasting through emitter semiconductor processing choices, layout design masking, utilizing interconnects and resistor elements, segmentation, and other layout and design techniques.

An ESD RF design practice in the design layout of heterojunction bipolar transistors can include the following concepts:

- Optimization of the emitter poly-silicon grain structure to provide inherent emitter resistance.

- Prevention of silicidation of the emitter poly-silicon structure either through semiconductor process or masking.

- Segmentation of emitter into minimum size emitters and utilization of single contacts.

- Maintenance of low emitter contact density and introduction of self-ballasting in the poly-silicon emitter film.

- Utilization of interconnect ballasting to each emitter region.

- Utilization of independent resistor ballast elements for each emitter finger.

- Utilization of independent resistor ballast element for each segmented region of the emitter structure.

In the utilization of emitter ballasting in ESD design, tradeoffs exist in the ESD protection network. These include the following:

- Triggering voltage.

- On-resistance.

- Current-to-failure.

In the integration of the emitter ballast resistor elements, improved current-to-failure can be achieved. Additionally, this can lead to a higher voltage across the HBT and emitter series resistor; this can be an advantage or disadvantage depending on the application. For example, if this is introduced into an ESD power clamp comprising an HBT and a series emitter resistor can impair the effectiveness of the ESD power clamp by increasing the total on-resistance and delaying the ability to discharge the amount of ESD current at a lower voltage.

Gao *et al.* [46] derived a relationship for the optimum emitter resistor for an HBT is as follows,

$$
R_{\mathrm{E}} = \frac{-V_{\mathrm{CE}}R_{\mathrm{TH}}\dfrac{\partial V_{\mathrm{BE}}}{\partial T} - \left(R_{\mathrm{EC}} + \dfrac{nkT}{qI_{\mathrm{E}}}\right)\left\{1 - (T_j - T_c)\dfrac{1}{\kappa(T)}\dfrac{d\kappa}{dT}\right\}}{1 + (T_j - T_c)\left\{\dfrac{1}{R_e}\dfrac{dR_e}{dT} - \dfrac{1}{\kappa(T)}\dfrac{d\kappa}{dT}\right\}}
$$

where R_{E} is the emitter ballast resistor, R_e is the emitter resistance, R_{EC} is the emitter contact resistance, and R_{TH} is the total thermal resistance of the device. From this development, there is a case resistance, T_{C}, and temperature, T_J.

$$
T_j = I_{\mathrm{C}}V_{\mathrm{CE}}R_{\mathrm{TH}} + T_{\mathrm{C}}
$$

5.9 ESD RF DESIGN LAYOUT: THERMAL SHUNTS AND THERMAL LENSES

In multiemitter structures, temperature differences in the heterojunction bipolar transistor (HBT) devices can lead to 'emitter collapse' at high temperatures and collector-to-emitter voltages. This thermal instability is initiated by having more than one emitter, where the emitter structures have different emitter temperatures. The temperature differential can be reduced in multiemitter structures with the introduction of a 'thermal shunt' element. A thermal shunt can be a shunt that is electrically insulating, but must be thermally conductive. For example, ceramic and electrically insulating materials can serve as a thermal shunt. Craft [47] proposed the use of insulating thermal shunts using thermally conductive ceramics.

Thermal shunts, also referred to as thermal lenses, are used to reduce the temperature differential between the emitter regions. Given a first and second emitter, in the case of no thermal coupling between the two emitters, the temperature can be expressed as

$$T_1 = T_0(1) + P_1(R_{TH})_1$$

$$T_2 = T_0(2) + P_2(R_{TH})_2$$

that can be expressed as the temperature at location and the product of the power in the transistor and the thermal resistance. In the case of thermal coupling between the first and second element, we can express this in tensor form and as a local temperature and the matrix term associated with the thermal resistance matrix and power product,

$$T_i = (T_0)_i + (R_{TH})_{ij}P_j$$

or,

$$\begin{bmatrix} T_1 \\ T_2 \end{bmatrix} = \begin{bmatrix} (T_0)_1 \\ (T_0)_2 \end{bmatrix} + \begin{bmatrix} (R_{TH})_{11} & (R_{TH})_{12} \\ (R_{TH})_{21} & (R_{TH})_{22} \end{bmatrix} \begin{bmatrix} P_1 \\ P_2 \end{bmatrix}$$

Using a thermal shunt, the temperature differences between the two elements can be reduced by a reduction of the extrinsic temperature differences or the thermal coupling and power differences in the two emitters.

Sewell et al. [49] introduced a thermal shunt into the emitter design of a heterojunction bipolar transistor. The heterojunction bipolar transistor design configuration was an EBCBE configuration. The thermal lens extended from the two emitter pads and was integrated over the base and collector regions. By extending the thermal shunt to the electrical pads reduces the temperature differential spatially across the region and also the external temperature differences. Sewell et al. [49] designed the emitter thermal shunt width significantly larger than the emitter metal width. By introducing a wide metal region of larger area, the region serves as a thermal reservoir, as well as a lower thermal conductance. Two types of emitter design styles exist in bipolar transistor design. Typically, the minimum emitter is a small square design (e.g., equal sides) that is able to be lithographically controlled; these minimum size structures are used for RF receiver networks and typically achieve the highest performance. These regions are typically connected with a single contact or bar contact. Moreover, the emitter design layout can be a rectangle, where one edge is the minimum design edge; for these structures there is typically a maximum length that is allowed for lithographic control (e.g., width/length ratio control). With segmentation of the emitter region into smaller segments, the current is distributed to more physical elements leading to a lower junction temperature. Moreover, with adequate emitter-to-emitter space, the thermal coupling between the emitters is reduced, again lowering the temperature. Additionally, with a design opening of the area between emitters, area can be used to introduce the thermal shunts, increasing the thermal shunt total width. Sewell *et al.* demonstrated that with the case of minimum size square emitter design, compared to the bar emitter, a superior result is achieved in the HBT stability for a given power level and equivalent total emitter area due to (1) spreading of the 'heat sources' and (2) larger thermal shunt width [49].

In RF applications, the RF performance can be impacted by the additional loading associated with the thermal shunt. Capacitance and reactance loading of the thermal shunt can impact the unity current gain cutoff frequency and the unity power gain cutoff frequency. In the work of Sewell *et al.*, the impact of the reactance loading effect was not regarded as significant in that specific device and application [49].

An ESD RF design practice in the design layout of heterojunction bipolar transistors can include the following concepts:

- Introduction of a thermal shunt element strapped between the emitter bond pads to lower emitter-to-emitter differential temperature.

- Extension of emitter thermal shunt from a first emitter to the second emitter bond pads.

- Significant enlargement of the emitter thermal shunt width than the emitter metal width (e.g., 3 to 10× greater).

- Segmentation of emitter into minimum size emitters to lower thermal resistance (e.g., instead of 'bar emitters').

- Maintenance of emitter-to-base spacing to avoid impacting the RF performance of the transistor element.

- Designing of emitter-to-emitter space to allow integration of the thermal shunt in-between in order to increase the width of the thermal shunt.

5.10 BASE-BALLASTING AND RF STABILITY

A design synthesis of the RF source stability and the ESD sensitivity of the base resistance can be understood from Smith Chart analysis [50,51]. Source stability requirements will determine a minimum resistance, R_{min}, for a source connected to a bipolar base to be unconditionally stable [52]. By modification of the base resistance, it is possible to achieve both unconditionally stable circuits and ESD robust circuitry. If the additional resistance added to a bipolar device for ESD robustness improvements exceeds the minimum base resistance, both ESD and the stability criteria can be achieved. The addition of resistance beyond r_{min} can exceed the minimum stability requirement with further improved ESD protection levels.

To optimize power amplifiers or other applications, this can be achieved by evaluating the unilateral constant gain and noise circles on a Smith chart. Plotting the source and load

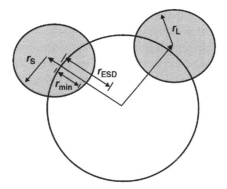

Figure 5.6 Unity radius Smith chart with source and load stability circles. Region outside the unity radius Smith chart is the unstable source. Stability and ESD improvement achieved with shift along axis toward radial center

circles on the normalized chart, stability of the circuit can be evaluated. For example, source and load stability circles can be defined by their radius and center (Figure 5.6). The radius, r, and center, C, are defined as follows:

$$r_s = \frac{|s_{12}s_{21}|}{|s_{11}|^2 - |\det S|^2}$$

and

$$C_s = \frac{(s_{11} - |\det S|s_{22}*)*}{|s_{11}|^2 - |\det s|^2}$$

If the source stability circle center is contained outside the Smith chart, there are conditions that unconditional stability is not achieved [52].

In this case, a minimum resistance r_{min} can be defined as that which moves the source stability circle within the boundary of the Smith Chart. In this case, adding resistance R_{min} to the base of an *npn* amplifier can place the amplifier into a metastable condition. Improved stability is achieved by adding a resistance R_{ESD} where $R_{ESD} = R_{min} + R_{ESD}*$, where the first term is for unconditional stability and the second term is the additional resistor ballasting for the base region. This method can also be applied to the load stability circles to provide a stable output and identifying a G_p (min) on the constant conductance circles.

From the Stern stability criteria, $K > 1$ achieves unconditionally stable two-port networks where

$$K = \frac{1 - |s_{11}|^2 - |s_{22}|^2 + |\det S|}{2|s_{12}s_{21}|}$$

To achieve stability and design margin, designers should know that $K > 1.2$. A circuit can be designed such that ESD elements are added to satisfy the Stern criteria and an additional ESD element can be added to provided the extra stability margin.

5.11 SUMMARY AND CLOSING COMMENTS

In this chapter we addressed radio frequency (RF) bipolar transistors. We have also discussed bipolar transistor physics, bipolar breakdown mechanisms, and bipolar transistor device parameters. In addtion we also discussed the very important Johnson Limit—this addresses the interrelationship of power and transistor frequency that governs the scaling of bipolar transistors and the choice of technology utilized for given applications. Because of the interrelationship between breakdown voltage and the transistor cutoff frequency, there is a significant interest in this issue for ESD protection. we also address the thermal issues of the bipolar transistor.

In Chapter 6, the text discusses ESD in silicon germanium and silicon germanium carbon technology. Chapter 6 will discuss HBM, MM, and TLP measurements of SiGe transistors and SiGe-derivative device elements and the interrelationship to the transistor layout and design. Self-aligned and nonself-aligned emitter–base SiGe transistor structures' results will be shown, highlighting the relationship of the spacings and how it influences ESD robustness. In addition, topics such as transistor ordering and multiple finger emitter transistors and their influence on ESD robustness will be discussed.

PROBLEMS

1. From the frequency response of the transistor

$$f_T = \frac{1}{2\pi}\frac{1}{\tau} = \frac{v_S}{2\pi L}$$

    assuming the transistor base width scales as scaling parameter, α, show the scaling of the cutoff frequency. Derive the frequency margin relationship between the frequency of the bipolar transistor and ESD frequency (e.g., CDM, HBM, MM).

2. From the Johnson Limit relationship, assuming a scaling parameter α, derive a relationship for the breakdown voltage scaling.

3. From the Johnson Limit relationship for current, maximum current, and reactive impedance, as

$$(I_m X_C) f_T = \frac{E_m v_s}{2\pi}$$

    where

$$X_C = \frac{1}{2\pi f_T C_0}$$

    Derive the scaling relationship for maximum current assuming a scaling parameter, α. Compare the maximum current scaling to the magnitude of ESD current magnitudes as a function of scaling relationship.

4. Compare the Johnson Limit maximum power relationship to the three different power-to-failure time regimes of the Wunsch–Bell model.

5. Show all possible current paths of the emitter, base, and collector in a semiconductor device structure within the intrinsic and extrinsic transistor regions.

6. Draw a transistor with a single emitter, base, and collector. Show the direction of current flow through different ESD events. Draw a transistor with a 'wrap-around' two-sided base region, single emitter, and single collector (BEBC). What are the advantages for performance? for ESD? Draw a transistor with a 'wrap-around base' and 'wrap-around collector' CBEBC configuration. What are the RF performance advantages? What are the ESD performance advantages?

7. Show all possible emitter–base–collector configurations for the case of single and multiple emitter structures. Draw possible current paths. Which versions are prone to parasitic undesired current paths. Discuss pros and cons of different implementations.

REFERENCES

1. Muller RS and Kamins TI. *Device electronics for integrated circuits*. New York: John Wiley and Sons; 1977.
2. Gray PE, Searle CL. *Electronic principles: physics, models, and circuits*. New York: John Wiley and Sons, Inc.; 1969.
3. Gray PE, DeWitt D, Boothroyd AR, Gibbons JF. *Physical electronics and circuit models of transistors*. Semiconductor Electronic Education Committee, vol. 2. New York: John Wiley and Sons; 1964.
4. Thornton RD, DeWitt D, Chenette ER, Gray PE. *Characteristics and limitations of transistors*. Semiconductor Electronic Education Committee, vol. 4. New York: John Wiley and Sons; 1964.
5. Sze SM. *Physics of semiconductor devices*. New York: John Wiley and Sons; 1981.
6. Cressler JD, Niu G. *Silicon germanium heterojunction bipolar transistors*. Boston: Artech House; 2003.
7. Zener C. A theory of electrical breakdown of solid dielectrics. *Proceedings of the Royal Society (London)* A 1933;**145**:523–9.
8. Wolff PA. Theory of electron multiplication in silicon and germanium. *Physical Review* 1954;**95**(6):1415–20.
9. Chynoweth AG. Ionization rates for electron and holes in silicon. *Physical Review* 1958;**109**(5):1537–40.
10. Moll JL, Van Overstraeten R. Charge multiplication in silicon p-n junctions. *Solid State Electronics* 1963;**6**:147–57.
11. Reisch M. On bistable behavior and open-base breakdown of bipolar transistors in the avalanche regime – modeling and applications. *IEEE Transactions on Electron Devices* 1992;**ED-39**:1398.
12. Hower PL, Reddi VGK. Avalanche injection and second breakdown in transistors. *IEEE Transactions on Electron Devices* 1970;**ED-17**:320–35.
13. Schafft HA, French JC. Second breakdown in transistors. *IEEE Transactions of Electron Devices* 1962;**ED-9**:129–36.
14. Scarlett RM, Schockley W, Haitz RH. Thermal instabilities and hot spots in junction transistors. In: Goldberg MF, Vaccaro J (eds.), *Physics of failure in electronics*. Baltimore: Spartan Books; 1963. p. 194–203.
15. Nienhuis RJ. Second breakdown in forward and reverse base current region. *IEEE Transactions on Electron Devices* 1966;**ED-13**:655–62.
16. Grutchfield HB, Moutoux TJ. Current mode second breakdown in epitaxial planar transistors. *IEEE Transactions on Electron Devices* 1966;**ED-13**:743.

17. Tasca DM, Peden JC, Miletta J. Non-destructive screening for thermal second breakdown. *IEEE Transactions of Nuclear Science* 1972;**NS-19**:57–67.
18. Dutton RW. Bipolar transistor modeling of avalanche generation for computer circuit simulation. *IEEE Transactions on Electron Devices* 1975;**ED-22**:334–8.
19. Ward A. An electrothermal model of second breakdown. *IEEE Transactions of Nuclear Science* 1976;**NS-23**:1679–84.
20. Gaur S, Navon D, Teerlinck R. Transistor design and thermal stability. *IEEE Transactions of Electron Devices* 1976;**ED-20**:527–34.
21. Ghandi SK. *Semiconductor power devices*. New York: John Wiley and Sons; 1977.
22. Koyanagi K, Hane K, Suzuki T. Boundary conditions between current mode and thermal mode second breakdown in epitaxial planar transistors. *IEEE Transactions of Electron Devices* 1977;**ED-24**:672–8.
23. Yee J, Orvis W, Martin L, Peterson J. Modeling of current and thermal mode second breakdown phenomenon. *Proceedings of the Electrical Overstress/Electrostatic Discharge (EOS/ESD) Symposium*, 1982. p. 76–81.
24. Orvis WJ, McConaghy CF, Yee JH, Khanaka GH, Martin LC, Lair DH. Modeling and testing for second breakdown phenomena. *Proceedings of the Electrical Overstress/Electrostatic Discharge (EOS/ESD) Symposium*, 1983. p. 108–17.
25. Domingos H. Basic considerations in electro-thermal overstress in electronic components. *Proceedings of the Electrical Overstress/Electrostatic Discharge (EOS/ESD) Symposium*, 1980. p. 206–12.
26. Ward AL. Calculations of second breakdown in silicon diodes at microwave frequencies. *Proceedings of the Electrical Overstress/Electrostatic Discharge (EOS/ESD) Symposium*, 1983. p. 102–7.
27. Hassan MMS, Domingos H. The double graded transistor and its beneficial effect on resistance to current mode second breakdown. *Proceedings of the Electrical Overstress/Electrostatic Discharge (EOS/ESD) Symposium*, 1989. p. 127–35.
28. Kirk CT. A theory of transistor cutoff frequency (f_t) falloff at high current densities. *IRE Transaction of Electronic Devices* 1962;**ED-9**:164–74.
29. Early JM. Maximum rapidly-switchable power density in junction triodes. *IRE Transactions on Electronic Devices* 1959;**ED-6**:322.
30. Early JM. Speed in semiconductor devices. *IRE National Convention*, March 1962.
31. Early JM. Structure-determined gain-band product of junction triode transistors. *Proceedings of the IRE*, December 1958. p. 1924.
32. Goldey JM, Ryder RM. Are transistors approaching their maximum capabilities? *Proceedings of the International Solid-State Circuits Conference (ISSCC)*, February 1963. p. 20.
33. Pritchard RL. Frequency response of grounded-base and grounded emitter junction transistors. *AIEE Winter Meeting*, January 1954.
34. Giacoletto LJ. Comparative high frequency operation of junction transistors made of different semiconductor materials. *RCA Review*, 16 March 1955. p. 34.
35. Johnson EO, Rose A. Simple general analysis of amplifier devices with emitter, control, and collector functions. *Proceedings of the IRE*, March 1959. p. 407.
36. Johnson EO. Whither the tunnel diode. *AIEE-IRE Electron Device Conference*, October 1962.
37. Johnson EO. Physical limitations on frequency and power parameters of transistors. *RCA Review*, June 1965. p. 163–77.
38. Liu W, Nelson S, Hill D, Khatibzadeh A. Current gain collapse in microwave multi-finger heterojunction bipolar transistors operated at very high power density. *IEEE Transactions on Electron Devices* 1993;**ED-40**:1917.
39. Liou LL, Bayraktaroglu B. Thermal stability analysis of AlGaAs/GaAs heterojunction bipolar transistors with multiple emitter fingers. *IEEE Transactions on Electron Devices* 1994;**ED-41**:629.

40. Liu W. The interdependence between the collapse phenomenon and the avalanche breakdown in AlGaAs/GaAs power heterojunction bipolar transistors. *IEEE Transactions on Electron Devices* 1995;**ED-42**:591.

41. Liu W, Khatibzadeh A. The collapse of current gain in multi-finger heterojunction bipolar transistors: its substrate temperature dependence, instability criteria, and modeling. *IEEE Transaction on Electron Devices* 1994;**ED-41**:1698.

42. Hower PL. High field phenomena and failure mechanisms in bipolar transistors. *Proceedings of the Electrical Overstress/Electrostatic Discharge (EOS/ESD) Symposium*, 1980. p. 112–6.

43. Minear RL, Dodson GA. Effects of electrostatic discharge on linear bipolar integrated circuits. *Proceedings of the International Reliability Physics Symposium (IRPS)*, 1974. p. 60.

44. Minear RL, Dodson GA. The phantom emitter – an ESD-resistant bipolar transistor design and its applications to linear integrated circuits. *Proceedings of the Electrical Overstress/Electrostatic Discharge (EOS/ESD) Symposium*, 1979. p. 188–92.

45. Wu WL, Chang CY, Ker MD. High-current characteristics of ESD devices in 0.35-μm silicon germanium RF BiCMOS process. *Proceedings of the Taiwan Electrostatic Discharge Conference (T-ESDC)*, 2003. p. 157–62.

46. Gao GB, Unlu MS, Morkoc H, Blackburn D. Emitter ballasting resistor design for, and current carrying capability of AlGaAs/GaAs power heterojunction bipolar transistors. *IEEE Transactions of Electron Devices* 1991;**ED-38**:185.

47. Craft S. Thermal shunt for electronic circuits. U.S. Patent No. 4,941,067 (July 10, 1990).

48. Bayraktaroglu B, Barrette J, Kehias L, Huang CI, Fitch R, Neikhard R, *et al.* Very high-power-density CW operation of GaAs/AlGaAs microwave heterojunction bipolar transistors. *IEEE Electron Device Letters* 1993;**EDL-40**:493.

49. Sewell J, Liou LL, Barlage D, Barrette J, Bozada C, Dettmer R, *et al.* Thermal characterization of thermally-shunted heterojunction bipolar transistor. *IEEE Transaction of Electron Devices* 1996;**ED-17**:19.

50. Voldman S, Lanzerotti LD, Johnson R. Emitter–base junction ESD reliability of an epitaxial base silicon germanium heterojunction transistor. *Proceedings of the International Physical and Failure Analysis of Integrated Circuits (IPFA)*, July 2001. p. 79–84.

51. Voldman S, Lanzerotti LD, Johnson RA. Influence of process and device design on ESD sensitivity of a silicon germanium heterojunction bipolar transistor. *Proceedings of the Electrical Overstress/ Electrostatic Discharge (EOS/ESD) Symposium*, 2001. p. 364–72.

52. Besser L. Avoiding RF oscillation. *Applied Microwave and Wireless Spring* 1995;44–55.

53. Liu W, Khatibzadeh A, Sweder J, Chau HF. The use of base ballasting to prevent collapse of current gain in AlGaAs/GaAs heterojunction bipolar transistors. *IEEE Transaction on Electron Devices* 1996;**ED-43**:245.

6 Silicon Germanium and ESD

6.1 HETEROJUNCTIONS AND SILICON GERMANIUM TECHNOLOGY

In today's RF technologies, the silicon (Si) homojunction bipolar transistor has been dominated by new heterojunction bipolar technologies from silicon germanium (SiGe), gallium arsenide (GaAs), to indium phosphide (InP) [1–5]. The heterojunction bipolar transistor (HBT) has outperformed the homojunction bipolar junction transistor (BJT). With the introduction of the silicon germanium HBT and presently the silicon germanium carbon HBT, these devices became the natural evolutionary step beyond the silicon BJT.

From the early days of the invention of the silicon bipolar transistor, as early as the 1950s and 1960s, the question of the limitations of the transistor speed was being addressed by early device researchers such as Early, Johnson, Webster, Kirk [6–14]. From the Johnson Limit relationships, it is clear that there was a tradeoff between the maximum power and the speed of the transistor. Additionally, the problem of scaling the base width and at the same time have a low base series resistance was a natural constraint to the scaling of the BJT. An independent variable was needed that allowed continuous scaling of the base region. The solution to faster transistors was the ability to modify the energy bandgap to improve device performance.

The work by Kroemer [17–21] on the heterojunction opened a new door on achieving higher speeds than anticipated with silicon homojunctions. Significant growth in both high-frequency wired and wireless markets has introduced new opportunities where compound semiconductors have unique application advantages over bulk CMOS technology [1–5]. With the introduction of the epitaxial-layer pseudo-morphic silicon germanium (SiGe) deposition processes by Meyerson [22], epitaxial-base SiGe HBTs have been integrated with mainstream advanced CMOS development processes, providing the advantages of SiGe technology for analog and RF circuitry while maintaining the full utilization of the advanced CMOS technology base for digital logic circuitry. Today, SiGe HBT and SiGeC HBT devices have replaced silicon bipolar junction (BJT) devices as the primary element in all analog and RF applications [23–65].

With the increased volume and growth in the RF applications that use SiGe HBT devices for external circuitry, the quantification of the SiGe HBT and SiGeC HBT devices during electrostatic overstress (EOS) and electrostatic discharge (ESD) sensitivities, as well as construction of ESD networks, has become more important [66–91]. In this chapter we focus on the ESD response of SiGe and SiGe HBT devices.

6.1.1 Silicon Germanium HBT Devices

With the rapid advancement of the epitaxial-layer pseudomorphic silicon germanium (SiGe) deposition processes, epitaxial-base SiGe HBTs have been integrated with mainstream advanced CMOS development processes, providing the advantages of SiGe technology for analog and RF circuitry while maintaining the full utilization of the advanced CMOS technology base for digital logic circuitry [23–44]. The advantages of forming an epitaxial layer SiGe HBT device in a standard CMOS process are as follows:

- Silicon germanium HBT devices naturally integrate with silicon semiconductor processing.

- SiGe HBT devices and circuits can be integrated with digital and analog CMOS on the same wafer reducing total cost.

- SiGe HBT devices and circuits integration with digital circuitry allow for high-level system integration and performance advantages.

- SiGe HBT devices can be developed in any CMOS foundry business.

- SiGe HBT devices can utilize mask levels from standard CMOS processing (e.g., substrate, n-wells, p-wells, shallow trench isolation [STI], polysilicon gate films, silicide films, contacts, vias, wiring and interlevel dielectric [ILD] films).

- High-speed SiGe HBT can be migrated across different CMOS technology generations because of the independence of the SiGe HBT film from the base CMOS technology.

Epitaxial deposition of silicon germanium allows for optimization of a bipolar transistor without the limitations and pitfalls of implanted dopants. The SiGe alloy film can achieve the following features:

- allow formation of atomically abrupt interface;

- avoid limitations of Gaussian profiles;

- avoid dopant-implant channeling;

- avoid hot-process deficiencies.

The ultra-high vacuum chemical vapor deposition (UHV/CVD) allows for tight control of dopant implants and low thermal budget. Because it can be formed after STI, or even after the completion of CMOS transistors, it can be well controlled.

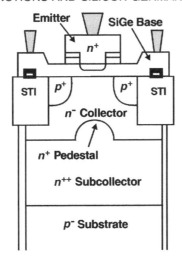

Figure 6.1   Cross section of a high performance SiGe heterojunction bipolar transitor (HBT)

6.1.2 Silicon Germanium Device Structure

Figure 6.1 shows a cross section of an epitaxial graded Ge-base SiGe HBT device [28,29]. The SiGe HBT device is formed on a p^- silicon substrate wafer. With the low-doped p^- substrate, good noise isolation between the analog and digital circuitry is achieved. The substrate doping concentration allows for low subcollector-to-substrate capacitance. In each technology generation, the substrate resistance will be increased to achieve lower capacitance and higher noise isolation.

The polysilicon deep trench (DT) isolation defines the subcollector region and also provides a low sidewall capacitance. The placement of the DT prevents out-diffusion of the subcollector, providing a smaller physical collector region and better spacing relative to adjacent devices. The DT also lowers the subcollector-to-substrate capacitance. The polysilicon DT isolation is used for high performance application as it provides a lower unity power cutoff frequency, f_{MAX}. Additionally, the DT isolation influences the subcollector-to-substrate breakdown voltage and eliminates parasitic devices in a SiGe HBT device. The DT extends beyond the subcollector lower edge. As a result, the DT structure is typically 4.5–5.5 μm deep. For low cost processes, the DT is replaced with either a low cost trench isolation (TI) or diffusion implants. The DT-defined collector contains a highly doped n^{++} subcollector, n^+ collector reach-through implant, and an optional n^{++} pedestal implant. The pedestal implant used for the high-frequency SiGe *npn*, provides a low-resistance collector, a high f_T, and an improved Kirk effect [16].

From the CMOS base technology, the STI is used for the definition of the collector contact and subcollector reach-through structure. Silicon germanium transistors can also be constructed with LOCOS isolation from older generation CMOS or BiCMOS base technology.

The deposition of the SiGe film has been demonstrated in a number of different chemical vapor depositions. Using UHV/CVD processing, SiGe is deposited in the base region over single-crystal silicon and the STI structure [22]. The germanium

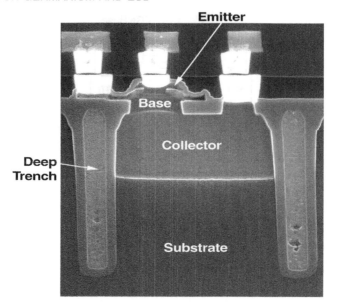

Figure 6.2   Silicon germanium HBT device

concentration is varied during the film deposition process to provide a position-dependent SiGe alloy film for profile and device optimization of the SiGe HBT base region. The epitaxial base region forms a single-crystal SiGe intrinsic base and an amorphous poly-SiGe extrinsic base region. A window is formed over the single-crystal intrinsic SiGe base region to form the n-type polysilicon emitter. After hot process steps, the n^+ dopants of the emitter diffuse into the SiGe intrinsic base region. Interconnection to the emitter, base, and collector is defined by a tungsten (W) local interconnect. Interlevel dielectrics (ILD) films, tungsten contacts, and aluminum interconnects are formed with the reactive ion etching (RIE) and chemical mechanical polishing (CMP) processes used in base CMOS technology. In today's technologies, copper interconnects have replaced aluminum interconnect films.

Figure 6.2 shows a cross sectional drawing of a silicon germanium HBT. In this structure, the DT structure borders the collector region. The collector region contains the n^{++} subcollector, the lightly doped region of the collector, and the n^+ pedestal implant. The STI is formed on the DT sidewall and serves as a region to define the base electrical contact. The collector contact is not shown in the drawing. The SiGe base region is formed over the single crystal silicon and the STI regions. Contacts to the base region are made over the STI regions to the polysilicon germanium film region. Dopants from the base diffuse into the single crystal region, forming the base–collector junction region. The emitter is formed in the single crystal silicon germanium region by diffusion.

Epitaxial-base profile optimization of the SiGe HBT device provides degrees of freedom for device designs. These design freedoms can also translate into the ability to design a heterojunction transistor with electrical characteristics suitable for both functional optimization and high-current pulse operation (e.g., ESD operation, overshoot/undershoot). Presently, SiGe HBT devices use a spatial-dependent Ge concentration to modulate the

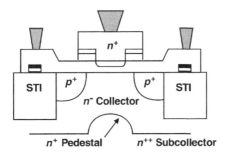

Figure 6.3   SiGe HBT device cross section

bandgap that provides a higher collector current density, base transit time reduction, low intrinsic base resistance, and an increased early voltage at cutoff frequencies. Silicon germanium HBT devices can be formed by varying the germanium concentration and the Si_xGe_{1-x} compound.

Figure 6.3 shows an expanded view of the emitter–base region. Over the single crystal silicon, a pseudomorphic single crystal SiGe film is formed. A facet occurs at the amorphous SiGe-to-SiGe single crystal interface. The $TiSi_x$ salicide is formed on the extrinsic-base region from the contact to the edge of the emitter polysilicon structure. A doped p^+ dopant is implanted into the extrinsic base region to reduce the base series resistance. A 'self-aligned' (SA) SiGe HBT device is a transistor where this extrinsic base implant is aligned to the edge of the emitter window opening using a disposable spacer forming a fixed extrinsic base-to-emitter space. A 'non-self aligned' (NSA) SiGe HBT device is a transistor where the extrinsic base implant is aligned to the edge of the polysilicon emitter shape.

6.2 SILICON GERMANIUM PHYSICS

The physics of a silicon germanium HBT device compared to the silicon BJT is different as a result of the bandgap engineering [28,29]. Epitaxial-base profile optimization of the SiGe HBTs provides degrees of freedom for transistor device design. Kroemer [17–21] pointed out in 1982 that heterojunctions can control electrons and holes separately and independently provide design freedom, that is not achievable in homojunction transistor junctions. These design freedoms can also translate into the ability to design a hetero-junction transistor with electrical characteristics suitable for both functional optimization and high-current pulse operation (e.g., ESD operation, overshoot/undershoot). Kroemer [17–21] noted that using wide-bandgap emitter structures was extremely advantageous for high base doping concentrations in the HBT. Presently, SiGe HBTs use a spatial-dependent germanium concentration to provide a higher collector current density, base transit time reduction, low intrinsic base resistance, and an increased early voltage at cutoff frequencies.

Figure 6.4 shows the SiGe bandgap highlighting the band modification from the germanium profile. Rectangular, triangular, and trapezoidal profiles are used to optimize the transistor transit time.

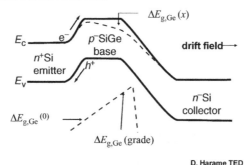

D. Harame TED

Figure 6.4 Silicon germanium HBT band diagram

With high base doping concentrations, the intrinsic temperature, T_i, of a SiGe HBT device can be significantly higher than a Si BJT device that provides a device less prone to thermal instability. The intrinsic temperature, T_i, is the temperature at which the intrinsic carrier concentration exceeds the dopant concentration. Thermal runaway or second breakdown is the point (I, V) at which a semiconductor device is at the onset of a negative resistance state because of thermal feedback initiated by Joule self-heating. The thermal instability is typically irreversible and leads to either leakage or destruction of the semiconductor structure. From Joule heating, the intrinsic carrier concentration increases with increasing temperature. It is well known that Si-BJT-device thermal instability typically occurs in the low-doped base region, where the intrinsic temperature is low due to the low base doping concentration. For SiGe devices, the position-dependent Ge concentration also plays a role. With a position-dependent Ge concentration, the intrinsic carrier concentration can be expressed as [28,29],

$$n_i^2(x) = \gamma n_{io}^2 e^{\Delta E(x)/kT}$$

and,

$$\Delta E_g(x) = \Delta E_{g,b}^{app} + \Delta E_{g,Ge}(\text{grade})(x/W_b kT) + \Delta E_{g,Ge}^{(0)}$$

where $\Delta E_g(x)$ is the position dependent bandgap, $\Delta E_{g,b}^{app}$ the heavy doping-induced apparent bandgap-narrowing, $\Delta E_{g,Ge}$ (grade) is the difference between the band-bending at the base–collector junction, ΔE_g ($x = W_b$), and the band-bending at the base–emitter junction, ΔE_g ($x = 0$). The effective density of states ratio of SiGe and Si is expressed as

$$\gamma = (N_C N_V)_{SiGe}/(N_C N_V)_{Si}$$

The position-dependent Ge concentration produces a position-dependent intrinsic temperature, $T_i(x)$. Hence, the intrinsic temperature for a HBT can be expressed as

$$T_i(x) = \{E_{g,Si} + \Delta E_{g,b}^{app} + \Delta E_{g,Ge}(\text{grade})(x/W_b)$$
$$+ \Delta E_{g,Ge}^{(0)}\}/k \ln\{[N_b(x)/n_{io}]^2 (N_C N_V)_{Si}/(N_C N_V)_{SiGe}\}$$

From an ESD perspective, a bipolar transistor with a high bipolar current gain is also advantageous for discharging current in a peripheral circuit or ESD network. From the position-dependent intrinsic carrier concentration and the Kroemer–Moll–Ross relationship [18–21], the collector current for a graded Ge concentration can be expressed as

$$I_C = A_C(q D_{nb} n_{io}^2 / W_b N_b) \{ e^{\Delta E_{g,b}^{app}/kT} e^{V_{be}/kT} - 1 \}$$
$$\{ \gamma v (\Delta E_g (\text{grade})/kT) / (1 - e^{-\Delta E_{g,Ge}(\text{grade})/kT}) \}$$

with γ and v position-averaged.

From the Kroemer–Moll–Ross collector current relationship, the ratio of the current gain of a SiGe HBT and Si BJT can be expressed as [28,29]

$$\frac{\beta_{SiGe}}{\beta_{Si}} = \frac{\gamma v (\Delta E_g (\text{grade})/kT) e^{\Delta E_{g,Ge}^{(0)}/kT}}{1 - e^{\Delta E_{g,Ge}(\text{grade})/kT}}$$

In a common-emitter mode, the collector current can be expressed as a function of the avalanche multiplication factor, M, and the current gain term, as

$$I_C = \frac{M \beta_{SiGe} I_B + M I_{CO}(1 + \beta_{SiGe})}{1 - \beta_{SiGe}(M - 1)}$$

where in an open-base condition,

$$I_C = \frac{M I_{CO}(1 + \beta_{SiGe})}{1 - \beta_{SiGe}(M - 1)}$$

From this expression, it is clear that the collector current is a strong function of both the avalanche multiplication factor, M, and bipolar current gain. The high bipolar current gain of SiGe HBT transistors allows for the ability to discharge significant current during ESD events. The avalanche multiplication term, M, also plays a role in the high-current conduction during ESD events.

With bandgap engineering, the SiGe transistor is designed so that the Ge concentration does not significantly penetrate into the Si collector region at the base–collector junction. Germanium has a lower high electric field effective ionization threshold energy compared to Si (e.g., Ge is approximately 1 eV compared to 3.6 eV in Si), leading to a higher impact ionization rate. Fortunately, as the penetration of the Ge concentration is a few tens of nanometers, there is not a significant accumulation of carrier energy leading to an enhancement of the avalanche multiplication term, M, at the base–collector junction.

For ESD phenomenon, it is important to have the bipolar transistor responsive on the time scale of the ESD model. The CDM model, the fastest of the ESD models, has a 250 ps rise time and an oscillatory waveform. The energy spectrum of a CDM event has a significant energy contained in the 1–5 GHz regime. The responsiveness of the SiGe HBT device is a function of the emitter, base, and collector transit times, whereas the base transit time is typically the frequency-limiting term. SiGe HBT devices use a graded Ge profile to establish a built-in electric field, that accelerates the electrons from the emitter to the collector, providing a base transit time significantly less than Si BJT devices.

From the double-integral expression of the base transit time for a heterojunction, the base transit time can be expressed as [28,29]

$$\tau = \left(W_b^2/D_{nb}\right)\left(kT/\Delta E_{g,Ge}(\text{grade})\right)\left[1 - \left(kT/\Delta E_{g,Ge}(\text{grade})\right)\left(1 - e^{-\Delta E_{g,Ge}(\text{grade})/kT}big\right)\right]$$

This expression comprises the base transit time in a uniform doped Si BJT, W_b^2/D_{nb}, multiplied by the terms associated with the base transit time enhancements due to the Ge bandgap modulation. The emitter transit time in a SiGe HBT device is proportional to the reciprocal of the bipolar current gain, so it has the advantage relative to the Si BJT for ESD phenomenon.

Breakdown voltages of SiGe HBT devices are important for both I/O and ESD applications. The collector-to-emitter breakdown voltage, BV_{CEO}, can be expressed as

$$BV_{CEO} = BV_{CBO}\left\{1 - \frac{(\beta_{SiGe}/\beta_{Si})}{\left[\dfrac{\beta_{SiGe}+1}{\beta_{Si}}\right]}\right\}$$

For SiGe HBT devices, the bipolar collector current gain improvement demonstrates itself as a low collector-to-emitter breakdown voltage. The collector-to-emitter breakdown voltage, BV_{CEO}, β tradeoff can be advantageous for both the performance and ESD protection by modulating the collector profile with a 'pedestal implant.' Using an optional pedestal implant, both the high frequency f_T and the high breakdown voltage SiGe *npn* can be built in a single process for I/O and ESD advantages.

6.3 SILICON GERMANIUM CARBON

Silicon germanium alloy films have enabled the reality of integration of advanced silicon technology with the HBT devices. In order to achieve higher speeds, introduction of new processes have emerged to combat the technology of scaling limitations of the silicon germanium HBT devices [39–44]. One of the scaling limitations of the SiGe *npn* HBT device is the *npn* transistor base width [39,41]. In order to reduce the transit time, the base width is required to scale to provide an increase in the unity current gain cutoff frequency, f_T. Moreover, the base doping concentration must be increased to achieve a high unity power gain cutoff frequency, f_{MAX}, to provide a low base sheet resistance as the base width is scaled. A limitation to achieve the RF performance is boron, which is used in the base region [45–65]. Boron diffusion will limit the ability to continually scale the SiGe HBT *npn* device base region. As a result, the silicon germanium carbon HBT was introduced in order to address this issue, allowing the continual scaling of the SiGe HBT device to higher cutoff frequencies [45,53,62–65].

To extend the SiGe HBT device to higher frequencies, carbon is incorporated into its base region. A key issue in the technology scaling of the HBT device is the dimensional scaling of the base width of the transistor to achieve smaller base transit times. A challenge in pseudomorphic epitaxial base SiGe HBT devices is to scale the narrow as-grown boron profile within the SiGe alloy film during post-epitaxial processing. The challenge comes in achieving a low sheet resistance by increasing doping concentration and at the same time limiting the boron diffusion process from thermal processes and transient enhanced diffusion (TED) processes [46–52].

In a pseudomorphic $Si_{1-x}Ge_x$ film, the film is compressed in the growth plane and expanded in the growth direction, causing the boron atom diffusion to be nonrandom. Boron will diffuse easier in the direction parallel to the silicon surface and diffuse slower in the film growth direction. The diffusion of boron in the $Si_{1-x}Ge_x$ HBT alloy film can be explained in a simple diffusion model that includes the modified strain, the effect of trapping between B and Ge, the drift field due to the bandgap narrowing, and the spatially dependent intrinsic carrier concentration. The universal diffusion equation for a generic impurity present in a structure is

$$\frac{\partial C_B}{\partial t} = -\nabla \bullet J_B$$

where C_B is the boron concentration and J_B is the boron impurity flux given by

$$J_B = -D_B\varphi\left(\frac{\partial C_B}{\partial x} + \frac{C_B}{Q}\frac{\partial N_T}{\partial x} + \frac{2C_B}{Q}\frac{n\partial \ln n_i}{\partial x}\right)$$

$$Q = \sqrt{N_T^2 + 4n_i^2} = n + p$$

$$\varphi = \frac{1 + \beta(p/n_i)}{1 + \beta}$$

and n and p are the electron and hole concentrations [49,61]. The factor ϕ allows for the inclusion of the concentration dependence of both B and Ge fractions (strain) dependence of the effective diffusion coefficient, β is a parameter, and N_T and n_i are the net doping and intrinsic concentration, respectively. In the boron impurity flux equation, the first term is the conventional Fickian diffusion term, the second is a term for the built-in electric field from the dopant distribution, and the third term is associated with the built-in electric field initiated by the spatial variation of the $Si_{1-x}Ge_x$ bandgap. The diffusion of boron in a $Si_{1-x}Ge_x$ system is influenced by the different Ge fractions, the boron density gradient leading to a segregation process, and the electrical active B percentage (which is approximately half of a silicon-only film). Germanium acts as a trap for the boron atoms that can be represented as a Ge-B clustering center for a boron clusterization process, which transfers B atoms into electrically inactive charge states.

$$B^- + Ge \rightarrow GeB^-$$

The diffusion of boron can be represented as the

$$D_B = \varphi D_{BO} \exp(-E_A/kT)$$

At high Ge contents, the strain due to the presence of Ge influences the boron diffusivity. The local strain, S, due to the Ge and B concentration is addressed through the modification of the activation energy,

$$D_B = \varphi D_{BO} \exp(-E_A/kT)^* \exp(-fS/kT)$$
$$E_A = E_{A_{mod}} + fS$$

The diffusion of boron in the presence of carbon is significantly altered in the $Si_{1-x}Ge_x$ film. Carbon diffusion in regions with significant carbon concentrations can cause localized undersaturation of the silicon self-interstitials. This C rich region results in the suppression of the boron diffusion process. The diffusion of C in silicon occurs as a result of a substitutional–interstitial mechanism. Silicon self-interstitials, I, react with immobile substitutional carbon, C_S, to form mobile interstitial carbon atoms, C_I, as represented in the reaction [49,61]

$$C_S + I \Leftrightarrow C_I$$

The boron diffusivity is modified by the concentration of Si interstitials C_I, and the equilibrium concentration, C_I^*

$$D_B = \left(\frac{C_I}{C_I^*}\right) \varphi D_{BO} \exp(-E_A/kT)$$

$$\varphi = \frac{1 + \beta(p/n_i)}{1 + \beta}$$

The boron activation energy in the equation above is a function of both the boron and the carbon. The activation is represented in the $Si_{1-x-y}Ge_xC_y$ system as $E_A = 3.62 + 0.4xy$, where x is the Ge and y is the C atom atomic fractional percentage, respectively [49,61]. Carbon, because of the atom size is small, relieves the strain in the $Si_{1-x}Ge_x$ film. From the relationship

$$D_B = \varphi D_{BO} \exp(-E_A/kT)^* \exp(-fS/kT)$$

the strain, S, can be represented as

$$S = |S_{Ge} - S_C|$$
$$S_{Ge} = \delta x$$
$$S_C = R\delta y$$

where δ is the lattice mismatch between Si and Ge, R is the strain compensation ratio, S_{Ge} and S_C are strain for Ge and C, respectively. The activation per unit strain, f', is a function of the diffusivity of the strained and unstrained materials,

$$f' = -\frac{kT}{S} \ln\left(\frac{\hat{D}}{D}\right)$$

6.4 SILICON GERMANIUM ESD MEASUREMENTS

ESD sensitivity of the silicon germanium HBTs needs to be evaluated in different bipolar configurations [66–79]. SiGe HBT device measurements of interest are collector-to-emitter,

emitter-to-base, and base–collector configurations. In the following sections, experimental results of SiGe HBT devices will be shown.

6.4.1 Silicon Germanium Collector-to-Emitter ESD Stress

In a collector–emitter mode, the emitter is grounded and an ESD pulse is applied to the collector region [69,70]. As the collector-to-emitter voltage increases, both the collector–base junction voltage and the maximum electric field in the junction increases, leading to the onset of avalanche multiplication. In SiGe HBT devices, the base doping concentration is significantly higher than for the Si BJT devices. With the high-doped base region, the depletion region extends into the lower doped collector region in both the pedestal and nonpedestal SiGe HBT devices. Avalanche multiplication occurs primarily in the Si collector region, where the holes are accelerated through the silicon collector into the SiGe base region and the electrons are accelerated into the collector. As the base region potential rises, the emitter–base junction forward-biases, leading to bipolar snapback of the SiGe HBT device.

As an example measurement configuration, a transmission line pulse (TLP) system is shown (Figure 6.5). A high-current pulse-testing I–V measurement of a SiGe HBT with a 0.32-µm wide by 20-µm-long emitter structure in a floating base common-emitter mode is shown in Figure 6.6. For this measurement, the pulse has 50-ns pulse width with 3-ns rise- and fall times. With single probes, no capacitive load exists on the base electrode.

As the TLP pulse current increases, the collector-to-emitter voltage increases across the SiGe HBT device until the first trigger voltage is reached. As the collector voltage increases, avalanche multiplication increases until avalanche breakdown occurs. From this I–V characteristic, the first trigger point of avalanche (I_{t1}, V_{t1}) is the onset of snapback. With increasing collector-to-emitter voltage, the current through the device increases, rising out of the low-voltage/high-current snapback state. The inverse slope of the I–V trace after the snapback point is the dynamic on-resistance of the SiGe HBT device. As the voltage increases, the current increases linearly with the increased pulse current. At this point, the voltage increases beyond the first trigger voltage, V_{t1}. Beyond the first trigger point, the current increases until a soft transition occurs. Then, a change in the dynamic resistance

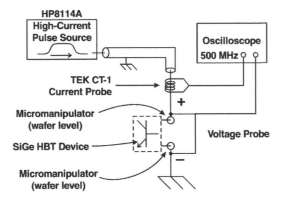

Figure 6.5   High current pulse test system

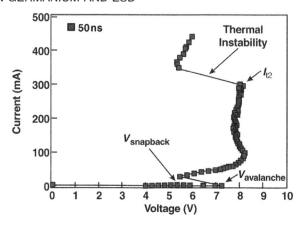

Figure 6.6   TLP *I–V* characteristic of SiGe HBT device (pulse width $\tau = 50$ ns)

occurs: the dynamic on-resistance is reduced, followed by an increasing current without a negative resistance transition or thermal instability. As the current pulse increases, the onset of thermal instability eventually occurs, where the SiGe HBT undergoes a negative resistance regime, and falls to a high-current/low-voltage state. The low-voltage/high-current state of the SiGe HBT is at a voltage value near or below the snapback voltage state. Typically, for most of the measurements taken in different configurations, the onset of thermal instability or the low voltage/current state is the point of increase in leakage current and is believed to be the point of device failure [69,70].

Thermal instability and power-to-failure in high-current pulse testing is a function of the applied pulse width as is evident from the Wunsch–Bell and Dwyer models. As shown in the Wunsch–Bell model, the critical current-to-failure and power-to-failure of a semiconductor device increases with decreasing pulse width. Figure 6.7 shows TLP *I–V* characteristics of a 0.32-μm × 20-μm SiGe HBT as a function of pulse width. For pulse widths significantly

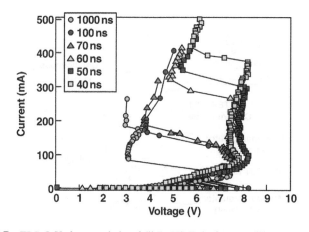

Figure 6.7   TLP *I–V* characteristic of SiGe HBT device ($\tau = 40$ ns to $\tau = 1000$ ns)

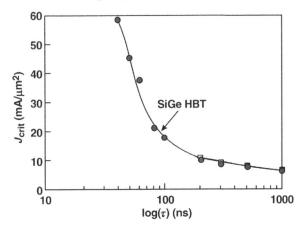

Figure 6.8   SiGe HBT critical current density, J_{crit}, versus pulse width, τ

shorter than the thermal diffusion time, the power-to-failure of a SiGe HBT will be inversely proportional to the pulse width and follow an adiabatic dependence. For pulse widths on the order of the thermal diffusion time, the critical current-to-failure will have a weaker dependence on pulse width due to the thermal diffusion of heat from the region of self-heating. As the pulse width increases above 50 ns, our measurements show that the onset of thermal instability decreases monotonically with increasing pulse width. The postthermal instability low-voltage/high-current I–V point also monotonically decreases in both current and voltage for the observed pulse-width range from 40 to 1000 ns. An additional observation is that, for very short pulse widths, a high-current state exists for the SiGe HBT prior to the onset of thermal instability whereas this is less evident in the longer pulse widths. For the dynamic on-resistance, the shorter pulse widths have a lower resistance than the longer pulse I–V measurements as the pulse width increases from 40 to 1000 ns [69,70].

From these results, the Wunsch–Bell curve can be plotted by mapping the point of the current-to-failure as a function of the pulse width. Figure 6.8 shows a linear log plot of the onset of thermal instability, J_{crit}, as a function of the pulse width for the DT-defined high-performance $0.32\,\mu m \times 20\,\mu m$ emitter SiGe HBT structure for the nine different pulse widths. The results show that J_{crit} decreases as the pulse width increases. The decrease in the onset of thermal instability and power-to-failure is consistent with the work of Wunsch–Bell and other established thermal physics failure models.

6.4.2 ESD Comparison of Silicon Germanium HBT and Silicon BJT

Silicon germanium transistors compared with silicon homojunction transistors have higher bipolar current gain characteristics, lower base transit time, lower emitter transit time, and lower intrinsic base resistance, which are advantageous for ESD protection networks. An experimental lot with two epitaxial base Si homojunction splits and the standard SiGe HBT device was constructed with identical masks in common structures. Figure 6.9 shows a TLP I–V characteristic of one of the Si homojunction BJT splits in the identical $0.32\,\mu m \times 20\,\mu m$

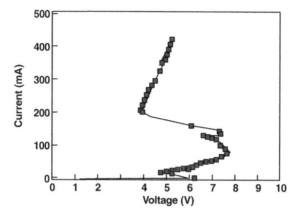

Figure 6.9   TLP *I–V* characteristic of silicon homojunction BJT ($\tau = 50$ ns)

emitter structure for a 50 ns pulse width compared with the SiGe HBT measurement. The SiGe HBT device remains in a high-current high-voltage state and does not undergo the transition into thermal instability, as is evident in the Si BJT device [69,70].

Further examination of this effect is evaluated by varying the pulse width from 40 to 1000 ns. Figure 6.10 shows the TLP *I–V* characteristics of the Si BJT homojunction. Comparing the response of the epitaxial base Si BJT homojunction *I–V* characteristics is similar to that of the SiGe HBT with one key qualitative distinction. It is evident that the Si homojunction BJT undergoes an earlier transition into thermal instability for shorter pulse widths.

Figure 6.11 shows a Wunsch–Bell J_{crit} versus pulse-width plot comparing the SiGe HBT and Si BJT device for a 0.32 μm × 20 μm transistor structure. For short pulse widths, the SiGe HBT has a higher J_{crit} than the Si BJT. For the longer pulse widths, the critical current density of SiGe HBT and Si BJT converges. Wunsch–Bell current-to-failure plots also demonstrate that the SiGe HBT ESD results are 30% superior to the Si BJTs at 100 ns. At shorter times, the difference is 200%. This comparison was also completed for other Si BJT experimental splits where, in all cases, the SiGe HBT showed superior results at shorter time constants [69,70].

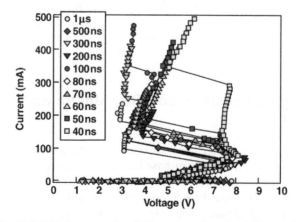

Figure 6.10   TLP *I–V* characteristic of silicon homojunction BJT ($\tau = 40$ ns to 1 μs)

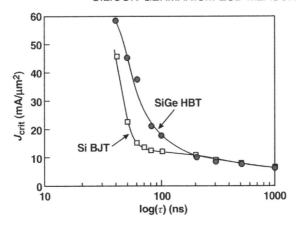

Figure 6.11 Wunsch–Bell J_{crit} versus τ plot SiGe HBT and Si BJT

The comparison of SiGe HBT and Si homojunction BJT low- and high-current characterization parameters are of considerable interest to circuit designers and device engineers who must optimize and tradeoff functional and ESD characteristics. In the Table 6.1, the electrical parameters and ESD-related parameters are shown. Key parameters are the current gain (β), the early voltage (V_A), breakdown voltages and second-breakdown values. Two key metrics of interest are the current gain–early voltage product, 'βV_A', and a new hybrid metric of the current gain–second breakdown, 'βI_{t2}' metric. The current gain–Early voltage metric provides a measure of the current drive and output impedance of a SiGe HBT. The current gain–second breakdown metric is a measure of the functional current-drive capability and the ESD robustness of the structure.

For these cases, where the SiGe-to-Si β_{SiGe}/β_{Si} ratio is 2.57 and 3.53, the SiGe HBT transistors provide both a higher βV_A and βI_{t2} compared with both epitaxial base Si homojunction npn BJT devices. The $(\beta I_{t2})_{SiGe}/(\beta I_{t2})_{Si}$ ratio can be expressed as

$$\frac{(\beta I_{t2})_{SiGe}}{(\beta I_{t2})_{Si}} = \eta \left(\Delta E_{g,Ge}(\text{graded})/kT \right) e^{\Delta E_{g,Ge}^{(0)}/kT} \left(1 - \exp^{-\Delta E_{g,Ge}(\text{graded})/kT} \right) \left\{ I_{t2_{SiGe}}/I_{t2_{Si}} \right\}$$

Table 6.1 Electrical parametrics of a SiGe HBT and Si BJT devices (e.g. two different sample wafers (a and (b))

Parameter	Si BJT (a)	Si BJT (b)	SiGe HBT
β	56.1	40.83	144
V_A (V)	41.63	53.12	103
I_{coll}@0.72 V (µA)	0.82	0.62	2.15
BV_{EBO} (V)	3.67	3.14	3.86
BV_{CEO} (V)	3.63	3.78	3.2
BV_{CES} (V)	11.16	11.1	11.26
I_{t2} (A)	0.08	0.08	0.12
P_f (W)	0.64	0.71	0.845
βV_A	2,335	2,169	14,832
βI_{t2}	4.48	3.38	16.58

From the Moll–Ross–Kroemer current relationships, the SiGe-to-Si β advantage is consistent with the correlation of improved I_{t2} with current gain, β [8,9].

A key advantage of the bandgap engineering for the SiGe base region allows for design optimization of the intrinsic base resistance, R_{bi}, and cutoff frequency, f_T. The intrinsic base resistance for a SiGe HBT device can be expressed as

$$R_{bi(SiGe)}/R_{bi(Si)} = (\beta_{SiGe}/\beta_{Si})\left\{e^{-E_g^{(0)}/kT}\left(kT/\Delta E_g(\text{grade})\right)\right\}$$

A lower zero bias intrinsic base resistance can be achieved for a given frequency f_T, allowing for higher doping concentration in the intrinsic and extrinsic base regions. By providing a higher doping concentration, a higher intrinsic temperature, $T_i(x)$, can be obtained that improves the second breakdown characteristics and provides a thermally stable base region. This is a significant advantage for ESD robustness of SiGe HBT devices. Additionally, the lower intrinsic base resistance reduces the impact of voltage-distribution drops in a SiGe npn stripe and lower resistance in a diode-configured base–collector npn transistor compared with a Si BJT so that they can be more suitable as ESD protection network elements. Successful usage of a SiGe HBT device in diode-like operation is also important to provide a low extrinsic base resistance and low collector resistance. The low extrinsic base resistance is minimized by design and device scaling whereas the low collector resistance is achieved with high subcollector doping concentrations and reach-through implants.

6.4.3 SiGe HBT Electrothermal Human Body Model (HBM) Simulation of Collector–Emitter Stress

Preliminary electrothermal simulation of the SiGe HBT device during an HBM ESD event was completed using a two-dimensional finite-element cross-section of the SiGe HBT device. A full SiGe semiconductor process was defined using the TSUPREME semiconductor process tool. Finite-element electrothermal simulation was completed using the FIELDAY III device simulator. The FIELDAY formulation is based on the three moments of the Boltzmann transport equations, conservation of charge equations, and lattice-energy conservation relationships. The constitutive equations for electron- and hole-current densities account for the current flow caused by thermal gradients. The constitutive equation for lattice energy flux represents the Fourier law of heat conduction. The lattice energy consists of the Joule heating and energy exchange with carriers and the lattice.

To simulate an HBM pulse, the Van Roozendaal double-exponential current waveform is used in the FIELDAY simulator [69,70]. Figure 6.12 shows the temperature field in the SiGe HBT device where the cross-section was sliced through the center of the emitter structure, using a reflecting thermal boundary condition ($dT/dy = 0$) with thermal contacts and proper thermal boundary conditions established at other interfaces. In the study, the simulation was completed with the test in common-emitter configuration. The emitter window is in the upper right corner of the device. In the common-emitter mode, the peak temperature can be observed in the collector region. In SiGe HBT devices, the base doping concentration is significantly higher than the collector region. As a result, the Joule heating in the collector region plays a significant role in the transient temperature field $T(x,y,t)$ in the SiGe HBT device. The darkest contour is the region of peak heating during the collector-to-emitter region.

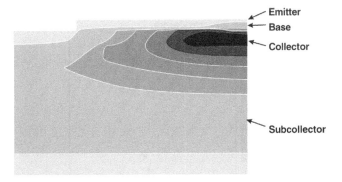

Emitter

Base

Collector

Subcollector

Figure 6.12 Transient electrothermal device simulation of SiGe HBT (HBM pulse)

6.5 SILICON GERMANIUM CARBON COLLECTOR–EMITTER ESD MEASUREMENTS

A SiGe system with carbon incorporation in the base region of a SiGe HBT device leads to influences in both the vertical and lateral transport of the boron in the intrinsic and extrinsic base region [45–65]. The reduction of the boron diffusion in the SiGe base regions allows for a narrower base width for a given base doping concentration. This provides a lower standard deviation of the base width and the base sheet resistance. The first case provides a higher unity current gain cutoff frequency, f_T, and the second case allows for a higher unity power gain cutoff frequency, f_{MAX}. Additionally, the retardation of the boron diffusion allows for the design re-optimization to increase the base doping concentration. The lowering of the base sheet resistance allows for a higher f_{MAX}. With the addition of carbon in the base region, the boron doping concentration can be increased to provide a lower sheet resistance. For ESD robustness, there are two advantages. A higher base doping concentration will also improve the thermal stability of the SiGe HBT device as a result of the increased intrinsic temperature, T_i, in the base region. Additionally, the higher boron concentration also will reduce the voltage drops in the base region leading to an improved current uniformity in the SiGe HBT during high current and ESD events [77].

In silicon bipolar transistors, statistical process variations lead to variations of the power-to-failure of transistor elements. Assuming a normal distribution, Pierce and Mason assumed that the power-to-failure probability distribution function has a form [66],

$$f_{P_f}(P_f) = \frac{1}{\sqrt{2\pi}S_p} \exp\left\{-1/2\left[\frac{P_f - \langle P_f \rangle}{S_p}\right]^2\right\}$$

where the power-to-failure, P_f, is the random variable and a mean power-to-failure $\langle P_f \rangle$, with standard deviation, S_p. The standard deviation can be expressed as

$$S_p = \left[\frac{1}{N-1}\sum_{i=1}^{N}\{P_{f_i} - \langle P_f \rangle\}^2\right]^{1/2}$$

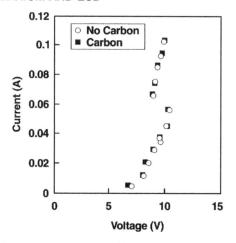

Figure 6.13 Silicon germanium versus silicon germanium carbon TLP characteristic (collector–emitter configuration)

Pierce and Mason showed that the probability of failure occurs when the ESD pulse power exceeds the device power-to-failure. This can be expressed as a cumulative distribution function [66].

$$\Pr\{P_f > P_f'\} = \int_{P_f'}^{\infty} f_{P_f}(P_f, \langle P_C \rangle / \alpha, \langle P_C \rangle / \alpha \lambda) \mathrm{d}P_f$$

In the Pierce–Mason model, the assumption in the analysis is that the parameter initiating the statistical variation in the failure sensitivity was the emitter junction depth. In our results, the power-to-failure variation will also be a function of the base width, doping distribution, sheet resistance and the thickness of the pseudomorphic $Si_{1-x}Ge_x$ film. Without carbon, the base width variation is dependent on the boron diffusion physics and implant statistical variations. The variations of the base width will be reflected in the collector-to-emitter breakdown voltage and the base sheet resistance [77].

Figure 6.13 shows a typical common-emitter TLP $I–V$ measurements comparing a $0.44\,\mu m \times 3.2\,\mu m$ SiGe and SiGeC HBT 45 GHz npn structure [77]. In this measurement, the pulse current is applied to the collector with the emitter grounded and the base floating. As the voltage increases in the collector region, avalanche multiplication occurs. In our results, current conduction is evident at the collector-to-emitter breakdown voltage BV_{CEO}. Snapback to a low current state is not evident after the breakdown voltage. This high voltage/high current regime is followed by a negative resistance state transition and ultimately an increase in the leakage current and device failure.

Figure 6.14 and 6.15 shows the comparison of the voltage-to-failure of a SiGeC and SiGe HBT devices [77]. A key conclusion is the SiGeC ESD results clearly shows a significantly tighter ESD distribution in all experimental splits performed. Parametric data shows that the base resistance control is superior when carbon is incorporated into the base of the SiGe npn HBT device, as anticipated because of improved control over the boron TED as well as improved Schuppen factor. This is evident from the d.c. electrical distribution of the base pinch resistance monitors. As anticipated, the SiGeC HBT provides improved ESD control as a result of improved base resistance distribution [77].

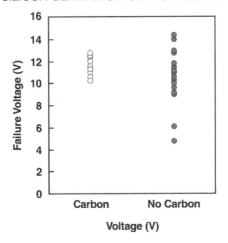

Figure 6.14   Voltage-to-failure of silicon germanium HBT with and without carbon

Figure 6.16 shows the comparison of the failure current histogram of the $Si_{1-x}Ge_x$ with and without carbon in the base region [77]. As was evident from the failure voltage histogram, the device with carbon is better controlled providing an improved standard deviation.

Experimental results show that the relationship for the power-of- failure follows a normal distribution. Experimental measurements were taken of the $Si_{1-x}Ge_x$ and $Si_{1-x-y}Ge_xC_y$ HBT devices as a function of the emitter width in a common-emitter configuration. Figures 6.17 and 6.18 show the TLP failure current and failure power as a function of the emitter width, respectively.

Our results show that for the $Si_{1-x}Ge_x$ and $Si_{1-x-y}Ge_xC_y$ HBT devices the ESD results increase linearly with increasing emitter width. At an emitter width of 0.35–0.45 µm, the TLP current-to-failure and power-to-failure begin to saturate [77].

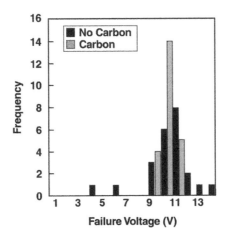

Figure 6.15   Histogram of silicon germanium voltage-to-failure with and without carbon incorporation

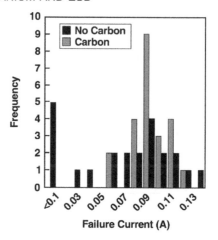

Figure 6.16   Histogram of silicon germanium current-to-failure with and without carbon incorporation

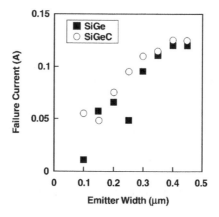

Figure 6.17   Silicon germanium and silicon germanium carbon HBT TLP current-to-failure as a function of emitter width

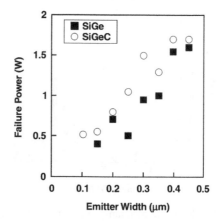

Figure 6.18   Silicon germanium and silicon germanium carbon HBT TLP power-to-failure as a function of emitter width

6.6 SILICON GERMANIUM TRANSISTOR EMITTER–BASE DESIGN

Emitter–base design of RF homojunction and heterojunction bipolar transistors has a significant influence on both the RF device performance and the ESD sensitivity. For RF applications, emitter–base design is important in order to achieve future RF performance objectives. An important device to achieve future RF objectives is the HBT. Heterojunction base–emitter design, bandgap engineering, and technology scaling, each will play a key role in the ability to achieve faster devices for the wired and wireless markets.

From an ESD perspective, the emitter–base region is typically the most ESD-sensitive regions of the bipolar transistor element [72,73]. As performance objectives increase, the size of the emitter region will scale to smaller physical dimensions. This is a concern for the ESD sensitivity of RF receiver circuits. The emitter and emitter–base region in RF receivers are sensitive to ESD failure of the emitter–base junction in both forward- and reverse-bias ESD stress.

Structural design, semiconductor process, and physical design layout influence the ESD robustness of the bipolar transistors. Structural and geometrical design features of the base region that influence both the ESD robustness and the HBT are as follows:

- epitaxial or nonepitaxial base region;
- raised extrinsic base or nonraised extrinsic base;
- placement of the extrinsic base over STI or LOCOS isolation;
- thickness of the base film;
- silicide or nonsilicide extrinsic base region;
- single contact or bar contact base region.

Structural design features of the emitter region that influence the RF performance and the ESD robustness of a homojunction and heterojunction bipolar transistor include the following:

- polysilicon or nonpolysilicon emitter film;
- polysilicon emitter film thickness;
- polysilicon film grain size;
- silicide or nonsilicide emitter structure;
- single contact or bar contact on the emitter;
- contact penetration into the emitter film.

The semiconductor process variables that influence the RF performance and the ESD robustness of the emitter–base region include the following:

- emitter region polysilicon grain structure;
- emitter region doping concentration;

- base region silicon concentration;

- base region germanium concentration and profile design;

- base region carbon concentration and profile placement.

Transistor layout and design of the emitter–base region have an influence on the RF performance and ESD robustness of a homojunction bipolar transistor or a heterojunction bipolar transistor. The transistor design layout features of interest can be as follows:

- SA or NSA emitter base design;

- emitter area;

- emitter aspect ratio (e.g., width to length ratio);

- emitter–base space;

- wrap-around or nonwrap around emitter base design layout;

- ordering: EBC, BEC, or ECB;

- single emitter or multiemitter;

- higher order design layouts: CBEBC, CEBEBEC, among others.

All the aforementioned features from a structural, semiconductor process, and design layout have an influence on the performance aspects as well as the ESD robustness of homojunctions and heterojunction bipolar transistors.

The evaluation of process variations and device design spacings on ESD robustness is evaluated for both positive and negative stress conditions as a function of the salicide location, emitter–base spacing, and collector opening. The evaluation of process variations and device design spacings on ESD robustness is evaluated for both positive and negative stress conditions as a function of the salicide location, emitter–base spacing, and collector opening.

6.6.1 Epitaxial-Base Heterojunction Bipolar Transistor (HBT) Emitter–Base Design

For forming a HBT, the epitaxial base region is formed on the wafer surface to allow for bandgap engineering of the base region. As an example of a heterojuction bipolar transistor, the structural aspects of an epitaxial base silicon germanium transistor will be discussed.

Figure 6.19 shows an expanded view of the emitter–base region. Over the single crystal silicon, a pseudomorphic single crystal SiGe film is formed. A window is opened on the wafer surface to allow formation of the epitaxial base region; the mask opening extends over a single crystal silicon region and the isolation area. Over the single crystal silicon, the SiGe film remains single crystal and forms the 'intrinsic' region of the transistor element. Over the isolation region, the silicon germanium film is amorphous forming a

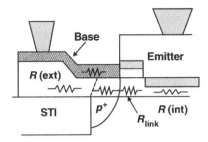

Figure 6.19   Emitter–base structure for a self-aligned epitaxial base transistor

polysilicon germanium region; this region is the 'extrinsic base' region used for electrical connection to the HBT base. In this fashion, a contact can be placed on the base region over the isolation. At the transition between the single crystal silicon wafer and the isolation, a facet occurs forming a transition in the SiGe film leading to an amorphous SiGe-to-SiGe single crystal interface. The direction of the growth of the facet, whether inward or outward toward the emitter is a function of the geometrical conditions at the single crystal-isolation transition. From an ESD perspective, the crystalline nature of the intrinsic and extrinsic base region has an influence on the electrical resistance and the thermal resistance of the base region. In the extrinsic base region, the nature of the polycrystalline film thickness, doping concentration, and grain structure will influence the ESD stability of the HBT [72,73].

In the extrinsic base region, a silicide can be formed on the polysilicon germanium surface. The silicide can be a refractory metal such as titanium, cobalt, tantalum, or tungsten. In the case of titanium, a $TiSi_x$ salicide is formed on the extrinsic-base region from the window opening edge (e.g., contact region) to the edge of the emitter polysilicon structure. A doped p^+ dopant is implanted into the extrinsic base region to reduce the base series resistance.

A SA SiGe HBT device is a transistor where this extrinsic base implant is aligned to the edge of the emitter window opening using a disposable spacer forming a fixed extrinsic base-to-emitter space. The emitter shapes forms a SA silicon germanium HBT,. After the emitter shape is etched and defined, a conformal film is deposited over the emitter region; this film is then etched to form a 'spacer' on the edge of the emitter structure. The formation of the spacer on the edge allows the ability to define the space between the emitter and the base implants. A p-type dopant is then implanted into the base region. In the SA HBT device, the spacer width defines the location of the space between the emitter and base implants. The implant is activated by a thermal process step, which can lead to the out-diffusion of the p^+ implant both vertically and laterally. In the region where there is no isolation, the implant diffuses below the epitaxial film depth and into the silicon wafer. In the extrinsic base region, there is no vertical diffusion over the STI. Silicide is deposited on the base structure to provide a low base series resistance. The silicide is formed from the edge of the polysilicon germanium film edge to the edge of the emitter structure.

For a NSA SiGe HBT device, the extrinsic base implant is aligned to the edge of the polysilicon emitter shape not to the edge of the spacer. In the case of a NSA structure, the emitter mask can extend beyond the region of the intrinsic device window. The polysilicon emitter shape can overhang the edges of the emitter insulators pushing out the extrinsic base implant. In this structure, the spacing between the extrinsic base implant and the emitter window is not a function of a spacer, but a result of the width of the emitter structure. As a result, in a NSA SiGe HBT device, the spacing between the extrinsic base p^+ implant and the emitter n^+ implant can be separated independently [72,73].

6.6.2 Emitter–Base Design RF Frequency Performance Metrics

Emitter–base design influences two key figures of the merit of a transistor. The two key figures of the merit for a bipolar transistor device are the unity current gain cutoff frequency, f_T, and unity power gain cutoff frequency, f_{MAX}. The unity current gain cutoff frequency f_T is inversely related to the emitter-to-collector transit time, τ_{EC}.

$$\frac{1}{2\pi f_T} = \tau_{EC} = \tau_E + \tau_B + \tau_{CSL} + \tau_C$$

where

$$\tau_E = \frac{C_{eb} + C_{bc}}{g_m}$$

$$\tau_B = \frac{W_B^2}{K D_B}$$

$$\tau_{CSL} = r_c(C_{csx} + C_{bc})$$

$$\tau_C = \frac{x_C}{v_{sat} L}$$

The emitter transit time is a function of the emitter–base capacitance, base–collector capacitance, and transconductance. The base transit time is a function of the base width, and diffusion coefficient and base grading factor, K. The collector terms consists of the collector capacitance, collector resistance, and velocity saturation term.

Minimizing the collector–base capacitance, C_{cb}, collector–substrate, C_{csx}, and base resistance, R_b improves f_{MAX}.

$$f_{MAX} = \sqrt{\frac{f_T}{8\pi R_b C_{bc}}}$$

6.6.3 SiGe HBT Emitter–Base Resistance Model

The base resistance is defined as an intrinsic base resistance and an extrinsic base resistance. The intrinsic base resistance is associated with the area under the n^+ emitter diffusion. The extrinsic base resistance consists of all base resistance components from the intrinsic base to the base contact. The extrinsic base resistance comprises of the amorphous SiGe region from the contact to the STI edge. At the STI–silicon transition, the SiGe film transitions from amorphous to single crystal SiGe. The first term is the resistance of the poly-SiGe. The second term is the resistance of the SiGe film from the facet to the edge of the extrinsic base implant. The third component of the extrinsic base implant is the single crystal SiGe base region between the emitter diffusion edge to the extrinsic base implant edge, known as the 'link' resistance [72,73].

$$R_{b,int} = \frac{1}{3} \frac{W_E}{L_E} \rho_B$$

$$R_{b,ext} = R_{polySiGe} + R_{SiGe} + R_{Link}$$

$$R_{Link} = \frac{L_{Link}}{L_E} \rho_{Link}$$

The first two regions of the external resistances are a function of the salicide location relative to the facet transition. The salicide is extends from the contact to the emitter polysilicon edge.

6.6.4 SiGe HBT Emitter–Base Design and Silicide Placement

In the emitter–base design, the placement of silicide influences both the emitter and the base resistance. The silicide influences the current density and the current distribution in the emitter and the extrinsic base structure. In the figure, the silicide is shown to extend to the full extent of the base region. With the silicide forming on the extrinsic polysilicon germanium film, the effective base resistance can be reduced. The ESD robustness of the transistor structure will be influenced by the silicide. The silicide influences the current density and the current distribution within the base structure. Optimization of the emitter–base region requires addressing the cutoff frequencies f_T and f_{MAX}, base resistance R_b, emitter–base breakdown voltage BV_{EBO}, and emitter–base leakage current IL_{EBO}. For SA SiGe HBT devices, there is an optimum spacer thickness that optimizes the emitter–base region [72,73]. As the spacer width decreases, the cutoff frequencies decrease because of to an increase of the emitter–base and collector–base capacitances. In addition the BV_{EBO} decreases and the IL_{EBO} increases. As the spacer width decreases, it has been shown that the leakage current transitions from a temperature sensitive temperature insensitive. This indicates transition from thermal leakage mechanism (e.g., Shockley Read Hall, Frenkel–Poole) avalanche multiplication to tunneling mechanisms (e.g., trap-to-band and band-to-band tunneling).

Refractory metal films are used in semiconductors in conjunction with silicon to form silicide films. Silicide are used in semiconductor technology to form low resistance regions where resistance is an impediment to design or chip performance. Silicide formation in the emitter base region of a selective epitaxial bipolar transistor influences the ESD robustness of the bipolar transistor. In the emitter–base design, the placement of silicide influences the base resistance. The silicide influences the current density and the current distribution in the extrinsic base structure.

Salicide formation of the extrinsic base region has no influence on the intrinsic region of the bipolar transistor. In the external base region, the link resistance is also not influenced by the salicide formation. The first two components of the external resistance expression are a function of the salicide location relative to the facet transition. The salicide extends from the contact to the emitter polysilicon edge. Hence, the salicide is formed over both the polysilicon and single crystal silicon regions of the external base.

The ESD robustness and transistor performance will be influenced by the silicide. The silicide influences the current density and the current distribution within the base structure. The introduction of the silicide has two effects. First, the current density is further increased as the current flows through the low- resistance thin silicide film leading to higher current densities. This introduces Joule heating in the base region. Secondly, the reduction in the lateral ballasting reduces the effective length of the bipolar transistor.

The silicide parameters that will influence the ESD robustness consist of the following:

- silicide material;

- silicide film stability;

- silicide phase state;

- silicide film thickness;

- silicide thickness uniformity;

- silicide temperature coefficient of resistance (TCR);

- silicide extent through the extrinsic base region (e.g., contacts to edge);

- silicide extension in the width of the base (e.g., continuous film or segmented).

The choice of the silicide material has significant effect on the performance as well as the ESD robustness of a structure [66]. The material choice influences the silicide film resistivity, the silicide film stability, the silicide phase states, the silicide film stability, the silicide TCR, the consumption of the underlying film (e.g., silicon or silicon germanium), and the required film thickness. The key reasons for the silicides being are very important issue for the ESD applications are its properties of low resistivity, film thickness, and the regions where silicides are formed. First, the low resistivity leads to an elimination of the natural ballasting resistance within the bipolar transistor base region, leading to current constriction and nonuniform current distribution; Moreover, the low resistivity lowers the electrical and thermal stability of devices. Second, the thin film thickness leads to high current density. High current densities lead to significant Joule heating in the silicide films. As the temperature in the silicide increases, the silicides can penetrate the metallurgical junction leading to failure, or undergo thermal instability and phase transformation. The third issue is that silicides are typically formed in the regions where device performance and low resistance are desired. These same regions, in many cases are also the regions of devices which are ESD sensitive. For example, MOSFET source drain regions use $TiSi_2$ to reduce the MOSFET source/drain region. In a bipolar transistor, silicides are used in the emitter and base region. In the bipolar transistor, the emitter–base junction region is the most ESD sensitive region of the transistor. Moreover, the design of how and where the silicide is formed can lead to nonuniform current distributions and regions of peak self-heating. With self-heating, the nature of the silicide films are modified. Hence, the nature of the film, and its thermal stability is important to provide ESD robust devices.

Titanium silicide has wide acceptance because of its characteristics of low resistivity, good thermal stability and self-alignment with silicon [66]. Titanium silicide is also very important to ESD applications because of its usage on MOSFET, diodes, resistors, and bipolar transistors [66]. For selective epitaxy deposited bipolar transistors, it is important that titanium silicide forms on both the single crystal silicon and polysilicon. The formation of $TiSi_2$ requires two silicon atoms for each deposited titanium atom. $TiSi_2$ is a polymorphic material and may exist in two states, a orthorhombic base-centered (C49) state with 12 atoms per unit cell when the formation temperature ranges from 550 to 700 °C or a thermo-dynamically favored orthorhombic face-center (C54) state with 24 atoms per unit cell when formation temperature is above 750 °C. The phase transformation kinetics that influence the state where the $TiSi_2$ resides is a function of the surface energy, film thickness, and microstructure. The phase transformation kinetics to the C54 state is limited by a low driving force from the small transformation enthalpy and high activation energy (e.g., greater than 5.0 eV). The meta-stable C49 phase has a higher resistance (on the order of 60–90 $\mu\Omega$ cm) than the thermodynamically favored C54 phase (12–20 $\mu\Omega$ cm). The phase transition is limited by the number of C54 nuclei that exist within the silicide area. This effect is known as 'fineline effect' where the phase transition is limited by the number of nuclei. As the line width decreases, this continues to become a problem. With technology scaling, the junction depths are scaled, requiring a scaling of the silicide film to avoid refractory metal atoms near the metallurgical junction. It is known that initially in processing the C49 phase of $TiSi_2$ forms first and its the transition to the C54 phase requires a transformation process. The transformation from the C49 to C54 phase is a function of the surface energy, film thickness, and microstructure [66]. The surface energy can be varied by adding impurities to the $TiSi_2$ or varying the substrate wafer materials. As thickness of the film decreases, the transformation from the C49 to C54 phase is more difficult with a smaller process window. If the time of formation is long, the $TiSi_2$ film undergoes undesirable agglomeration. As the silicide is formed, it transits gradually from an initial high resistance to a low resistance where at times only a fraction of the film undergoes the transformation,

$$\zeta = \frac{\rho_O - \rho(t)}{\rho_O - \rho_f}$$

The process of transition is similar to nucleation and growth mechanisms of incubation, induction, rapid growth, and then slow completion. For thin films this can be explained by the Johnson–Mehl–Avrami equation,

$$\ln(1 - \zeta) = \pi\delta \int_0^t T^2 N(t - \tau)^2 \, d\tau$$

where the N is the nucleation rate, T the growth rate, t the time of the process, and τ the induction period [66]. From this relationship, by knowing the process time, the percentage of the transformed film can be calculated and from this knowledge, the film $TiSi_2$ resistivity can be estimated.

In a $TiSi_2$ film, a seed of the C54 phase must be present in order to initiate the nucleation process. As the line width is decreased, the probability of a C54 phase crystal structure decreases. Experimental results show that the C54 grain size is influenced by the diffusion area on which the salicide is formed. As RF bipolar transistors and RF MOSFETs are scaled, the ability to undergo the low resistance state is less probable, with a decreasing process window between the C49 phase and agglomeration. Additionally, it was found that the transition from the C49 to C54 phase is also more difficult in n^+ diffusions relative to a p^+ diffusion. In the early 1990s this was believed to be due to Arsenic atoms in the n^+ diffusion influenced the transformation. In mid-1990s it was shown that the molybdenum atom from the implanters assisted the transition from the C49 to C54 phase.

For ESD applications, silicide block masks are used on diffusions to prevent $TiSi_2$ formation in some regions of the design. Note that the reduction of the area can influence the transformation of a region to the low resistivity C54 state.

In the development of the film, if the heating process continues, a rapid transition can occur between the transformation and the agglomeration process. After periods of heating, the $TiSi_2$ C54 phase microstructure at the grain boundaries become thin and begin to physically separate. As the heating continues, the interior of the $TiSi_2$ C54 atoms separate into independent regions and lead to increased series resistance. During an ESD event, the self-heating in the salicide film in the extrinsic base region can lead to agglomeration. The C54-phase microstructure separation can lead to nonuniform current distribution and instability [66]. As the ESD current increases, the likelihood of agglomeration in the base region also increases. The resistance change can manifest itself as a latent defect or failure. The bipolar transistor base resistance change can lead to changes in the bipolar transistor RF parameters; these can include the bipolar transistor unity current gain cutoff frequency, f_T, unity power gain cutoff frequency, f_{MAX}, and S-parameters.

In the formation of the STI, the objective is to achieve a planar interface between the insulator regions and the silicon areas. Etch and masking processes, and salicide formation volumetric changes, however, cause a nonplanar intersection of the isolation and silicon areas. In formation of STI regions, they are exposed during the etch process, leading to nonplanar STI edges where the silicon region extends above the isolation edges; this is known as the STI pull-down effect. Geiss showed that the nonplanar region at the STI-silicon edge changes the direction of the growth of the facet. Additionally, the nonplanar topography leads to variations of the silicide thickness near the emitter mask edge. As a result, the uniformity of the silicide in this region can be uniform or separated; with titanium silicide, this can prevent the transformation from the high resistance C49 state to the low resistance C54 state.

Various approaches have been taken to initiate the full transition of the C49 to C54 state, such as rapid thermal annealing (RTA), doing with antimony (Sb), and germanium preamorphization implants (PAI). It was shown by Mann *et al.* that molybdenum (Mo) implanted prior to titanium deposition can eliminate the fineline structure issue that occurs in the titanium disilicide [66]. The molybdenum implant depth as to be consumed in the silicon required to form the titanium disilicide. According to Mann *et al.*, the molybdenum enhances the nucleation of C54 in the C49 matrix by increasing the number of active grain triple junctions in the C49 where the C54 nucleation occurs.

The molybdenum segregates into the C49 grain triple point, lowering the nucleation energy for C54 phase transformation [66].

The importance of the Mo-implanted $TiSi_2$ salicide is two-folded. First, the ability to eliminate the fine-line effect from the titanium disilicide allows extension of titanium disilicide into smaller line-widths in advanced technologies. The second issue is that the Mo changes the physical stability of the titanium disilicide, leading to a different sensitivity to self-heating during an ESD event. During an ESD event, even prior to thermal runaway, the temperatures in the silicide films can exceed the activation energies of the titanium disilicide film and can lead to shifts in the morphology and film stability [66].

Although titanium silicide is of significant importance in the semiconductor industry from 1.0 to 0.25 μm technology because of the sheet resistance line-width dependence, cobalt silicide is used today in advanced technologies [66]. Titanium silicide had wide acceptance because of its characteristics of low resistivity, good thermal stability, and self-alignment with silicon but the problem of the transformation from the high resistivity orthorhombic base-centered (C49) state to the thermodynamically favored orthorhombic face-center (C54) state at small line-width used in advanced technologies is limited. This is critical in bipolar structures and MOSFET gates whose physical polysilicon gate width is below 1 μm. Cobalt salicides vary from the formation of $CoSi_x$ to $CoSi_2$. Cobalt salicide formation of $CoSi_2$ undergoes transitions from [66].

$$Co \rightarrow Co_2Si \rightarrow CoSi \rightarrow CoSi_2$$

Using cobalt salicide, the physical dimensions of devices can be scaled. It is anticipated that as the linewidth becomes small compared to the cobalt salicide grain size, nucleation of $CoSi_2$ at line edges may influence the rate of transformation.

With the low resistivity achieved in the cobalt salicide film, structures with salicide films will be sensitive to ESD events [66]. Cobalt salicide can be an advantage or disadvantage for ESD elements depending on the semiconductor device. A disadvantage of cobalt salicide($CoSi_2$) is that its melting temperature is 1326 °C compared to the $TiSi_2$ melting temperature of 1500 °C. Hence, ESD elements or circuit elements whose current stressing leads to temperature near the melting temperature is more likely to fail at a lower critical current with cobalt salicide. Additionally, RF bipolar transistor base regions whose local temperatures initiate agglomeration will also lead to salicide-induced degradation mechanism. Cobalt stability and spiking has been demonstrated to decrease with germanium PAI prior to the cobalt deposition. With the introduction of cobalt salicide, another concern is leakage mechanisms. Early leakage measurements by Goto pointed out that the cobalt salicide leakage problem was driven by an area effect as opposed to perimeter effects [66]. Goto claimed that cobalt salicide leakage issues occurred at the phase transition temperature from Co_2Si to $CoSi$ as a result of $CoSi$ spikes through the metallurgical junction area. It is well known that leakage current is a concern of the stress-induced $Co_2Si/CoSi/Si$ triple points leading to area spiking [66]. But RF bipolar transistors place the polysilicon germanium film over the STI structure; in this region of the extrinsic base, there is no concern on leakage current because of the lack of a metallurgical junction.

6.6.5 Self-Aligned (SA) Emitter–Base Design

For the SA HBT device, as the emitter polysilicon width increases, the salicide is moved away from the emitter diffusion area while the extrinsic base implant is fixed. The extrinsic base implant is aligned to the emitter window. For RF device performance, the link resistance region (e.g., space between the extrinsic base implant and the emitter window) must scale with the emitter junction scaling. Hence, to continue to achieve improved RF device performance, the space must be reduced. Scaling of the spacing between the extrinsic base implant and the emitter diffusion can be achieved by reduction of the spacer (e.g., which defines the implanted edge of the extrinsic base implant). Additionally, the self-alignment of the emitter and the extrinsic base improves the statistical control. This is important in achieving performance objectives. This is distinct from the NSA transistor where the space between the extrinsic base implant and the emitter junction can be varied and is a function of NSA masking steps. As a result, the RF performance and statistical control are superior in the SA HBT.

From an ESD perspective, the SA bipolar transistor has a fixed spacing between the extrinsic base implant and the emitter n^+ diffusion; this has both advantages and disadvantages. One of the advantages of the SA transistor is low link resistance. The low link resistance allows for a low resistance in a forward bias mode of operation. In a diodic mode of operation, the p^+ extrinsic base implant, p^- link resistance region and n^+ emitter form a $p^+/p^-/n^+$ diode structure. Hence, using the SA transistor in a forward-bias mode of operation will allow for a graded low anode resistance. Second, this is a well controlled region, hence will have good statistical control. Third, this region will be scaled in future devices. In a reverse breakdown mode of operation, the $p^+/p^-/n^+$ region in the HBT is well controlled. The disadvantage is that the failure will most likely occur at a lower BV_{EBO} breakdown voltage as the transistor is scaled [72,73].

In this SA design, the salicide can be spatially separated from the edge of the extrinsic base implant. This feature allows for the salicide edge to be separated from the emitter–base metallurgical junction region. As the current flows through the salicide, the current density increases leading to high current density. The self-heating in the film can increase leading to high temperature near the metallurgical junction. The truncation of the salicide film also allows for the re-distribution of the current into the extrinsic polysilicon base regions as it in approaches the metallurgic junction; this allows for a lower current density and lowers the self-heating in the link resistance region. From an ESD perspective, the ability to move the salicide region provides the ability to move the region of peak self-heating away from the metallurgical junction.

Figure 6.20 shows HBM ESD results of a 0.44 μm × 10 μm SA SiGe HBT device where the salicide-to-emitter space was decreased below nominal spacing while the extrinsic base implant was kept fixed (in this case, the extrinsic base implant is self-aligned to the emitter

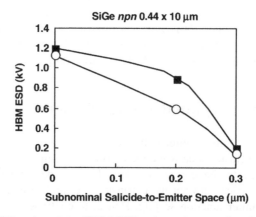

Figure 6.20 SiGe HBT emitter–base HBM ESD robustness as a function of salicide-to-emitter spacing

window). This can be achieved using the natural masks of the SA transistor, or by the addition of a silicide block mask over the extrinsic base region [72,73].

For a SA SiGe HBT device, it was found that the base resistance was modulated by removal of the salicide film leading to an increase in the positive stress base–emitter HBM ESD results. With removal of the salicide film, current must flow through the polysilicon SiGe film instead of the salicide film itself. In a SA SiGe HBT, the increase in the size of the emitter polysilicon increases the base series resistance. The link resistance remains constant while the other two extrinsic terms are modified depending on whether the salicide extends to the single-crystal Si region or over the isolation (e.g., the facet transition region). As the silicide is pushed out beyond the facet, the base resistance will increase significantly exposing the amorphous SiGe film.

RF measurements were also taken to evaluate the implications of the salicide-to-emitter spacing. From the unity current gain cutoff frequency–collector current plot (f_T versus I_C plot), unity power gain cutoff frequency–collector current plot (f_{MAX} versus I_C) plot RF measurements, and the peak f_{MAX} versus I_C characteristics show that the peak f_{MAX} decreases with increasing spacing of the salicide from the silicon emitter. This effect can be demonstrated by increasing the polysilicon emitter width or by using a salicide block mask.

Using a salicide block mask level, RF measurements show an insignificant change in the peak f_T when the salicide is removed from within 0.4 μm of the emitter polysilicon edge. With the increase of the base series resistance, the peak f_{MAX} decreases from 45 to 42 GHz. This can be expressed as

$$\frac{\partial f_{MAX}}{\partial R_b} = -\frac{f_{MAX}}{2R_b}$$

This effect was studied for two different collector openings. Figure 6.21 shows the HBM ESD results as a function of two collector spacings where the higher ESD results are with a larger collector opening. For the large opening, the HBM results increase to 1.2 kV HBM levels. As the spacing was decreased, the HBM ESD results rapidly decreased. As the salicide approached the emitter, the HBM ESD results decreased to 200 V HBM levels. In the case of the smaller collector opening, the HBM failure occurred at 600 V HBM. In this specific design, the HBM failure levels decreased to 200 V HBM. In both cases, the

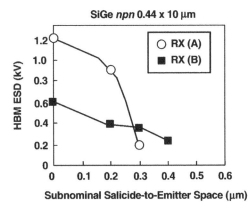

Figure 6.21   SiGe HBT emitter–base HBM ESD robustness for different collector openings

experimental data indicates that as the salicide approaches the emitter–base junction, the HBM ESD results decrease; this also indicates that as the SA transistor is scaled, salicide block masks may be valuable in preserving the ESD robustness. Additionally, these HBM results are indicators of the potential geometrical and design scaling issues that will occur in epitaxial HBTs [72,73].

In conclusion, in the SA design, the forward-bias ESD results improve with larger spacing between the emitter polysilicon edge and the emitter window. In the reverse characteristic, no ESD variation has been observed as the extrinsic-base implant position does not vary relative to the emitter. As the base link resistance is decreased, the forward-bias ESD result improves with the better f_{MAX} but the reverse-bias ESD results degrades. These results help improve our understanding of the optimization of the emitter–base junction and the role of the link resistance, Si/SiGe extrinsic base, salicide, and emitter base profiles. As devices are scaled, new structures design and new materials may influence both the transistor performance and the ESD sensitivity.

6.6.6 Non-self aligned (NSA) Emitter–Base Design

For a NSA transistor, the space between the p^+ extrinsic base implant and the n^+ emitter implant dopants can be modified by the change in the emitter mask edge. In the emitter design, the first shape is formed to define the emitter area. Insulator films are placed to avoid electrical shorting between the emitter and the base regions. A small 'emitter window' is formed in the emitter region, opening an area in the insulator stack. A polysilicon film (e.g., n^+ doped or undoped) is deposited to form the emitter region. This emitter polysilicon conformal film fills the emitter window and extends over the insulator stack. An n^+ implant is performed in the emitter polysilicon to allow its usage as a dopant source. The edges of the emitter polysilicon film and the stack is eliminated to define an emitter region. After a thermal process step, the n^+ dopants diffuse out of the emitter polysilicon region, forming the emitter. Over the insulator region, the emitter polysilicon overhangs on its edges. The emitter polysilicon film width can be increased with a fixed emitter area. In this case, as the polysilicon extends outward, the space between the metallurgical junction (formed between the n^+ emitter dopants and the p^- base region) and the emitter polysilicon edge is increased. Given that the extrinsic base $p^+ +$ implant is done after the emitter formation, the emitter polysilicon serves as a mask. As the polysilicon region extension is increased, the spacing between the p^+ extrinsic base implant and the emitter–base metallurgical junction increases. In this fashion, a $p^+/p^-/n^+$ diode region is formed at the base–emitter junction. As the spacing increases, the region of p^- dopant increases (e.g., link resistance area); this serves three purposes from an ESD perspective. First, as the p^- region increases, the base series resistance of the $p^+/p^-/n^+$ region increases. In the NSA HBT device, the link resistance increases as the emitter polysilicon width increases and the salicide is moved farther from the emitter. Second, in a reverse mode of operation, the heavily doped regions are being spaced farther apart; this leads to a higher metallurgical breakdown voltage as well as lower reverse bias leakage current. The increase in the emitter polysilicon region space (e.g., the link resistance region) has a significant influence on the emitter-to-base breakdown voltage BV_{EBO} and the emitter–base leakage current IL_{EBO}. During an ESD event, as the reverse bias is applied, the emitter–base depletion width increases [72,73]. As the emitter is heavily doped, the depletion region extends laterally through the p-doped SiGe link resistance region. With increased reverse bias, the depletion region extends into the p^+ doped SiGe extrinsic base implant. Hence, the ESD robustness of NSA SiGe HBT device will be a strong function of the extrinsic base to emitter spacing. Third, as the emitter overlap region is increased, the silicide film is moved physically farther from the emitter region. In the NSA epitaxial HBT, the extrinsic base implant and the silicide moves together with the emitter polysilicon width. Hence, the emitter shape acts as a 'silicide block mask' in that it moves the silicide film away from the metallurgical junction. Given the current density is highest in the silicide film, this can be the region of peak Joule heating; hence, the peak temperature will be at this location. As the peak temperature is moved away from the emitter–base metallurgical junction, the temperature at the metallurgical junction will be lower (during forward bias operation).

6.6.6.1 NSA Human body model (HBM) step stress

To evaluate from the NSA HBT to ESD events, the transistor was stressed on an HBM commercial test system. The HBM ESD step-stress is performed in 30 V stress increments of the NSA 47 GHz SiGe HBT device [72,73].

For negative polarity HBM stress across the emitter–base junction, experimental results show that as the emitter–base spacing is decreased, there is a decrease in the ESD robustness of the emitter–base junction (Figure 6.22). A 0.44 μm × 10 μm emitter npn SiGe HBT device was tested with variations in the emitter design size. For negative polarity HBM stress, the emitter–base junction is reverse-biased, leading to avalanche breakdown. In the commercial HBM test systems, the ESD failure criteria is established by a change in the I–V characteristic (e.g., the failure is associated with an increase in the emitter–base leakage current, IL_{EBO}).

The second HBM step-stress is performed on a second set of HBTs. A 0.44 μm × 3 μm emitter npn SiGe HBT device was tested with variations in the emitter design size. For positive polarity HBM pulse stress, the ESD results degrade as the link resistance is increased. Figure 6.23 shows the HBM ESD results of the forward bias as a function of the extrinsic-base encroachment. At very small extrinsic base–emitter spacing, enhanced

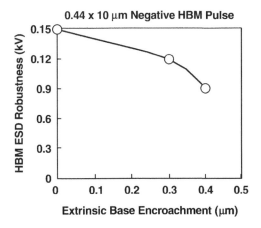

Figure 6.22   HBM ESD failure level as a function of the extrinsic-base implant encroachment in a NSA SiGe HBT device (for a negative polarity HBM pulse)

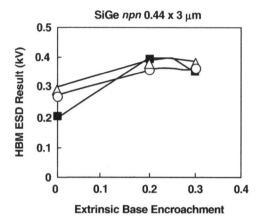

Figure 6.23 NSA SiGe HBT device HBM measurement in a forward bias voltage stress as a function of extrinsic base encroachment

leakage current is evident, impacting the low current gain parameters. In Figure 6.23, it can be seen that as the link resistance increases, the ESD robustness of the base–emitter junction decreases.

6.6.6.2 Transmission line pulse (TLP) step stress

Transmission line pulse (TLP) testing provides significant insight into the emitter–base physics during stress. A TLP system with a commercial-based high current pulse source (Hewlett Packard HP8114A), an oscilloscope, and a wafer probe system is used to test the SiGe HBT devices. The pulse source is connected to a 50 Ω matching resistor to ground, and a second 50 Ω resistor in series with the device under test (DUT). A Tektronix TEK CT-1 current probe evaluates the current flowing through the emitter–base junction during the test. The shield of the cable was locally grounded near the wafer-probe station.

From TLP measurements, it can be seen that the failure point and failure mechanism is associated with a significant increase in the current at a breakdown voltage. The TLP measurement of a NSA SiGe HBT device under reverse-bias emitter–base stress is shown in Figure 6.24. As the TLP applied current pulse increases, the emitter–base voltage increases. The increase in the emitter–base current is also associated with a small increase in the leakage current. The leakage current saturates at some value prior to the breakdown of the emitter–base junction [72,73].

At some value, the emitter–base junction undergoes a sudden increase in the reverse bias emitter–base current as well as a significant jump in the leakage current (the final leakage current prior to failure of the 0.44 μm $\times$ 3 μm SiGe *npn* is 19 pA). A key result is that both leakage current as well as the emitter–base current increase after approximately 4–5 V across the emitter–base junction; this condition is beyond the safe operating area (SOA) of the SiGe HBT device. After 10 V, the leakage current saturates prior to the final leakage increase. From the results in Figure 6.24, the emitter–base leakage current IL_{EBO} increases with the emitter–base current after the voltage exceeds the SOA voltage followed by a saturation phenomenon. A significant jump in leakage occurs till the avalanche breakdown voltage point [72,73].

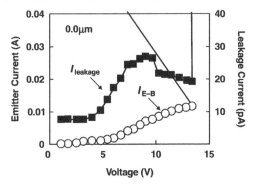

Figure 6.24 TLP Measurement of NSA SiGe HBT emitter–base for large emitter diffusion-to-extrinsic base implant spacing

In Figure 6.25, the extrinsic base encroachment is increased by 0.3 μm. In this case, the SiGe HBT NSA device shows an increase in the emitter–base current with the TLP pulse increase. The leakage current (IL_{EBO}) again increases after 4 V followed by a saturation at 6 V. The IL_{EBO} has a sudden increase at the emitter–base current avalanche point.

In the NSA SiGe HBT, as the extrinsic-base implant to emitter window edge is reduced, the breakdown voltage reduces. In Figure 6.26, TLP measurements of the minimum extrinsic base implant to emitter window space NSA SiGe HBT device is shown

In Figure 6.26, for the 0.4 μm encroachment case, the failure voltage reduces to 5.05 V. In Table 6.2, the emitter–base encroachment and corresponding TLP failure voltage is shown with the leakage value prior to the failure of the structure [72,73].

Figure 6.27 shows the TLP failure voltage has a monotonically decreasing value with emitter–base encroachment. As can be observed, the failure voltage is a strong function of the extrinsic base implant spacing relative to the emitter–base metallurgical junction [72,73].

6.6.6.3 RF testing of SiGe HBT emitter–base configuration

In the case of a SiGe HBT and SiGeC HBT devices, the emitter–base region is a sensitive region to high current stress. A test method for ESD testing of transistors can use the following procedure [76,77]:

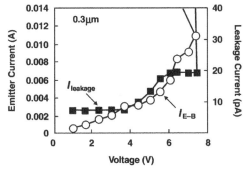

Figure 6.25 NSA SiGe HBT TLP measurement

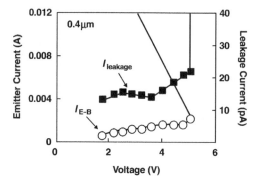

Figure 6.26 TLP measurement of NSA SiGe HBT emitter–base for minimum emitter-to-extrinsic base spacing

- RF characterization of the unity current gain cutoff frequency;
- RF characterization of the unity power gain cutoff frequency;
- RF characterization of the Gummel characteristics (e.g., base and collector currents);
- RF characterization of the *S*-parameters of the transistor element;
- ESD Pre-evaluation of the emitter–base current voltage forward bias characteristic;
- Single ESD stress at a given stress level and a given polarity;
- ESD Post-evaluation of the emitter–base current voltage forward bias characteristic;
- RF characterization of the transistor element.

For determination of the ESD degradation and non-in situ RF characterization, pre- and post-RF characterization is completed on different transistor elements and at different ESD stress levels.

As an example of the emitter–base stress study, HBM testing is performed on ten different sites at different current stress levels [76]. Tables 6.3 and 6.4 contain the results of a SiGeC HBT for the forward bias voltage shifts and RF degradation for positive and negative HBM pulse stress, respectively. Table 6.5 shows the relationship of the normalized percent change in the unity current gain cutoff frequency versus the normalized percent change in

Table 6.2 TLP measurement summary of the non-self-aligned (NSA) SiGe HBT emitter–base for minimum emitter-to-extrinsic base spacing

Emitter–base encroachment (μm)	TLP leakage (pA)	TLP failure voltage (V)
0.00	19.3	13.487
0.20	12	10.453
0.30	19.83	7.41
0.40	18.0	5.046

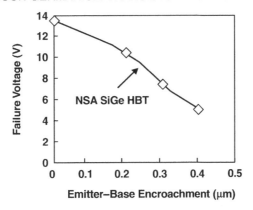

Figure 6.27   TLP failure voltage as a function of emitter–base encroachment in a NSA SiGe HBT device

Table 6.3   SiGeC HBT forward d.c. voltage shift and RF degradation versus positive HBM pulse stress

HBM pulse (V)	Forward voltage (V)	Peak unity current gain cutoff frequency (GHz)
+50	0.78	46
+100	0.78	44
+200	0.78	44
+240	0.30	36
+250	0.18	31

Table 6.4   SiGeC HBT forward d.c. voltage shift and RF degradation versus negative HBM pulse stress

HBM pulse (V)	Forward voltage (V)	Peak cutoff frequency (GHz)
−50	0.72	46.49
−75	0.42	42.59
−100	0.38	34.78
−150	0.05	32.47

Table 6.5   Percent change in the peak f_T and forward bias voltage

$\Delta f_T/f_T$ (V)	$\Delta V_f/V_f$ (V)
0.0227	0.0
0.0860	0.416
0.217	0.384
0.258	0.472
0.301	0.935
0.326	0.769

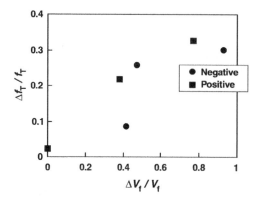

Figure 6.28 Plot of the percent change in the unity current gain cutoff frequency as a function percent change in the forward bias voltage (SiGeC HBT device)

the forward bias voltage. Additionally, the forward-bias voltage maintained an anticipated forward-bias voltage of 0.78 V. As the HBM voltage stress was increased on additional sites, the peak unity current cutoff frequency value decreased from 46 to 36 GHz as well as an observed post-stress V_{BE} of 0.3 V (Table 6.4). As the stress level continued, the d.c. characterization shows continued degradation of the I–V characteristic with an additional decrease in the peak unity current gain cutoff frequency (Figure 6.28) [76].

From these experimental results, the percent change in the peak f_T and the percent change in the forward-bias voltage can be compared. Evaluation of these results indicates that there is a relationship between the percent change in the peak f_T and the percent d.c. shift in the forward-bias voltage.

It can be observed that forward bias shifts greater than 30% can lead to RF degradation of the peak f_T greater than 10%. Forward bias shifts greater than 50% can lead to RF peak f_T degradation greater than 20% [76].

6.6.6.4 Unity current gain cutoff frequency–collector current plots

From the evaluation of the current gain cutoff frequency–collector current plot (f_T versus I_C), significant insight into the ESD degradation can be understood. As the d.c. shift occurs in the I–V characteristic, the peak f_T degrades. The f_T versus I_C plot shows three characteristics effects (Figure 6.29). First, for current levels below the peak f_T, the f_T is decreased with a larger slope in the df_T/dI_C term. Second, the peak f_T is decreased. Third, the current gain after the peak f_T is a weak variation with a convergence of the df_T/dI_C term.

The cutoff frequency f_T is inversely related to the emitter-to-collector transit time, τ_{EC}.

$$\frac{1}{2\pi f_T} = \tau_{EC} = \tau_E + \tau_B + \tau_{CSL} + \tau_C$$

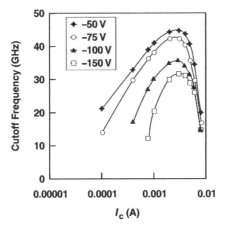

Figure 6.29 Unity current gain cutoff frequency versus collector current degradation plots with increased ESD stress in a SiGeC HBT device [76]

where

$$\tau_E = \frac{C_{eb} + C_{bc}}{g_m}$$

$$\tau_B = \frac{W_B^2}{KD_B}$$

$$\tau_{CSL} = r_C(C_{csx} + C_{bc})$$

$$\tau_C = \frac{x_C}{v_{sat}L}$$

From these experimental results, the transconductance term, g_m, is a function of current I_C and is the source of the degradation mechanism. Hence the first region is explainable as transconductance degradation. The last region is a function of the Kirk effect at the collector junction and is not influenced by the emitter–base ESD stress. For the unity power gain cutoff frequency, ESD testing also leads to degradation in the peak f_{MAX}. This is defined as:

$$f_{MAX} = \sqrt{\frac{fT}{8\pi R_b C_{bc}}}$$

with collector–base capacitance, C_{cb}, collector–substrate, C_{csx}, and base resistance, R_b. With the degradation of f_T, the f_{MAX} also decreases. ESD testing showed greater signs of S-parameter degradation in the S_{11}, S_{12}, and S_{21} terms compared to the S_{22}. The S-parameter degradation is significantly more evident in the parameters associated with the emitter–base junction.

The Gummel plots show the base current I_B and the collector current I_C as a function of V_{BE}. As the d.c. degradation increases, the base current increases, and the base ideality factor (slope) decreases. The collector current does not shift significantly with the d.c. degradation. From the Gummel plots, the d.c. current gain, β ($\beta = I_C/I_B$) decreases with the increased ESD stress level.

From this study, some significant conclusions can be made on the relationship of the d.c. shifts, the degradation mechanism, the RF parameter degradation and the condition of final HBM failure. Our work shows that the f_T parameter does not significantly shift without some shift in the d.c. I–V characteristic. Changes in f_{MAX} are evident when changes occur in the base resistance when f_T is fixed. The failure criteria based on a d.c. shift is coupled to the level of acceptable degradation of the unity current gain cutoff frequency, unity power gain cutoff frequency, and S-parameters of a device.

6.6.7 Silicon Germanium Carbon – ESD-Induced S-Parameter Degradation

Ronan and Voldman established the RF ESD test procedure for the evaluation of a silicon germanium carbon (SiGeC) HBT pre- and post-ESD stress utilizing a wafer level S-parameter extraction on a wafer-level HBM tester, and wafer-level RF functional test system [76]. As discussed in the prior section on RF ESD testing of single components, a test procedure was established for an evaluation of the S-parameter degradation.

RF ESD test methods can be established as follows [76]:

- Pre-stress evaluation: Evaluation of all S-parameter terms of the RF component using a TDR methodology as a function of frequency.

- S-parameter matrix: Formulation of the S-parameter matrix terms (e.g., two port matrix terms are S_{11}, S_{12}, S_{21}, S_{22}).

- ESD step-stress: Apply an ESD stress between two terminals of the RF component; one of the terminals has an ESD stress applied, whereas a second terminal is reference ground. A given ESD model waveform shape, waveform polarity, and pulse magnitude is chosen (e.g., HBM, MM, CDM, CDE, TLP, or VF-TLP).

- Post-stress S-parameter evaluation: Evaluation of all S-parameter parameters terms of the RF component using a TDR methodology as a function of frequency.

- Post-stress S-parameter matrix formulation: Formulation of the S-parameter matrix terms (e.g., two-port matrix terms are S_{11}, S_{12}, S_{21}, S_{22}) post-ESD stress.

- ESD step-stress: The pulse magnitude of the defined ESD stress is increased on the same RF component, and the above procedure is repeated, where the S-parameters are re-evaluated.

A silicon germanium carbon (SiGeC) HBT placed in an S-parameter wafer-level pad set. Short and open wafer level pad set structures were also created for de-embedding and S-parameter extraction. The RF SiGeC HBT device was configured in a common-emitter configuration where the two-port input terminals are between the base and the emitter, and the two-port output terminals are between the collector and the emitter.

In the test procedure, the S-parameters were extracted on a full RF functional test system and the measurements were performed on a wafer level. The HBM stress was applied using a wafer-level HBM test system source, single probes, and an oscilloscope to evaluate the I–V

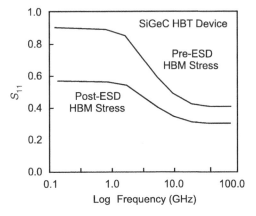

Figure 6.30   *S*-parameter S_{11} comparison as a function of frequency for a SiGeC HBT device before and after ESD HBM stress

degradation magnitude in the d.c. forward-bias *I–V* characteristic. From an earlier study of the ESD-induced forward-bias degradation, the magnitude of the HBM pulse to cause ESD failure was known. A single HBM ESD pulse of a magnitude that would induce ESD failure was applied between the SiGeC HBT base and emitter pads. Post-ESD stress, the test wafer was re-evaluated for *S*-parameter extraction.

In the Figures 6.30–6.33, the four two-port *S*-parameters (e.g., S_{11}, S_{12}, S_{21}, and S_{22}) are plotted as a function of frequency before and after HBM ESD stress, respectively.

In the Figure 6.30, the *S*-parameter S_{11} results are shown pre- and post-HBM ESD stress; it can be observed that the S_{11} parameter magnitude is lower for all frequencies post-ESD stress. After ESD stress, the largest degradation is observed.

In the Figure 6.31, the *S*-parameter S_{12} results are shown pre- and post-HBM ESD stress. In this parameter, post-ESD stress measurements were higher for lower frequencies but lower for the higher frequency regime.

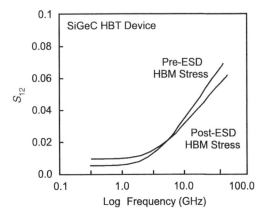

Figure 6.31   *S*-parameter S_{12} comparison as a function of frequency for a SiGeC HBT device before and after ESD HBM stress

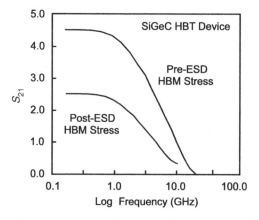

Figure 6.32 S-parameter S_{21} comparison as a function of frequency for a SiGeC HBT device before and after ESD HBM stress

In the Figure 6.32, the S-parameter S_{21} results are shown pre- and post-HBM ESD stress; it can be observed that the S_{21} parameter magnitude is lower for all frequencies post-ESD stress. Prior to HBM ESD stress, the S_{21} parameter magnitude of the SiGeC HBT device is 4.5; after, the ESD stress, S_{21} is approximately 2.5. This reduction is related to the decrease in the bipolar current gain (e.g., reduction in the peak unity current gain cutoff frequency). In the last figure, Figure 6.33, the S-parameter S_{22} results are shown pre- and post-HBM ESD stress; it can be observed that the S_{22} parameter magnitude is also lower for all frequencies in the SiGeC HBT device post-HBM ESD stress.

6.6.8 Electrothermal Simulation of Emitter–Base Stress

Electrothermal simulation can be used to assist understanding of the device shifting; simulation shows the peak self-heating occurs in the extrinsic base region [72,73,76,77].

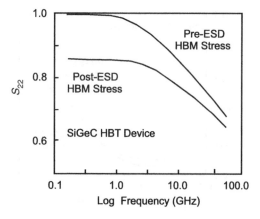

Figure 6.33 S-parameter S_{22} comparison as a function of frequency for a SiGeC HBT device before and after ESD HBM stress

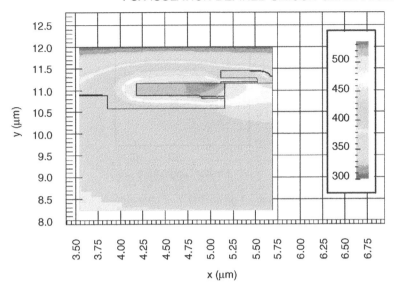

Figure 6.34   Electrothermal simulation as a function of emitter–base stress. A positive HBM pulse is applied to the base region

Figures 6.34 and 6.35 are simulation results emulating the RF pre- and post-stress ESD test of a SiGe HBT in emitter–base configuration. Simulation results show that for a positive stress, the peak self-heating occurs in the extrinsic base regions.

6.7 FIELD-OXIDE (FOX) ISOLATION DEFINED SILICON GERMANIUM HETEROJUNCTION BIPOLAR TRANSISTOR HBM DATA

Transistor structural design and implantation of HBTs have a significant influence on both the RF device performance and ESD sensitivity. Homojunction and heterojunction bipolar

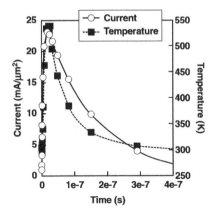

Figure 6.35   Current density and peak lattice temperature as a function of time

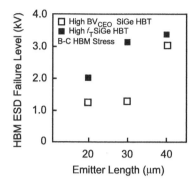

Figure 6.36 HBM ESD failure levels of SiGe HBT *npn* transistors as a function of emitter length (base–collector forward-bias stress)

transistors can be constructed in LOCOS isolation or STI. Additionally, the HBT characteristics can be modified using implants in the base–collector region. Implantation into the collector region can modulate the breakdown voltage and the unity current gain cutoff frequency, f_T. Placement of the 'pedestal' implant in the intrinsic sections of a HBT influences the collector-to-emitter breakdown voltage, BV_{CEO}.

Wu *et al.* [88] measured the ESD robustness of a field oxide (FOX)-defined silicon germanium HBT in a 0.35-µm SiGe BiCMOS technology. In the physical structure, the epitaxial silicon germanium region was deposited on a field oxide (FOX) isolation structure. Electrical contacts were placed on the polycrystalline SiGe film in the extrinsic base region. The emitter was formed by establishing a window within the FOX region, and placing over the single crystal SiGe film region. Different transistors were analyzed to evaluate the HBM sensitivity of SiGe HBT devices. The high voltage BV_{CEO} SiGe *npn* HBT is a 30 GHz/6 V f_T/BV_{CEO} *npn* HBT device, and the high f_T SiGe *npn* HBT is a 70 GHz/2.5 V f_T/BV_{CEO} *npn* HBT device. HBM stressing in the emitter–base configuration and the collector-to-emitter configuration of the SiGe HBT *npn* transistor elements showed that the ESD robustness was a weak function of emitter length. Evaluation of the HBM stress for the base–collector showed that as emitter length increased (e.g., the base length dimension scales with the emitter length), the HBM ESD robustness increased. At emitter lengths of 20 µm, the HBM ESD robustness was above 1 kV HBM for both the high BV_{CEO} and the high f_T SiGe HBT *npn* devices. As the emitter length increased to 40 µm, the ESD results increased above 3.0 kV HBM levels [88]. These experimental results demonstrate that the utilization of the SiGe film as an anode structure and the subcollector as a cathode structure could be utilized as an ESD protection element (Figure 6.36).

6.8 SILICON GERMANIUM HBT MULTIPLE-EMITTER STUDY

In transistor structures, many different configurations can be established with the emitter, base, and collector region. The way the current flows from the collector through the multiple emitter and multiple base region can strongly influence the ESD robustness. As was observed by Minear and Dodson, even in the case of a single emitter and a single base region, there exists multiple current paths from the subcollector region to the base or emitter region. In this case, the current path can strongly influence ESD failure in an epitaxial bipolar transistor.

Wu *et al.* [88] measured a multifinger emitter and multiple base structure, and observed dependencies on which base region was grounded. Wu *et al.* [88] evaluated the ESD robustness of a four emitter, five base stripe CBEBEBEBEBC-configured field oxide (FOX)-defined silicon germanium HBT in a 0.35-μm SiGe BiCMOS technology. In the physical structure, the epitaxial silicon germanium region was deposited on a field oxide (FOX) isolation structure. Electrical contacts were placed on the polycrystalline SiGe film in the extrinsic base region. The emitter was formed by establishing a window within the FOX region and placing over the single crystal SiGe film region. Different transistors were analyzed to evaluate the HBM sensitivity of SiGe HBT devices. The high voltage BV_{CEO} SiGe *npn* HBT is a 30 GHz/6 V f_T/BV_{CEO} *npn* HBT device and the high f_T SiGe *npn* HBT is a 70 GHz/2.5 V f_T/BV_{CEO} *npn* HBT device.

HBM ESD stressing of the CBEBEBEBEBC configured SiGe *npn* HBT structure were taken in the collector-to-emitter configuration where all four emitters were at ground potential. In the study, the structure was designed to allow independent biasing or floating of the five base contacts. Figure 6.37 shows the plot of the HBM ESD robustness as a function of the cases of (a) all the base regions grounded; (b) base region B1, B3, and B5; and (c) the first end base region B1, and center region B3. From the experimental results, in the first (a) and the second (b) case showed low ESD results. In the third case (c), when the first region, B1, and the center region B3 were grounded, the ESD results were the highest. Although these results seem counter-intuitive, it is postulated that the floating of the region B5 node adjacent to the collector region lead to a reduction of current crowding as well as elimination of a local current path to the node B5 and failure [88].

An RF ESD design practices can be as follows:

- In multiple emitter and base structures in a bipolar transistor, nonuniform current flow can exist in the emitter structures because of the nonequal spacing of the collector structure relative to each emitter and base region.

- The nonuniformities of the collector-to-emitter spacings and the collector-to-base spacings may be addressed by the floating or ballasting of specific nodes (e.g., adjacent base regions) in the structure.

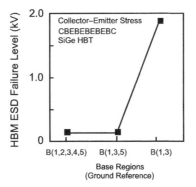

Figure 6.37   HBM collector-to-emitter stress ESD failure levels of CBEBEBEBEBC SiGe HBT *npn* transistors as a function of referenced base regions

6.9 SUMMARY AND CLOSING COMMENTS

In Chapter 6, we have discussed ESD in silicon germanium and silicon germanium carbon technology. HBM, MM, and TLP measurements of SiGe and SiGeC HBTs were shown. Additionally, SiGe and SiGeC HBT results were compared demonstrating an improvement in the SiGeC HBT as a result of better statistical control of the transistor base width; this leads to a better control of the breakdown characteristics and hence second breakdown results. Self-aligned and non-self aligned emitter–base SiGe transistor structures results will be shown, highlighting the relationship of the spacings and the way it influences ESD robustness; the critical information is that the scaling of the physical dimensions of the emitter and base regions lead to a lower emitter-breakdown voltage, followed by second breakdown. Hence, the design of the emitter base spacings and film dimensions may play a critical role in the performance and the ESD robustness of the transistor. This may also influence the ESD robustness scaling with each technology generation. In addition, transistor ordering and multiple finger emitter transistors highly influence on ESD robustness through the physical path the ESD current undergoes within the structure (e.g., in the intrinsic and in the extrinsic regions of the transistor).

In Chapter 7, we will discuss ESD in gallium arsenide-based technology. ESD robustness of GaAs MESFETs devices to GaAs HBT devices will be shown. Early GaAs work in GaAs MESFETs to present ESD measurements in GaAs HBT will be discussed. ESD testing of the different transistor regions and junctions will be reviewed. In addition, ESD failure mechanisms in gallium arsenide, and indium gallium arsenide will be discussed and tabulated.

PROBLEMS

1. From the Johnson Limit relationship for maximum voltage–unity current cutoff frequency product,

$$V_m f_T = E_m v_S / 2\pi$$

   show the magnitude of the right hand side of the equation for silicon germanium.

2. From the equation in problem 1, assume the material is silicon, and evaluate the right hand side of the Johnson Limit equation.

3. From the comparison of silicon germanium to silicon, derive the application breakdown voltages that are possible given the silicon BJT and SiGe HBT are the same unity current gain cutoff frequency.

4. From the Johnson Limit power formulation,

$$(P_m X_C)^{1/2} f_T = E_m v_S / 2\pi$$

   where the reactive impedance X_C is defined as

$$X_C = \frac{1}{2\pi f_T C_O}$$

derive the ratio of the power of the maximum power of a SiGe device and silicon bipolar homojunction device. Assuming the same unity current gain cutoff frequency, derive the relationship between the two power values.

5. Derive a relationship of a SiGe HBT in a diode configuration used in a diode string. Assume N SiGe HBT devices in series.

6. Derive a relationship of a trigger network consisting of N SiGe HBT diode-configured elements in series with an open-base SiGe HBT in a collector-to-emitter configuration.

7. Given a SiGe HBT power amplifier, SiGe HBT diode strings can be used on the RF output. Derive the relationship of a 'diode string' network consisting of SiGe HBT with M elements in one direction in parallel with a SiGe HBT RF output. Assume a SiGe HBT with an open-base connections and the SiGe HBT is in a 'common-emitter' configuration. Derive the number of diodes needed to avoid collector-to-emitter breakdown of the SiGe HBT.

8. Given a SiGe HBT power amplifier, SiGe HBT diode strings can be used on the RF output. Derive the relationship of a 'diode string' network consisting of SiGe HBT with M elements in one direction in parallel with a SiGe HBT RF output. Assume a SiGe HBT has a resistor load on its base input and the SiGe HBT is in a 'common-emitter' configuration. Derive the number of diodes needed to avoid collector-to-emitter breakdown of the SiGe HBT (hint: BVCER is required).

9. Given a SiGe HBT power amplifier, assume a SiGe HBT devices are used as ESD elements on the input in a grounded-base common-emitter-configuration, derive the unity current cutoff frequency and the base resistance required to prevent an over-voltage of the input of the SiGe HBT power amplifier.

10. Given a SiGe HBT power amplifier with a breakdown voltage of $(BVCEO)_{PA}$ on its output, using a grounded-base SiGe HBT ESD element in a common-emitter configuration, what should the frequency be of the SiGe HBT ESD element?

REFERENCES

1. Konig U. SiGe and GaAs as competitive technologies for RF-applications. *Proceedings of the Bipolar/BiCMOS Circuits and Technology Meeting (BCTM)*, 1998. p. 87–92.
2. Yuan JS. *SiGe, GaAs, and InP heterojunction bipolar transistors*. New York: John Wiley and Sons, Inc.; 1999.
3. Liu W. *Fundamentals of III–V devices: HBTs, MESFETs, and HFETs/HEMTs*. New York: John Wiley and Sons, Inc.; 1999.
4. Cressler JD, Niu G. *Silicon–germanium heterojunction bipolar transistors*. Boston: Artech House, Inc.; 2003.
5. Singh R, Harame D, Oprysko M. *Silicon germanium: technology, modeling and design*. New York: John Wiley and Sons; 2004.
6. Early JM. Maximum rapidly-switchable power density in junction triodes. *IRE Transactions on Electronic Devices* 1959;**ED-6**:322.
7. Early JM. Speed in semiconductor devices. *IRE National Convention*, March 1962.

8. Early JM. Structure-determined gain-band product of junction triode transistors. *Proceedings of the IRE*, December 1958. p. 1924.

9. Goldey JM, Ryder RM. Are transistors approaching their maximum capabilities? *Proceedings of the International Solid-State Circuits Conference*, February 1963.

10. Pritchard RL. Frequency response of grounded-base and grounded emitter junction transistors. *AIEE Winter Meeting*, January 1954.

11. Webster WM. On variation of junction transistor current amplification with emitter current. *Proceeding of the IRE* 1954;**42**:914–20.

12. Giacoletto LJ. Comparative high frequency operation of junction transistors made of different semiconductor materials. *RCA Review* 1955;**16**:34.

13. Johnson EO, Rose A. Simple general analysis of amplifier devices with emitter, control, and collector functions. *Proceedings of the IRE*, March 1959. p. 407.

14. Johnson EO. Whither the tunnel diode. *AIEE-IRE Electron Device Conference*, October 1962.

15. Johnson EO. Physical limitations on frequency and power parameters of transistors. *RCA Review* 1965;**26**:163–77.

16. Kirk CT. A theory of transistor cutoff frequency (f_T) falloff at high current densities. *IRE Transactions on Electron Devices* 1962;**ED-9**:164–74.

17. Kroemer H. Zur theorie des diffusions und des drifttransistors, Part III. *Archiv für Elektrotechnik Ubertrangung* 1954;**8**:499–504.

18. Kroemer H. Theory of wide-gap emitter for transistors. *Proceedings of the IRE* 1957;45: 1535–7.

19. Kroemer H. Two integral relations pertaining to the electron transport through a bipolar transistor with a non-uniform energy gap in the base region. *Solid-State Electron* 1958;**28**:1101–3.

20. Kroemer H. Heterostructure bipolar transistors and integrated circuits. *Proceedings of the IEEE* 1982;**70**(1):13–25.

21. Kroemer H. Heterojunction bipolar transistors: what should we build?. *Journal of Vacuum Science and Technology* 1983;**B1**(2):112–30.

22. Meyerson B. Low temperature silicon epitaxy by ultrahigh vacuum/chemical vapor deposition. *Applied Physics Letters* 1986;**48**:797–9.

23. Patton GL, Iyer SS, Delage SL, Tiwari S, Stork JMC. Silicon germanium base heterojunction bipolar transistors by molecular beam epitaxy. *IEEE Electron Device Letters* 1988;**9**(4):165–7.

24. Harame DL, Stork JM, Meyerson BS, Nguyen N, Scilla GJ. Epitaxial-base transistors with ultra-high vacuum chemical vapor deposition (UHV-CVD) epitaxial/enhanced profile control for greater flexibility in device design. *IEEE Electron Device Letters* 1989;**10**(4):156–8.

25. Burghartz JN, Comfort JH, Patton GL, Meyerson BS, Sun JY-C, Stork JMC, *et al.* Self-aligned SiGe-base heterojunction bipolar transistor by selective epitaxy emitter window (SEEW) technology. *IEEE Electron Device Letters* 1990;**11**(7):288–90.

26. Harame DL, Crabbe EF, Cressler JD, Comfort JH, Sun JYC, Stiffler SR, *et al.* A high performance epitaxial SiGe-base ECL BiCMOS technology. *International Electron Device Meeting (IEDM) Technical Digest* 1992;19–22.

27. Harame DL, Stork JMC, Meyerson BS, Hsu KY, Cotte J, Jenkins KA, *et al.* Optimization of SiGe HBT technology for high speed analog and mixed signal applications. *International Electron Device Meeting (IEDM) Technical Digest* 1993;874–6.

28. Harame D, Comfort J, Cressler J, Crabbe E, Sun JY-C, Meyerson B. Si/SiGe epitaxial-base transistors – Part I: materials, physics, and circuits. *IEEE Transactions on Electron Devices* 1995; **ED-42**(3):455–68.

29. Harame D, Comfort J, Cressler J, Crabbe E, Sun JY-C, Meyerson B. Si/SiGe epitaxial-base transistors – Part II: process integration and analog applications. *IEEE Transactions on Electron Devices* 1995;**42**(3):469–82.

30. Burghartz JN, Ginsberg BJ, Mader SR, Chen TC, Harame DL. Selective epitaxy base transistor (SEBT). *IEEE Electron Device Letters* 1998;**9**(5):259–61.

31. Prinz EJ, Garone PM, Schwartz PV, Xiao X, Sturm JC. The effects of base dopant out-diffusion and undoped SiGe junction space layers in Si/SiGe/Si heterojunction bipolar transistors. *IEEE Transactions on Electronic Devices Letters* 1991;**12**:42–4.

32. Prinz EJ, Sturm JC. Analytical modeling of current gain–Early voltage products in Si/Si(1 – x)–Ge(x) heterojunction bipolar transistors. *International Electron Device Meeting (IEDM) Technical Digest* 1991;853–6.

33. Crabbe EF, Cressler JD, Patton GL, Stork JMC, Comfort JH, Sun JYC. Current gain rolloff in graded-base SiGe heterojunction bipolar transistors. *IEEE Electron Device Letter* 1993; **14**(4):193–5.

34. Crabbe EF, Stork JMC, Baccarani G, Fischetti MV, Laux SE. The impact of nonequilibrium transport on breakdown and transit time in bipolar transistors. *International Electron Device Meeting (IEDM) Technical Digest* 1990;463–6.

35. Zanoni E, Crabbe EF, Stork JM, Pavan P, Verzellesi G, Vendrame L, *et al.* Extension of impact ionization multiplication coefficient measurements to high electric fields in advanced Si BJTs. *IEEE Electron Device Letters* 1993;**14**(2):69–71.

36. Slotboom JW, Streutker G, Pruijmboom A, Gravesteijn DJ. Parasitic energy barriers in SiGe HBT's. *IEEE Electron Device Letters* 1991;**12**(9):486–8.

37. Tiwari S. A new effect at high currents in hetero-structure bipolar transistors. *IEEE Electron Device Letters* 1989;**10**:2105.

38. Comfort JH, Crabbe EF, Patton GL, Stork JMC, Sun JYC, Meyerson BS. Impact ionization reduction in SiGe HBT's. *Device Research Conference* 1990;21–4.

39. Jeng SJ, Greenberg DR, Longstreet M, Hueckel G, Harame DL, Jadus D. Lateral scaling of the self-aligned extrinsic base in SiGe HBTs. *Proceedings of the Bipolar/BiCMOS Circuits and Technology Meeting (BCTM) Symposium*, 1996. p. 15–8.

40. Jang J, Kan E, Dutton R, Arnborg T. Improved performance and thermal stability of interdigitated power RF bipolar transistor with nonlinear base ballasting. *Proceedings of the Bipolar/BiCMOS Circuits and Technology Meeting (BCTM) Symposium*, 1997. p. 143–6.

41. Walkey D, Schroter M, Voinigescu S. Predictive modeling of lateral scaling in bipolar transistors. *Proceedings of the Bipolar/BiCMOS Circuits and Technology Meeting (BCTM) Symposium, IEEE Bipolar Circuit Technology Meeting*, 1995. p. 74–7.

42. Richey D, Cressler J. Scaling issues and Ge profile optimization in advanced UHV/CVD SiGe HBTs. *Proceedings of the Bipolar/BiCMOS Circuits and Technology Meeting (BCTM) Symposium*, 1996. p. 19–23.

43. Tran H, Schroter M, Walkey D, Marchesan D, Smy TJ. Simultaneous extraction of thermal and emitter series resistances in bipolar transistors. *Proceedings of the Bipolar/BiCMOS Circuits and Technology Meeting (BCTM) Symposium*, 1997. p. 170–4.

44. Larson LE. Silicon bipolar transistor design and modeling for microwave integrated circuits applications. *Proceedings of the Bipolar/BiCMOS Circuits and Technology Meeting (BCTM) Symposium*, 1997. p. 142–8.

45. Lanzerotti L. Suppression of boron out-diffusion in silicon germanium HBT by carbon incorporation. *International Electron Device Meeting (IEDM) Technical Digest*, 1996. p. 249–52.

46. Cowern NEB, Zalm PC, Sluis PVD, Graresteijn DJ, Groer WBD. Diffusion in strained SiGe. *Physical Review Letters* 1994;**72**:2585–8.

47. Poate JM, Eaglesham DJ, Gilmer GH, Gossmann HJ, Jaraiz M, Rafferty CS, *et al.* Ion implantation and transient enhanced diffusion. *International Electron Device Meeting (IEDM) Technical Digest*, 1995. p. 77–80.

48. Lever RF, Bonar JM, Willoughby AFW. Boron diffusion across silicon-silicon germanium boundaries. *Journal of Applied Physics* 1999;**83**:1998–4.

49. Rajendran K, Schoenmaker W. Studies of boron diffusivity in strained $Si_{1-x}Ge_x$ epitaxial layers. *Journal of Applied Physics* 2001;**89**(2):980–7.

50. Eberl K, Iyer SS, Zalner S, Tsand JE, Lehover FK. Growth and strain compensation effects in the ternary $Si_{1-x-y}Ge_xC_y$ alloy system. *Applied Physics Letters* 1992;**60**:3033–5.

51. Nishikawa S, Takeda A, Yamaji T. Reduction of transient boron diffusion in pre-amorphized Si by carbon incorporation. *Applied Physics Letters* 1992;**60**:2270–2.

52. Stolk PA, Eaglesham DJ, Gossmann H-J, Poate JM. Carbon incorporation in silicon for suppressing interstitial-enhanced boron diffusion. *Applied Physics Letters* 1995;**66**(11):1370–2.

53. Lanzerotti L, Sturm J, Stach E, Hull R, Buyuklimanli T, Magee C. Suppression of boron out-diffusion in silicon germanium HBT by carbon incorporation. *International Electron Device Meeting (IEDM) Technical Digest*, 1996. p. 249–52.

54. Osten HJ, Lippert G, Knoll G, Barth R, Heinemann B, Rucker H, *et al*. The effect of carbon incorporation on SiGe heterojunction bipolar transistor performance and process margin. *International Electron Device Meeting (IEDM) Technical Digest*, 1997. p. 803–6.

55. Scholtz R, Gosele U, Huh JY, Tan TY. Carbon-induced saturation of silicon self-interstitials. *Applied Physics Letters* 1998;**72**:200–2.

56. Rucker H, Heinemann B, Ropke W, Kurps R, Kruger D, Lippert G, *et al*. Suppressed diffusion of boron and carbon in carbon-rich silicon. *Applied Physics Letters* 1998;**73**:1682–4.

57. Scholz R, Werner P, Gosele U, Tan TY. The contribution of vacancies to carbon out-diffusion in silicon. *Applied Physics Letters* 1999;**74**:392–4.

58. Rucker H, Heinemann B, Bolze D, Kurps R, Kruger D, Lippert G, *et al*. The impact of supersaturated carbon on transient enhanced diffusion. *Applied Physics Letters* 1999;**74**: 3377–9.

59. Kurata H, Suzuki K, Futasugi T, Yokoyama N. Shallow *p*-type SiGeC layers synthesized by ion implantation of Ge, C, and B in Si. *Applied Physics Letters* 1999;**75**:1568–70.

60. Rucker H, Heinemann B. Tailoring dopant diffusion for advanced SiGe:C heterojunction bipolar transistor. *Solid State Electronics* 2000;**44**:783–9.

61. Rajendran K, Schoenmaker W. Modeling of complete suppression of boron out-diffusion in $Si_{1-x}Ge_x$ by carbon incorporation. *Solid State Electronics* 2001;**45**:229–33.

62. Joseph A, Coolbaugh D, Zierak M, Wuthrich R, Geiss P, He Z, *et al*. A $0.18\,\mu m$ BiCMOS technology featuring a $120/100\,GHz$ (f_T/f_{MAX}) HBT and ASIC compatible CMOS using copper interconnects. *Proceedings of the Bipolar/BiCMOS Circuits and Technology Meeting (BCTM) Symposium*, 2001. p. 143–46.

63. Gruhle A. Prospects for $200\,GHz$ on silicon with silicon germanium heterojunction bipolar transistors. *Proceedings of the Bipolar/BiCMOS Circuits and Technology Meeting (BCTM) Symposium*, 2001. p. 19–25.

64. van den Oever LCM, Nanver LK, Slotboom JW, *et al*. Design of 200 GHz SiGe HBTs. *Proceedings of the Bipolar/BiCMOS Circuits and Technology Meeting (BCTM) Symposium*, 2001. p. 78–81.

65. Kirchgessner J, Bigelow S, Chai Fk, *et al*. A 0.18 μm SiGe:C RF BiCMOS technology for wireless and gigabit optical communication applications. *Proceedings of the Bipolar/BiCMOS Circuits and Technology Meeting (BCTM) Symposium*, 2001. p. 151–54.

66. Voldman S. *ESD: physics and devices*. Chichester: John Wiley and Sons; 2004.

67. Voldman S. *ESD: circuits and devices*. Chichester: John Wiley and Sons; 2005.

68. Voldman S. The state of the art of electrostatic discharge protection: Physics, technology, circuits, designs, simulation and scaling. *Proceedings of the Bipolar/BiCMOS Circuits and Technology Meeting (BCTM) Symposium*, September 27–29, 1998. p. 19–31.

69. Voldman S, Juliano P, Johnson R, Schmidt N, Joseph A, Furkay S, *et al*. Electrostatic discharge and high current pulse characterization of epitaxial base silicon germanium heterojunction bipolar transistors. *Proceedings of the International Reliability Physics Symposium (IRPS)*, March 2000. p. 310–6.

70. Voldman S, Schmidt N, Johnson R, Lanzerotti L, Joseph A, Brennan C, *et al*. Electrostatic discharge characterization of epitaxial base silicon germanium heterojunction bipolar transistors.

Proceedings of the Electrical Overstress/Electrostatic Discharge (EOS/ESD) Symposium, September 2000. p. 239–51.

71. Voldman S, Juliano P, Schmidt N, Botula A, Johnson R, Lanzeratti L, *et al.* ESD robustness of a silicon germanium BiCMOS technology. *Proceedings of the Bipolar/BiCMOS Circuits and Technology Meeting (BCTM) Symposium,* September, 2000. p. 214–17.

72. Voldman S, Lanzerotti LD, Johnson R. Emitter–base junction ESD reliability of an epitaxial base silicon germanium hetero-junction transistor. *Proceedings of the International Physical and Failure Analysis of Integrated Circuits (IPFA),* July 2001. p. 79–84.

73. Voldman S, Lanzerotti LD, Johnson RA. Influence of process and device design on ESD sensitivity of a silicon germanium hetero-junction bipolar transistor. *Proceedings of the Electrical Overstress/Electrostatic Discharge (EOS/ESD) Symposium,* 2001. p. 364–72.

74. Voldman S, Botula A, Hui D, Juliano P. Silicon germanium hetero-junction bipolar transistor ESD power clamps and the Johnson Limit. *Proceedings of the Electrical Overstress/Electrostatic Discharge (EOS/ESD) Symposium,* 2001. p. 326–36.

75. Voldman S. Variable trigger-voltage ESD power clamps for mixed voltage applications using a 120 GHz/100 GHz (f_T/f_{MAX}) silicon germanium hetero-junction bipolar transistor with carbon incorporation. *Proceedings of the Electrical Overstress/Electrostatic Discharge (EOS/ESD) Symposium,* October 2002. p. 52–61.

76. Voldman S, Ronan B, Ames S, Van Laecke A, Rascoe J, Lanzerotti L, *et al.* Test methods, test techniques and failure criteria for evaluation of ESD degradation of analog and radio frequency (RF) technology. *Proceedings of the Electrical Overstress/Electrostatic Discharge (EOS/ESD) Symposium,* October 2002. p. 92–100.

77. Ronan B, Voldman S, Lanzerotti L, Rascoe J, Sheridan D, Rajendran K. High current transmission line pulse (TLP) and ESD characterization of a silicon germanium hetero-junction bipolar transistor with carbon incorporation. *Proceedings of the International Reliability Physics Symposium (IRPS),* 2002. p. 175–83.

78. Voldman S. Variable trigger-voltage ESD power clamps for mixed voltage applications using a 120 GHz/100 GHz (f_T/f_{MAX}) silicon germanium heterojunction bipolar transistor with carbon incorporation. *Proceedings of the Electrical Overstress/Electrostatic Discharge (EOS/ESD) Symposium,* October 2002. p. 52–61.

79. Voldman S, Gebreselasie E. Low-voltage diode-configured SiGe:C HBT triggered ESD power clamps using a raised extrinsic base 200/285 GHz (f_T/f_{MAX}) SiGe:C HBT device. *Proceedings of the Electrical Overstress/Electrostatic Discharge (EOS/ESD) Symposium,* 2004. p. 57–66.

80. Voldman S, Strang S, Jordan D. A design system for auto-generation of ESD circuits. *Proceedings of the International Cadence Users Group,* September, 2002.

81. Voldman S, Strang S, Jordan D. An automated electrostatic discharge computer-aided design system with the incorporation of hierarchical parameterized cells in BiCMOS analog and RF technology for mixed signal applications. *Proceedings of the Electrical Overstress/Electrostatic Discharge (EOS/ESD) Symposium,* October 2002. p. 296–305.

82. Voldman S, Perez CN, Watson A. Guard rings: theory, experimental quantification and design. *Proceedings of the Electrical Overstress/Electrostatic Discharge (EOS/ESD) Symposium,* 2005. p. 131–40.

83. Perez CN, Voldman S. Method of forming a guard ring parameterized cell structure in a hierarchical parameterized cell design, checking and verification system. U.S. Patent Application 20040268284 (December 30, 2004).

84. Gebreselasie E, Sauter W, St. Onge S, Voldman S. ESD structures and circuits under bond pads for RF BiCMOS silicon germanium and RF CMOS technology. *Proceedings of the Taiwan Electrostatic Discharge Conference (T-ESDC),* 2005. p. 73–8.

85. Voldman S, Lanzerotti LD, Ronan B, St. Onge S, Dunn J. The influence of process and design of sub-collectors on the ESD robustness of ESD structures and silicon germanium heterojunction

bipolar transistors in a BiCMOS SiGe technology. *Proceedings of the International Reliability Physics Symposium (IRPS)*, 2003. p. 347–56.

86. Voldman S. The effect of deep trench isolation, trench isolation and sub-collector doping on the electrostatic discharge (ESD) robustness of radio frequency (RF) ESD STI-bound p^+/n-well diodes in BiCMOS silicon germanium technology. *Proceedings of the Electrical Overstress/Electrostatic Discharge (EOS/ESD) Symposium*, 2003. p. 214–23.

87. Chen SS. Design of low-leakage deep-trench diode string for ESD application in 0.18-μm SiGe BiCMOS process. *Proceedings of the Taiwan Electrostatic Discharge Conference (T-ESDC)*, 2003. p. 151–6.

88. Wu WL, Chang CY, Ker MD. High current characteristics of ESD devices in 0.35-μm silicon germanium RF BiCMOS process. *Proceedings of the Taiwan Electrostatic Discharge Conference (T-ESDC)*, 2003. p. 157–63.

89. Voldman S, Lanzerotti L, Morris W, Rubin L. The influence of heavily doped buried layer implants on electrostatic discharge (ESD), latchup, and a silicon germanium heterojunction bipolar transistor in a BiCMOS SiGe technology. *Proceedings of the International Reliability Physics Symposium (IRPS)*, 2004. p. 143–51.

90. Voldman S. A review of CMOS latchup and electrostatic discharge (ESD) in bipolar complimentary MOSFET (BiCMOS) silicon germanium technologies: Part I – ESD. *Journal of Microelectronics and Reliability* 2005;**45**:323–40.

91. Voldman S. A review of CMOS latchup and electrostatic discharge (ESD) in bipolar complimentary MOSFET (BiCMOS) silicon germanium technologies: Part II – Latchup. *Journal of Microelectronics and Reliability* 2005;**45**:437–55.

7 Gallium Arsenide and ESD

7.1 GALLIUM ARSENIDE TECHNOLOGY AND ESD

Gallium arsenide (GaAs) technology has been pervasive in semiconductor technology where power, speed, and reliability are necessary [1–4]. GaAs has advantages over other competing technologies, such as RF CMOS and RF silicon germanium because of its performance [5–8]. In recent times, GaAs technology has been dominating telecommunications, optical interconnect systems, cell phones, as well as space and military applications in the sectors where performance is needed [6–9]. Because reliability is important in these applications, electrostatic discharge (ESD) has been an issue since the early usage of GaAs semiconductor products [9–39]. Additionally, gallium-based and indium-based technologies [40–42] also have growing interest in the area of reliability and ESD protection [40–50]. Presently, the ESD protection is being pursued from on-chip ESD power clamps [50,51], on-chip spark gaps and field emission devices (FEDs) [35], separate GaAs ESD chips [9] to off-chip protection surge protection devices [52,53].

7.2 GALLIUM ARSENIDE ENERGY-TO-FAILURE AND POWER-TO-FAILURE

The power-to-failure of RF GaAs MESFETs over a wide range of pulse widths was of interest for microwave and military applications. Early works on the power-to-failure models of GaAs MESFETs focused on the power-to-failure as a function of the pulse width from 1 to 10 ns time scales [10–15]. The power-to-failure of the GaAs MESFETs under these pulse conditions can be related to single pulse very fast transmission line pulse (VF-TLP) ESD testing.

In 1979, Whalen, Thorn, Rastefano, and Calcatera evaluated 1 µm gate length (and a 3.5 µm channel length); GaAs MESFETs used for transmitter–receiver radar systems under pulsed power stress [13–15]. The 9.3 GHz X-band GaAs MESFETs were constructed using different metallurgy for the MESFET gate and electrical contacts to the source/drain regions. A first GaAs MESFET structure consisted of titanium/chromium/platinum/gold

ESD: RF Technology and Circuits   Steven H. Voldman
© 2006 John Wiley & Sons, Ltd

(Ti/Cr/Pt/Au) gate structure; a second GaAs MESFET was processed with an aluminum (Al) gate structure.

In the evaluation of the GaAs MESFET structures, to evaluate the RF devices in the nanosecond regime utilized a 9.3 GHz CW source. A diode modulator element was initiated by a pulse generator; the diode modulator switch became a pulse amplitude modulated (PAM) signal with a 1.5 ns rise time and 2.0 ns fall time. The nanosecond pulse was amplified by a preamplifier and a 100 W output power traveling wave tube (TWT) power amplifier. The RF pulse system measured the incident power applied to the GaAs MESFET as well as the reflected power using directional coupling elements [10,11]. Crystal detectors for both the incident and the reflected power signal were transmitted to the oscilloscope; in this manner, the absorbed power and the absorbed energy in the GaAs MESFET device under test (DUT) could be evaluated. Using an integration process, the integration over the power, the energy absorbed by the GaAs MESFET was evaluated. Integration of the incident power over the pulse width gives the incident energy and can be expressed as

$$E_I = \int P_I(t) dt$$

and integration of the reflected power gives the reflected energy and can be expressed as

$$E_R = \int P_R(t) dt$$

The absorbed energy is then obtained from the difference between the two terms, which can be expressed as

$$E_A = E_I - E_R = \int \{P_I(t) - P_R(t)\} dt$$

Figures 7.1 and 7.2 show the experimental results of RF testing of the GaAs MESFET structures for RF pulses from 1 to 10 ns. In Figure 7.1, the incident RF pulse power-to-failure

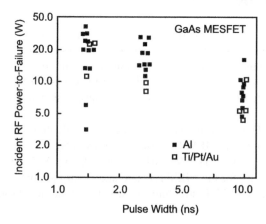

Figure 7.1 RF incident RF pulse power-to-failure versus RF pulse width is shown for the Ti/Pt/Au and Al gate GaAs MESFET structures

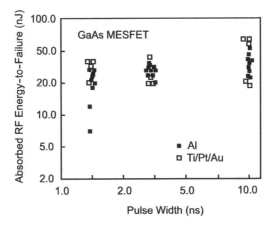

Figure 7.2   Absorbed RF pulse energy-to-failure versus RF pulse width is shown for the Ti/Pt/Au and Al gate GaAs MESFET structures

versus RF pulse width is shown for the Ti/Pt/Au and Al gate GaAs MESFET structures. As the applied pulse width increases, the incident power-to-failure decreases. From the experimental results, Whalen [10] assumed a model of incident power-to-failure versus pulse width,

$$P_1 = \frac{K_1}{\tau^n}$$

where P_1 is the incident power (in W), K_1 is the constant for the incident power, τ is the applied pulse width (in ns), and n is a constant. Experimental fitting of the relationship obtained is expressed as

$$P_1(\text{W}) = \frac{25}{\tau^{1/2}}$$

In Figure 7.2, the absorbed energy-to-failure is shown as a function of the pulse width. As the RF pulse width increases, the absorbed energy-to-failure increases. From the work of Whalen et al., the relationship of the absorbed energy can be expressed as

$$E_A = K_E \tau^n$$

where E_A is the absorbed energy-to-failure, K_E is a constant, and τ is the applied pulse width. Whalen et al. showed the experimental results fit the following relationship, where

$$E_A(\text{nJ}) = 22\tau^{0.3}$$

where the absorbed energy is in units of nJ and the pulse width is in units of ns. In the experimental results, the GaAs MESFET with the Ti/Pt/Au gate structure achieved higher energy-to-failure compared to the Al gate structures.

From this relationship, knowing the pulse width τ, we can define an average absorbed power, where

$$\langle P_{\mathrm{A}} \rangle \tau = E_{\mathrm{A}}$$
$$\langle P_{\mathrm{A}} \rangle \tau \approx E_{\mathrm{A}} = K_{\mathrm{E}} \tau^{n}$$

or expressing as a function of an averaged absorbed power-to-failure

$$\langle P_{\mathrm{A}} \rangle = \frac{K_{\mathrm{E}} \tau^{n}}{\tau} = \frac{K_{\mathrm{E}}}{\tau^{1-n}}$$

From the experimental studies, the classification of the GaAs MESFET failure mechanisms due to the short pulse width phenomenon included the following [10–15]:

- MESFET gate-to-source gold metal filaments.

- MESFET gold metal short.

- MESFET gold metallurgy whisker growth.

- MESFET GaAs material gate-to-source damage.

- MESFET GaAs material gate-to-drain damage.

- MESFET gate metallurgy damage.

The RF ESD design practices are as follows:

- Using a nanosecond RF pulse system, the RF device energy-to-failure can be obtained by evaluation of the difference between the integral of the incident power and the integral of the reflected power (e.g., integration over the pulse time).

- A physical model can be established for the relationship between the absorbed power and the pulse width, in the form of

$$E_{\mathrm{A}} = K_{\mathrm{E}} \tau^{n}$$

7.3 GALLIUM ARSENIDE ESD FAILURES IN ACTIVE AND PASSIVE ELEMENTS

Early works on GaAs devices were based on GaAs MESFET devices. Rubalcava *et al.* [16] completed extensive studies on the ESD robustness of GaAs MESFET technology. The evaluation quantified the ESD robustness of the transistors, the passives and the metallurgy used in GaAs technology.

In GaAs MESFET-based technologies, the ESD failure mechanisms occurred in the following structures [16]:

- interconnects;

- thin film nichrome resistors;

- metal–insulator–metal (MIM) capacitors.

Interconnects. Interconnect failures were observed by Rubalcava *et al.* in both the first and second level metal films. In the first level metal, a titanium/palladium/gold (TiPdAu) failed from an ESD pulse of 1331 V (HBM level); the thickness of the film was 0.44 μm and the width was 2 μm. In this interconnect structure, the TiPdAu film material was blown 'open,' leaving residual TiPdAu on the wafer surface; the ESD damage pattern was not dissimilar to the HBM testing of modern tantalum nitride/tantalum/copper (TaN/Ta/Cu) damascene interconnects. Based on scaling from aluminum interconnects, a 0.5 μm Ti/Al/Ti film would fail at approximately 500 V HBM/μm of width; hence estimating the result, it would be anticipated that ESD failure of the TiPdAu would exceed 800 V HBM level (note that a TaN/Ta/Cu interconnect would achieve approximately 1600 V HBM). Rubalcava *et al.* [16] noted that in the HBM ESD stressing of the film, there was little indication of resistance shifts in the structure. In the second metal level film, the interconnect consisted of titanium/gold (Ti/Au) materials; the thickness of the second level metal film was 1.9 μm. For a 4 μm wide Ti/Au film, the HBM ESD stress of 2000 V did not lead to ESD failure.

Thin Film Metal Resistors. Rubalcava *et al.* [16] completed HBM ESD stress on nichrome (NiCr) resistors with Ti/Pd/Au contacts. The ESD robustness of the NiCr resistor was linear with the resistor width. The ESD HBM robustness of the NiCr resistors was 24–29 mA/μm of resistor width. The ESD stress failure mechanism in the NiCr resistor occurred at the contact of metal interfaces and the center of the resistor element (note: the center point of the resistor is the location of maximum peak temperature).

Metal–Insulator–Metal (MIM) Capcacitors. Metal–insulator–metal capacitor elements demonstrated failure from ESD HBM stress. These capacitors were formed using the TiPdAu metal level as the lower plate and the TiAu metal level for the upper plate, with a 0.2 μm silicon nitride (SiN) dielectric. Leakage current was used as the failure criteria for these capacitor elements. The ESD HBM failures were 'pin-hole' like failures in the central region of the capacitor element (at HBM levels of 300 to 500 V). In one instance, there was a capacitor corner failure at 600 V HBM and an edge capacitor edge failure at 1500 V (e.g., both associated with enhanced electric fields).

7.4 GALLIUM ARSENIDE HBT DEVICES AND ESD

In the present day GaAs power amplifier circuits, ESD failure can occur in both the GaAs heterojunction bipolar transistor (HBT) as well as the passive elements. Circuit elements that are prone to failure during GaAs ESD stress are:

- GaAs HBT;
- resistor element;
- capacitor element;
- bond pad.

From HBM testing of GaAs circuits, the failure levels can range from 200 to 2000 V levels. The GaAs failure mechanisms from HBM ESD stress consist of GaAs HBT emitter–base failure at test increments from 200 to 500 V HBM levels (e.g., wide band applications). From the RF functional testing, it can be observed that there is degradation in the gain and output power. In some narrow band GaAs applications, ESD HBM levels of 500–2000 V HBM are achievable; in this case, when the GaAs transistor does not fail, the passive elements such as capacitor elements lead to failure. In RF circuits, capacitor elements can exist between the RF input signals and ground potential. During ESD testing for HBM frequencies, these capacitor elements can lead to HBM ESD failures.

From machine model (MM) stress of GaAs devices, typical MM ESD failure levels occur from 30 to 100 V MM levels. In an MM stress, GaAs failure mechanism occurs in the GaAs HBT device as well as the in the passive elements. As in the HBM stress condition, GaAs HBT emitter–base junction damage is evident in the levels of 50–100 V. Additionally, passive elements such as bias resistors fail under MM stress. In the MM stress, the observed RF degradation mechanism included the GSM output power.

GaAs failures manifested themselves from the following RF functional tests:

- output power;

- gain;

- GSM maximum power.

In the GaAs HBM and MM stress failure, ESD failure occurred in the RF input and RF output. The failure paths consist of the following cases:

- RF input to ground.

- RF output to ground.

- Voltage bias circuit to ground.

7.4.1 Gallium Arsenide HBT Device ESD Results

Gallium arsenide (GaAs) HBT devices are designed for both power and performance. GaAs HBT devices have the following advantages for ESD robustness:

- *Power Design Point*: GaAs HBT regions are optimized for power and performance.

- *Low Resistance Cathode*: GaAs HBT collector regions have low resistance because of the formation of the subcollector and reach-through structures.

- *Low Capacitance Cathode*: GaAs HBT subcollectors have low subcollector-to-substrate junction capacitance.

- *Performance Optimized Anode*: GaAs HBT base regions are optimized for both resistance and capacitance to achieve a high unity power current gain cutoff frequency.

- *Interconnect Structures*: Interconnects are suitable for high current because of the electromigration and power requirements.

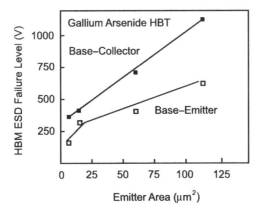

Figure 7.3   GaAs HBT ESD robustness as a function of the emitter area for the base–collector and the base–emitter region

Unfortunately, the ESD disadvantages of the GaAs HBT device are the size of the emitter, the emitter–base sensitivity, and the interconnect metallurgical penetration.

Figure 7.3 shows a plot of the HBM ESD robustness of a GaAs HBT. As the GaAs HBT emitter area increases from 0 to 100 μm^2, the GaAs base–collector HBM ESD results show an linear increase from 300 to 1250 V HBM; concurrently, the GaAs emitter–base junction HBM ESD level increases from 200 to 600 V HBM. From the plot, it is evident that the GaAs HBT emitter–base junction metric of ESD robustness per unit area is less than the base–collector region. These results show that it is possible to utilize the GaAs HBT base–collector region to provide RF ESD protection. Additionally, these results also show that as the GaAs HBT emitter is scaled to small dimensions, the ESD sensitivity of the GaAs HBT is below 250 V HBM levels, where the first failure level will occur due to GaAs HBT emitter–base ESD stress.

An RF ESD design practice is as follows:

- GaAs HBT devices used in a base–collector forward bias operation are suitable for ESD protection elements because of the GaAs HBT power design point: low resistance heavily doped collector region, low subcollector-to-substrate capacitance, high base doping concentration, and good interconnect structures.

- As GaAs HBT devices are scaled to small emitter area of less than 5 μm^2, the ESD robustness of the GaAs HBT device will decrease below 250 V HBM levels, with the first failure mechanism being the failure of the GaAs HBT emitter–base region.

7.4.2 Gallium Arsenide HBT Diode Strings

In the integration of GaAs applications, GaAs HBT structures can be used in back-to-back diode string configurations.

Figure 7.4 shows an example of a back-to-back RF GaAs diode string circuit formed from GaAs HBT element in an emitter–base diode configuration (e.g., base–collector junction is

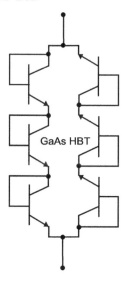

Figure 7.4 GaAs HBT-based ESD back-to-back emitter–base diode string network (e.g., base–collector junction shorted)

shorted). In this application, the GaAs HBT-based diode back-to-back diode string is configured with the emitter–base junction.

The advantages of an emitter–base HBT back-to-back diode string are the following:

- *Low Capacitance*: Small emitter–base region allows for low capacitive coupling between power supplies (e.g., when used between two ground rails such as RF V_{SS} and digital V_{SS}.

- *Noise Isolation*: GaAs HBT base–emitter junction isolates from the substrate when the collector regions are electrically connected (note: in the present figure, the base–collector is shown to be electrically connected).

In Figure 7.5, the transmission line pulse (TLP) I–V characteristic is shown for a GaAs HBT-based emitter–base diode string network. The TLP test system was a 50 Ω commercial TLP system and the pulse waveform was 100 ns pulse width. From the TLP I–V characteristic, it was observed that the turn-on of the GaAs HBT E–B diode string occurred after the sum of the GaAs emitter–base forward voltages. The GaAs HBT device has a linear dynamic on-resistance until current saturation occurs at high current; this is followed by a negative resistance region at avalanche breakdown. From the leakage measurements, it was noted that the failure of the GaAs HBT diode string occurred at the negative resistance transition.

The RF ESD design practices are as follows:

- GaAs HBT-based emitter–base diode string structures can be utilized for back-to-back diode string implementations between RF, analog, and digital sections to minimize capacitance coupling between the rails.

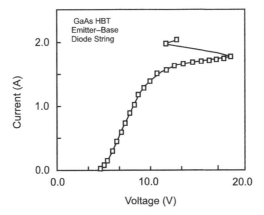

Figure 7.5 GaAs HBT emitter–base (E–B) back-to-back ESD diode string transmission line pulse (TLP) *I–V* characteristic

- GaAs HBT E–B diode strings TLP *I–V* characteristics demonstrate a turn-on voltage, a linear dynamic on-resistance region, a saturation region, followed by avalanche breakdown and a reverse resistance region.

- GaAs HBT structures using emitter–base (E–B) diode strings with an isolated collector regions can provide noise isolation from the substrate region.

7.5 GALLIUM ARSENIDE HBT-BASED PASSIVE ELEMENTS

Gallium arsenide (GaAs) HBT can be used for GaAs derivative passive elements. The GaAs HBT derivative elements can be utilized for ESD protection.

7.5.1 GaAs HBT Base–Collector Varactor

In GaAs technology, the base–collector region of a GaAs HBT device can be used as a base–collector varactor element. Varactor elements provide variable capacitance as a function of the d.c. bias condition. GaAs HBT-based varactor structures can be used in a forward bias mode of operation for RF ESD elements. The RF ESD advantages of utilization of GaAs HBT derivative elements are as follows:

- *Power Design Point*: GaAs HBT regions are optimized for power and performance.

- *Low Resistance Cathode*: GaAs HBT collector regions have low resistance because of the formation of the subcollector and reach-through structures.

- *Low Capacitance Cathode*: GaAs HBT subcollectors have low subcollector-to-substrate junction capacitance.

- *Performance Optimized Anode*: GaAs HBT base regions are optimized for both resistance and capacitance to achieve a high unity power current gain cutoff frequency.

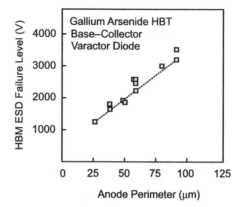

Figure 7.6 GaAs HBT derivative base–collector varactor HBM ESD robustness as a function of the anode perimeter

- *Interconnect Structures*: Interconnects suitable for high current because of the electromigration and power requirements.

Figure 7.6 shows a plot of the HBM ESD robustness of a GaAs HBT base–collector varactor as a function of the anode perimeter. As the GaAs HBT base–collector varactor increases from 25 to 100 μm, the HBM ESD level increases from 1 to 3 kV. The HBM ESD robustness per unit perimeter is greater than 25 V/μm.

Using the GaAs HBT derivate base–collector structure, different types of ESD protection networks can be synthesized:

- *Rail-to-Rail Back-to-Back ESD Diode Strings*: GaAs HBT base–collector diode strings between RF, analog, and digital ground rails.

- *V_{CC} to V_{SS} ESD Power Clamps*: GaAs HBT diode strings can be formed from the base–collector GaAs HBT varactor structure between the V_{CC} and V_{SS} ground.

Figure 7.7 shows an example of a GaAs HBT base–collector back-to-back diode string. In the case of using a GaAs HBT device, the emitter–base structure will be shorted. Using a GaAs HBT device, the emitter region interferes with the base–collector area. Hence, it is advantageous to utilize the GaAs HBT derivative element without the emitter structure.

The RF ESD design practices are as follows:

- GaAs HBT-based derivatives are suitable for ESD protection elements because of the GaAs HBT power design point: low resistance heavily doped collector region, low subcollector-to-substrate capacitance, high base doping concentration, and good interconnect structures.

- GaAs HBT-based base–collector varactors can be utilized to form diode string circuits between RF, analog, and digital sections for ESD protection between the power rails.

- GaAs HBT B–C varactor structures are advantageous over the usage of GaAs HBT devices because of the increased base area under the emitter region.

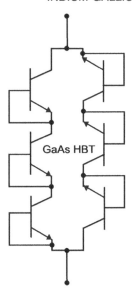

Figure 7.7 GaAs HBT-based ESD back-to-back base–collector diode string network (e.g., base–emitter junction shorted)

7.6 GALLIUM ARSENIDE TECHNOLOGY TABLE OF FAILURE MECHANISMS

In GaAs technology, the failure mechanisms are different for the different transistor structure [10–24]. Table 7.1 shows a summary of GaAs MESFET, HEMT, and HBT devices. In GaAs MESFET structures, the common failure mechanism was gate-to-source failure. In GaAs HEMT devicesalso the gate-to-source failures were evident. In GaAs HBT devices, the lowest failure level is the emitter–base metallurgical region. In ESD networks, construction of GaAs E–B diode emitter–base junction, the failure level was a function of the emitter area. Additionally, GaAs-based B–C varactor structure ESD robustness was a function of the base, with the base–collector metallurgical junction as the limiting failure mechanism.

7.7 INDIUM GALLIUM ARSENIDE AND ESD

In the III–V semiconductor arena, there are many combinations of indium and GaAs-based semiconductor devices (Table 7.2) [35–37]. In electrical and optical electronic industry, there are a wide variety of structures for photodiodes, light emitting diodes (LED), and lasers to transistor devices that are sensitive to ESD phenomenon:

- GaInAs/GaAlAs MODFETs;
- InGaAs lasers;
- InGaAs PIN diodes;

Table 7.1 HBM ESD failures in GaAs-based devices

GaAs Device	Length (μm)	Width (μm)	HBM failure level (V)	Failure mechanism
MESFET	0.8 (Gate)	1500	+100	Gate-to-source
	1.0 (Gate)	2800	+350	Gate-to-source
	2.0 (Gate)	220	+50	Gate-to-source
HEMT	0.5 (Gate)	280	+90	Gate-to-source
HBT		20	+120	Emitter–base metallurgical junction
		6×20	+480	Emitter–base metallurgical junction
HBT Diode string (E–B)				Emitter–base metallurgical junction
HBT B–C varactor				Base–collector metallurgical junction

- InGaAs/InP photodetectors;

- InP heterojunction bipolar transistors.

In these structures, changes in the functional response, whether electrical or optical response, become modified as a result of ESD stress [33–37]. In the case of lasers,

Table 7.2 HBM ESD failures in indium GaAs-based devices

InGaAs and InP Device	Structure	Width	HBM failure level (V)	Failure mechanism	Reference
InGaAs Laser	GaAs substrate	10 mil	−1500 V		Bock [35]
		20 mil	−4000 V		
		30 mil	−5500 V		
GaInAs/GaAlAs MODFET	GaAs substrate InGaAs channel			Gate Burnout	Vashenko [36]
InGaAs/InP photodetectors		50–80 μm	700 V	Photo-luminescence degradation	Neitzert [37]
InGaAs PIN	InGaAsP substrate		−200 V		Bock [35]

ESD-induced damage can influence the laser gain characteristics [33]. In the case of photodiode elements, dark current and photoluminescence can be influenced by ESD-induced stress [37]. ESD stress can induce physical damage to the heterostructure, which can introduce degradation in quantum efficiency, gain characteristics, and lead to latent mechanisms in the electrical and optical response.

7.8 INDIUM PHOSPHIDE (InP) AND ESD

In recent times, significant interest exists in InP HBT devices for high-speed wire communication and optical communication systems [40–46]. Design and reliability of the InP HBT devices have been evaluated and it demonstrated high unity current gain cutoff frequencies. At that time, there had been no reported measurements of ESD robustness of InP HBT devices.

7.9 SUMMARY AND CLOSING COMMENTS

Electrostatic discharge (ESD) in GaAs technology is important with the growth of cell phone and wireless industry. In this chapter we showed early GaAs ESD works in GaAs MESFETs; ESD measurements from Rubalcava, Whalen, and Bock are also discussed. ESD failure mechanisms associated with the GaAs technology metallurgy played a key role in the failure levels. ESD measurements have demonstrated that the emitter–base region is the most sensitive region of GaAs HBT devices.

In the next chapter, we will discuss the ESD protection of bipolar receiver and transmitter circuits. Additionally, ESD bipolar power clamps that utilize Zener diode breakdown, bipolar junction transistor breakdown, and forward bias trigger networks will be discussed. Moreover, capacitive-coupled bipolar ESD power clamps will be shown. Silicon bipolar, silicon germanium, silicon germanium carbon, gallium arsenide, and indium gallium arsenide ESD power clamp networks will also be shown.

PROBLEMS

1. From the Johnson Limit relationship for maximum voltage–unity current cutoff frequency product,

$$V_m f_T = \frac{E_m v_S}{2\pi}$$

   show the magnitude of the right hand side of the equation for gallium arsenide.

2. From the Johnson Limit relationship, compare Silicon Germanium technology.

3. From the comparison of gallium arsenide to silicon germanium, derive the application breakdown voltages that are possible, given the SiGe HBT and GaAs HBT are the same unity current gain cutoff frequency.

4. From the Johnson Limit power formulation,

$$(P_m X_C)^{1/2} f_T = \frac{E_m v_S}{2\pi}$$

where the reactive impedance X_c is defined as

$$X_C = \frac{1}{2\pi f_T C_0}$$

derive the ratio of the power of the maximum power of a gallium arsenide device and silicon germanium device. Assuming the same unity current gain cutoff frequency, derive the relationship between the two power values.

5. Derive a relationship of a GaAs HBT in a diode configuration used in a diode string. Assume N GaAs HBT devices in series.

6. Derive a relationship of a trigger network consisting of N GaAs HBT diode-configured elements in series with an open-base GaAs HBT in a collector-to-emitter configuration. Compare this to a trigger network of M SiGe HBT diode-configured elements in series with an open-base SiGe HBT (in a common-emitter configuration). Compare the two trigger voltages. Show the ratio of the two devices. Assuming the same trigger voltage condition, derive the relationship between the N GaAs HBT devices and the M SiGe HBT devices.

7. Given a GaAs HBT amplifier, GaAs HBT diode strings can be used on the RF output. Derive the relationship of a 'diode string' network consisting of GaAs HBT with M elements in one direction in parallel with a GaAs HBT RF output. Assume a GaAs HBT with an open-base connections and the GaAs HBT is in a 'common-emitter' configuration. Derive the number of diodes needed to avoid collector-to-emitter breakdown of the GaAs HBT.

8. Given a GaAs HBT amplifier, GaAs HBT diode strings can be used on the RF output. Derive the relationship of a 'diode string' network consisting of GaAs HBT with M elements in one direction in parallel with a GaAs HBT RF output. Assume that the GaAs HBT has a resistor load on its base input and the GaAs HBT is in a 'common-emitter' configuration. Derive the number of diodes needed to avoid collector-to-emitter breakdown of the GaAs HBT (hint: BV_{CER} is required).

REFERENCES

1. Yuan JS. *SiGe, GaAs, and InP heterojunction bipolar transistors*. New York: John Wiley and Sons, Inc.; 1999.
2. Liu W. *Fundamentals of III-V devices: HBTs, MESFETs, and HFETs/HEMTs*. New York: John Wiley and Sons, Inc.; 1999.
3. Williams RE. *Gallium arsenide processing technologies*. Dedham, MA: Artech House; 1985.
4. Christou A. *Reliability of gallium arsenide MMICs*. Chichester: John Wiley and Sons, Inc.; 1992.
5. Konig U. SiGe and GaAs as competitive technologies for RF-applications. *Bipolar/BiCMOS Circuit Technology Meeting (BCTM)*, 1998. p. 87–92.
6. Otsuji T. Present and future of high-speed compound semiconductors IC's. In: *Compound semiconductor integrated circuits*. London: World Scientific Publishing Co. Pte. Ltd. p. 1–26. April 2003.

7. Voinigescu SP, McPherson DS, Pera F, Szilagyi S, Tazlauana M, Tran H. A comparison of silicon and III–V technology performance and building blocks for 10 and 40 Gb/s optical networking IC's. *Compound semiconductor integrated circuits*. London: World Scientific Publishing Co. Pte. Ltd. p. 27–58. April 2003.

8. LeBlanc R, Gasmi A, Zahzouh M, Smith D, Auvray F, Moron J, *et al.* GaAs PHEMT chip sets and IC processes for high-end fiber optic applications. In: *Compound semiconductor integrated circuits*. London: World Scientific Publishing Co. Pte. Ltd. p. 91–110. April 2003.

9. Berger O. GaAs HBT for power amplifier applications. *Bipolar/BiCMOS Circuit Technology Meeting (BCTM)*, 2004. p. 52–5.

10. Whalen J. The RF pulse susceptibility of UHF transistors. *IEEE Transaction of Electromagnetic Compatibility* 1975;**EMC-17**;220–5.

11. Whalen J, Domingos H. Square pulse and RF pulse overstressing of UHF transistors. *Proceedings of the Electrical Overstress/Electrostatic Discharge (EOS/ESD) Symposium*, 1979. p. 140–6.

12. Whalen JJ, Calcatera MC, Thorn M. Microwave nanosecond pulse burnout properties of GaAs MESFETs. *Proceedings of the IEEE MTT-S International Microwave Symposium*, May 1979. p. 443–5.

13. Whalen JJ, Calcatera MC, Thorn M. Microwave nanosecond pulse burnout properties of GaAs MESFETs. *IEEE Transactions on Microwave Theory and Techniques* 1979;**MTT-27**;1026–31.

14. Thorn ML. *Failure diagnostics of gallium arsenide metal semiconductor field effect transistors overstressed with high power microwave pulses.* M.S. Thesis, Department of Electrical Engineering, State University of New York at Buffalo; September 1979.

15. Whalen JJ, Thorn ML, Rastefano E, Calcatera MC. Microwave nanosecond pulse burnout properties of one micron MESFETs. *Proceedings of the Electrical Overstress/Electrostatic Discharge (EOS/ESD) Symposium*, 1979. p. 147–57.

16. Rubalcava AL, Stunkard D, Roesch WJ. Electrostatic discharge effects on gallium arsenide integrated circuits. *Proceedings of the Electrical Overstress/Electrostatic Discharge (EOS/ESD) Symposium*, 1986. p. 159–65.

17. Anderson Jr WT, Chase EW. Electrostatic discharge (ESD) thresholds for GaAs FETs, gallium arsenide reliability workshop. *Proceedings of the Gallium Arsenide Integrated Circuits Symposium*, Grenelefe, Florida, 1986.

18. Anderson WT, Buot FA, Christou A, Anand Y. High power pulse reliability of GaAs Power FETs. *Proceedings of the International Reliability Physics Symposium (IRPS)*, 1986. p. 144–9.

19. Wurfl J, Hartnagel HL. Field and temperature dependent life-time limiting effects of metal GaAs interfaces of device structures studied by XPS and electrical measurements. *Proceedings of the International Reliability Physics Symposium (IRPS)*, 1986. p. 138–43.

20. Buot FA, Anderson WT, Christou A, Sleger KJ, Chase EW. Theoretical and experimental study of sub-surface burnout and ESD in GaAs FETs and HEMTs. *Proceedings of the International Reliability Physics Symposium (IRPS)*, 1987.

21. Wurfl J, Hartnagel HL, Gupta RP. *Proceedings from the SPIE Symposium on Advances in Semiconductor Structures*, 1987. p. 218–22.

22. Anderson Jr WT, Chase EW. Electrostatic discharge effects in GaAs FETs and MODFETS. *Proceedings of the Electrical Overstress/Electrostatic Discharge (EOS/ESD) Symposium*, 1987. p. 205–7.

23. Buot FA, Sleger KJ. Numerical simulation of hot electron effects on source-drain burnout characteristics of GaAs power FETs. *Solid State Electronics* 1984;**27**:1067–81.

24. Rubalcava AL, Roesch WJ. Lack of latent and cumulative ESD effects on MESFET-based GaAs ICs. *Proceedings of the Electrical Overstress/Electrostatic Discharge (EOS/ESD) Symposium*, 1988. p. 62–4.

25. Chase E. Theoretical Study of expected EOS/ESD sensitivity of III–V compound semiconductor devices. *Proceedings of the Electrical Overstress/Electrostatic Discharge (EOS/ESD) Symposium*, 1988. p. 65–9.

26. Grovenor CRM. *Microelectronic materials*. Bristol: Adam Hilger; 1989.

27. Bock K, Fricke K, Krozer V, Hartnagel HL. *Proceedings of the International Workshop on Solid State Power Amplifiers for Space Applications*, ESA/ESTEC, Noordwijk, the Netherlands, November 15–16, 1989.

28. Bock K, Fricke K, Hartnagel HL. *Conference on ESD of the Fachkreis ESD, ESD Forum*, Grainau FRG, February 6–7, 1990.

29. Bock K, Fricke K, Krozer V, Hartnagel HL. Improved ESD protection of GaAs FET microwave devices by new metallization strategy. *Proceedings of the Electrical Overstress/Electrostatic Discharge (EOS/ESD) Symposium*, 1990. p. 193–6.

30. Bilodeau TM. The electrostatic discharge sensitivity of GaAs MMIC amplifiers. *Proceedings of the Electrical Overstress/Electrostatic Discharge (EOS/ESD) Symposium*, 1990. p. 131–6.

31. Domingos H. Input protection design. *Tutorial Notes of the Electrical Overstress/Electrostatic Discharge (EOS/ESD) Symposium*, 1990.

32. Bock K, Lipka KM, Hartnagel HL. *Proceedings of the ESREF Symposium*, Bordeaux, France, October 7–10, 1991.

33. DeChairo LF, Unger BA. Degradation in InGaAsP semiconductor lasers resulting from human body model ESD. *Journal of Electrostatics* 1993;**29**:227–50.

34. Ragle D, Decker K, Loy M. ESD effects on GaAs MESFET lifetime. *Proceedings of the International Reliability Physics Symposium (IRPS)*, March 1993. p. 352–6.

35. Bock K. ESD issues in compound semiconductor high-frequency devices and circuits. *Proceedings of the Electrical Overstress/Electrostatic Discharge (EOS/ESD) Symposium*, 1997. p. 1–12.

36. Vashenko V, Sinkevitch V, Martynov J. Gate burnout of small signal MODFETs at TLP stress. *Proceedings of the Electrical Overstress/Electrostatic Discharge (EOS/ESD) Symposium*, 1997. p. 13–8.

37. Neitzert HC, Cappa V, Crovato R. Influence of device geometry and inhomogeneity on the electrostatic discharge sensitivity of InGaAs/InP avalanche photo-detectors. *Proceedings of the Electrical Overstress/Electrostatic Discharge (EOS/ESD) Symposium*, 1997. p. 18–26.

38. Lipka KM, Schmid P, Birk M, Slingart B, Kohn E, Schneider J, *et al.* Novel concept for high level overdrive tolerance of GaAs based FETs. *Proceedings of the Electrical Overstress/Electrostatic Discharge (EOS/ESD) Symposium*, 1997. p. 27–32.

39. Yamada FM, Oki AK, Kaneshiro EN, Lammert MD, Gutierrez-Aitken AL, Hyde JD. ESD Sensitivity of various diode protection circuits implemented in a production 1 μm GaAs HBT technology. *Proceedings of the Gallium Arsenide Reliability Workshop*, 1999. 139–46.

40. Raghavan G, Sokolich M, Stanchina E. Indium phosphide IC's unleash the high frequency spectrum. *IEEE Spectrum Magazine* 2000;**37**(10):47–52.

41. Jalali B, Pearton SJ. *InP HBTs: growth, processing and applications*. Norwood, MA: Artech House; 1995.

42. Katz A. *Indium phosphide and related materials: processing, technology, and devices*. Norwood, MA: Artech House; 1992.

43. Hanson AW, Stockman SA, Stillman GE. InP/In(0.53)Ga(0.47)As heterojunction bipolar transistors with a carbon-doped base grown by MOCVD. *IEEE Electron Device Letters* 1992;**EDL-13**:504.

44. Gee RC, Chin TP, Tu CW, Asbeck PN, Lin CL, Kirchner PD, *et al.* InP/InGaAs heterojunction bipolar transistors grown by gas-source molecular beam epitaxy with carbon-doped base. *IEEE Electron Device Letters* 1992;**EDL-13**:247.

45. Van der Zanden K, Schreurs DMMP, Menozzi R, Borgaarino M. Reliability testing of InP HEMT's using electrical stress methods. *IEEE Transactions on Electron Devices* 1999;**ED-46**(8):1570–6.

46. Kiziloglu K, Thomas III S, Williams Jr F, Paine BM. Reliability and failure criteria for AlInAs/GaInAs/InP HBTs. *Proceedings of the Indium Phosphide Related Materials Conference*, May 2000. p. 294–7.

47. Sun M, Xie K, Lu Y. Robust PIN photodiode with a guard ring protection structure. *IEEE Transactions of Electron Devices* 2004;**ED-51**(6):833–8.

48. Sun M, Lu Y. A new ESD protection structure for high-speed GaAs RF ICs. *IEEE Electron Device Letters* 2005;**EDL-26**(3):133–5.

49. DeChairo LF, Unger BA. Degradation in InGaAsP semiconductor lasers resulting from human body model ESD. *Journal of Electrostatics* 1993;**29**:227–50.

50. Ma Y, Li GP. A novel on-chip ESD protection circuit for GaAs HBT RF power amplifiers. *Proceedings of the Electrical Overstress/Electrostatic Discharge (EOS/ESD) Symposium*, 2002. p. 83–91; *Journal of Electrostatics* 2003;**59**:211–27.

51. Ma Y, Li GP. InGaP/GaAs HBT DC-20 GHz distributed amplifier with compact ESD protection circuits. *Proceedings of the Electrical Overstress/Electrostatic Discharge (EOS/ESD) Symposium*, 2004. p. 50–4.

52. Shrier K, Truong T, Felps J. Transmission line pulse test methods, test techniques, and characterization of low capacitance voltage suppression device for system level electrostatic discharge compliance. *Proceedings of the Electrical Overstress/Electrostatic Discharge (EOS/ESD) Symposium*, 2004. p. 88–97.

53. Shrier K, Jiaa C. ESD enhancement of power amplifier with polymer voltage suppressor. *Proceedings of the Taiwan Electrostatic Discharge Conference (T-ESDC)*, 2005. p. 110–5.

8 Bipolar Receiver Circuits and Bipolar ESD Networks

In this chapter, bipolar receiver networks and bipolar-based ESD elements, as well as bipolar ESD networks appropriate for RF applications are discussed. In the discussion of bipolar receiver networks, focus will not be to address the RF circuit design, but instead the interest will be to use them as case studies of ESD failure mechanisms in RF bipolar circuits. The chapter will then shift toward discussing ESD elements that are suitable for RF applications. It will close with the discussion of RF bipolar-based ESD power clamps for both homojunction and heterojunction technologies.

8.1 BIPOLAR RECEIVERS AND ESD

Receiver circuits are very important in RF ESD design because of the ESD sensitivity of these networks. Typically, the receiver circuits are the most sensitive circuits in a chip application. Receiver performance has a critical role in the semiconductor chip performance. The primary reasons for this are as follows:

- RF receiver circuits are small in physical area.

- Receiver performance requirements limit the ESD loading allowed on the receiver. MOSFET gate area, bipolar emitter area, and electrical interconnect wiring widths impact the receiver performance.

- RF receiver input are electrically connected to either the MOSFET gate (in a CMOS receiver) where the MOSFET gate dielectric region is the most ESD sensitive region in RF MOSFET receiver networks. RF MOSFET gate dielectric thickness and channel length is scaled to achieve RF performance objectives.

- RF receiver inputs are electrically connected to the bipolar base region (in a bipolar receiver), where the bipolar transistor emitter-base metallurgical junction is the most ESD sensitive

ESD: RF Technology and Circuits   Steven H. Voldman
© 2006 John Wiley & Sons, Ltd

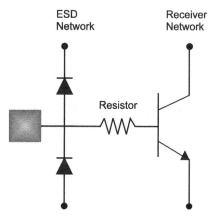

Figure 8.1 Bipolar receiver network

region of the bipolar transistor. The pipolar transistor emitter-base region is scaled to achieve RF performance objectives.

- Both the MOSFET gate dielectric region and the bipolar transistor base region are the most sensitive regions of the structures.

- RF receivers require low series resistance.

8.2 SINGLE ENDED COMMON-EMITTER RECEIVER CIRCUITS

Receiver circuits are a common ESD sensitive circuit in bipolar and BiCMOS technologies. Bipolar receiver circuits typically consist of *npn* bipolar transistor configured in a common-emitter configuration (Figure 8.1). For bipolar receivers, the input pad is electrically connected to the base contact of the *npn* transistor, with the collector connected to V_{CC} either directly or through additional circuitry. The *npn* bipolar transistor emitter is electrically connected to V_{SS}, or through a emitter resistor element, or additional circuitry.

In bipolar receiver networks, for a positive polarity human body model (HBM) ESD event, as the base voltage increases, the base-to-emitter voltage increases leading to a forward biasing of the base–emitter junction. The base–emitter junction becomes forward active, allowing current to flow from the base to the emitter region. Typically in bipolar receiver networks, the physical size of the emitter regions are small. When the ESD current exceeds the safe operation area (SOA), degradation effects occur in the bipolar transistor. The bipolar device degradation is observed as a change in the transconductance of the bipolar transistor. From the electrical parametrics, the unity current cutoff frequency, f_T, decreases with increased ESD current levels. From a f_T–I_C plot, the f_T magnitude decreases with ESD pulse events, leading to a decrease in the peak f_T.

For a negative pulse event, the base–emitter region is reverse biased. As the voltage on the signal pad decreases, the base–emitter reverse-bias voltage across the base–emitter metallurgical junction increases. Avalanche breakdown occurs in the emitter–base metallurgical junction, leading to an increase in the current that flow through the emitter and base regions; this leads to thermal runaway and bipolar second breakdown in the bipolar transistor. The

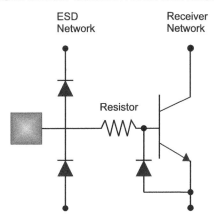

Figure 8.2 Bipolar receiver network with ESD diode

experimental results show that the negative polarity failure level has a lower magnitude compared to the positive polarity failure level.

One common ESD design solution used to provide improved ESD results in a single-ended bipolar receiver network is to place a p–n diode element in parallel with the npn bipolar transistor emitter–base junction (Figure 8.2). Using a parallel element, the p–n junction is placed in such a manner that the anode is electrically connected to the npn emitter and the cathode is electrically connected to the npn base region; this ESD element serves as a bypass element avoiding avalanche breakdown of the npn base–emitter junction. The diode element is placed local to the npn transistor element to avoid substrate resistance from preventing early turn-on of the ESD diode element. Note that this element is analogous to the CDM solution used in CMOS receiver networks. For a bipolar transistor, it is serving for events from both the signal pad and potential events from the emitter electrode.

8.2.1 Single-Ended Bipolar Receiver with D.C. Blocking Capacitors

In RF bipolar receivers, metal–insulator–metal (MIM) capacitors are used between the signal pad and the base electrode. For positive or negative mode polarity events, the MIM capacitor can fail because of dielectric degradation. Without ESD protection on the receiver network, the ESD failure levels of the receiver network will be limited by the MIM capacitor element. An ESD solution to prevent ESD failure in these RF bipolar receivers is to use a p–n diode element in parallel with the MIM capacitor element. The p–n diode element can be in a reverse configuration so that it serves as a parallel capacitor element and does not allow a d.c. voltage to be transmitted between the signal pad and the bipolar receiver base element. The functional disadvantage of the p–n element is the impact of the effective quality factor 'Q' of the capacitor element.

In high-performance RF bipolar receiver networks, capacitors are placed in series between the RF (in) input pad structure and the RF bipolar receiver network. Figure 8.3 shows an example of an input signal pad RF (in), a d.c. blocking capacitor, and the input of a RF npn bipolar receiver in a common-emitter configuration. The capacitor structure can be a MIM, a metal–ILD–metal, or a vertical parallel plate (VPP) capacitor structure.

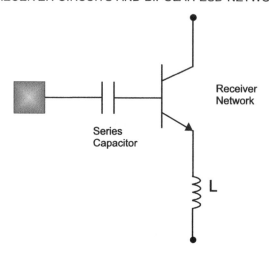

Figure 8.3 Bipolar receiver with d.c. blocking capacitor

In the case of no ESD protection network, when an ESD event occurs on the RF (input) signal pad, the voltage differential of the ESD event is across the d.c. blocking capacitor and the base electrode of the RF *npn* transistor. When the substrate is grounded and the RF signal pad is pulsed with a positive potential, the voltage is across the capacitor structure and the forward-biased *p–n* metallurgical junction of the RF *npn* receiver network. For a negative polarity ESD event, a negative voltage is across the capacitor structure and reverse-biased RF *npn* base electrode. In both cases, a voltage stress occurs across the d. c. blocking capacitor.

Experimental results show that HBM ESD failure levels for positive polarity HBM events will occur from 200 to 500 V HBM for typical MIM capacitor structures using a typical silicon germanium *npn* receiver network. For negative polarity HBM events, the ESD failure level will occur at −100 V HBM ESD levels.

The ESD design practices to improve the ESD robustness of the RF bipolar receiver are as follows:

- placement of a ESD network between the RF signal pad and the d.c. blocking capacitor;

- placement of a ESD element in parallel with the d.c. blocking capacitor element;

- placement of a decoupling element between the d.c. blocking capacitor and the RF *npn* input network to avoid 'pinning' across the capacitor element.

8.2.2 Single-Ended Bipolar Receiver with D.C. Blocking Capacitors and ESD Protection

Direct current blocking capacitors can lead to ESD failure in RF bipolar networks when placed between the RF (input) signal pad and the RF bipolar receiver network. As discussed in the last section, ESD solutions can be instituted that provide ESD protection.

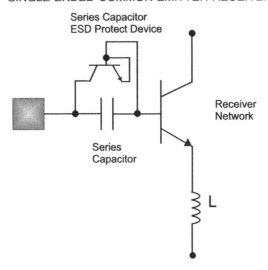

Figure 8.4   Bipolar receiver with d.c. blocking capacitor and parallel ESD element

As an example of an ESD solution to avoid failure of the d.c. blocking element, a parallel element can be placed in parallel with the capacitor. In this solution, the parallel element must provide a breakdown voltage lower than the d.c. blocking capacitor.

Figure 8.4 shows an example of utilizing an RF *npn* bipolar transistor in parallel with the d.c. blocking capacitor. The RF *npn* bipolar transistor is placed in a manner that it acts as a capacitor element. When transistor breakdown voltage occurs, the bypass transistor element avoids dielectric breakdown of the capacitor element.

8.2.3 Bipolar Single-Ended Common-Emitter Receiver Circuit with Feedback Circuit

Bipolar receivers utilize feedback elements between the output and input electrical nodes to improve electrical stability.

8.2.3.1 Bipolar single-ended common-emitter circuit with resistor feedback element

In bipolar single-ended common-emitter receivers, resistor elements are integrated between the collector and base element, as shown in Figure 8.5. Figure 8.5 shows the feedback resistor element is electrically connected between the transistor collector and base.

With the introduction of the resistor element, the feedback element is in parallel with base–collector junction. In testing between the input node and the power supply voltage, the feedback resistor, and the base–collector metallurgical junction are in parallel. During a positive pulse event in the *npn* base region, *p–n* base–collector junction will forward bias, when a forward-bias V_{BE} occurs across the resistor element. At this point, the *npn* base–collector junction allows current to flow to the power supply.

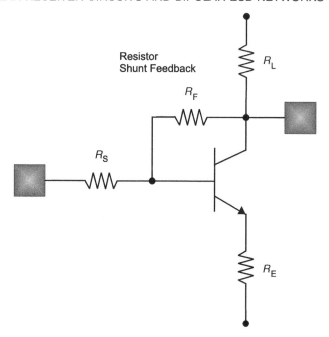

Figure 8.5   Bipolar single-ended common-emitter receiver with resistor feedback element

In testing between the collector and emitter, the feedback resistor also is involved in base-coupling and influencing the ESD response. In this configuration, without the resistor element, the current conduction will flow from the collector to emitter region at the BV_{CEO} avalanche breakdown voltage. During a positive pulse event into the *npn* collector region, *p–n* base–collector junction will forward bias, when a forward-bias V_{BE} occurs across the resistor element. At this point, the *npn* base–collector junction allows current to flow to the power supply.

With an ESD stress between the RF (out) and RF (in), the parallel combination of the feedback resistor and base–collector junction are between the input and output pins. In a reverse biased state, two possible failure mechanisms can occur; the failure of the feedback resistor element, or the failure of the base–collector junction.

8.2.3.2 Bipolar single-ended common-emitter receiver circuit with resistor–capacitor feedback element

Bipolar receiver utilize feedback elements between the output and input electrical nodes can influence the biasing conditions of the base region. To avoid d.c. biasing isolation between the input and output, a capacitor is placed in series with the feedback resistor (Figure 8.6).

With the introduction of the capacitor in series with the resistor element, the resistor–capacitor network becomes vulnerable to ESD events. With an ESD stress between the RF (out) and RF (in), the feedback capacitor–resistor, and base–collector junction are between

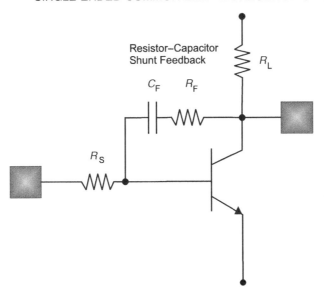

Figure 8.6   Bipolar single-ended common-emitter receiver with capacitor–resistor feedback element

the input and output pins. In a reverse biased state, two possible failure mechanisms can occur; the failure of the capacitor in the feedback element, or failure of the reverse-biased base–collector junction.

8.2.4 Bipolar Single-Ended Common-Emitter Receiver Circuit with Emitter Degeneration

Bipolar receiver utilize degeneration elements in series with the emitter structure to improve both electrical and thermal stability. Emitter degeneration can be achieved using either resistor elements or inductor elements.

Figure 8.7 shows an example of a single-ended common-emitter bipolar receiver with emitter degeneration using a resistor element. With the introduction of the resistor element, the emitter degeneration element is in series with the current flow through the transistor. From an ESD perspective, the emitter resistor limits the current flow from collector-to-emitter (akin to resistor ballasting techniques). Observing the pulse response, the resistor element increases the bipolar transistor 'on-resistance.' Additionally, for ESD events on the RF (input), the resistor is in series with the base–emitter junction limiting the current flow from the RF (input) to the reference ground. Given that the ESD networks exist on the RF receiver pad, the emitter degeneration element buffers the current flow through the receiver element, hence diverting the ESD current through the ESD network.

Note that for a machine model (MM) pulse waveform, the addition of the emitter resistance further dampens the waveform response.

Figure 8.8 shows another example of a single-ended common-emitter receiver network with an emitter degeneration resistor, a resistor shunt feedback element, and a series base

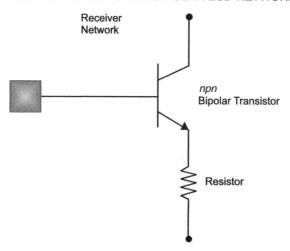

Figure 8.7 Bipolar single-ended common-emitter receiver with emitter degeneration resistor element

resistor. In this implementation, during an RF (input) stress, the series base resistor and the emitter degeneration resistor limit the current flow through the base–emitter junction, protecting the base–emitter region from electrical overstress. As in the last implementation, the emitter degeneration resistor adds to the dynamic on-resistance during ESD testing.

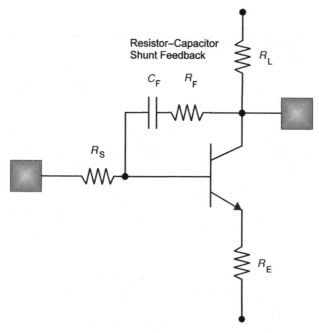

Figure 8.8 Bipolar single-ended common-emitter receiver with emitter degeneration resistor element, shunt feedback resistor, and series resistor element

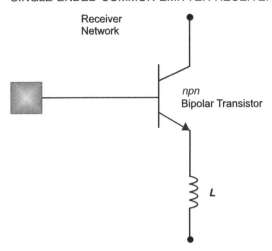

Receiver
Network

npn
Bipolar Transistor

L

Figure 8.9 Bipolar single-ended common-emitter receiver with emitter degeneration inductor element

Additionally, a concern in this network is ESD stress from the RF (output) to the RF (input). In this case, the electrical stress occurs across the resistor shunt feedback element and the collector–base junction.

Although emitter degeneration is desired for RF circuits, the use of resistors for emitter degeneration is not desirable because of the noise generation from the resistor element. Inductors can also serve to provide emitter degeneration without introducing lossy or noise generation. Figure 8.9 shows a circuit schematic of a single-ended common-emitter RF receiver network with inductor emitter degeneration.

In the case of a positive polarity ESD event on the RF (input) pad, the ESD current will flow through the base–emitter junction. The degeneration inductor provides a voltage drop in series, $L(di/dt)$, limiting the current flow through the emitter–base junction. Given that ESD networks exist on the RF receiver pad, the emitter degeneration current buffers the current flow through the receiver element. This diverts the ESD current through the ESD network and the alternative current loop through the power supplies. In the case of positive polarity events on the RF (output) pad, the inductor provides a voltage drop during current transients, limiting the peak current through the *npn* bipolar transistor.

In the case of a HBM pulse, after the switch is closed, note that the combination of the ESD source term and the circuit introduces an RCL response. Hence, the inductor adds to the oscillation of the incoming waveform.

Figure 8.10 shows an example of a single-ended common-emitter RF receiver with both inductor emitter degeneration and resistor–capacitor feedback network. As discussed in the other implementations, the emitter degeneration inductor reduces the overcurrent condition in the RF base–emitter junction. But, the RC feedback network is vulnerable during ESD stress from RF (in) to RF (out) pin-to-pin testing.

In addition to the above implementations, the d.c. blocking capacitors can be added to the RF network, as well as d.c. bias inductors. Figure 8.11 shows the single-ended common-emitter network with inductor emitter degeneration, RC feedback element, and d.c. blocking

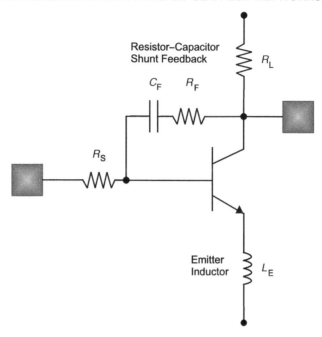

Figure 8.10 Bipolar single-ended common-emitter receiver with emitter degeneration inductor element and RC feedback element

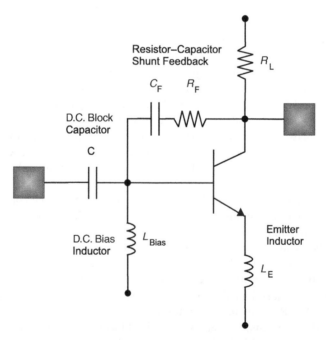

Figure 8.11 Bipolar single-ended common-emitter receiver with emitter degeneration inductor element, RC feedback element, d.c. blocking capacitor, and inductor-bias element

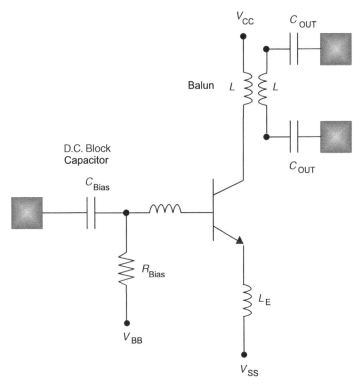

Figure 8.12   Bipolar single-ended common-emitter receiver with Balun output

capacitor as well as the base region d.c. bias inductor element. In this network, the d.c. blocking capacitor and the feedback capacitor are vulnerable for pin-to-pin testing.

8.2.5 Bipolar Single-Ended Common Emitter Circuit with Balun Output

In RF receiver networks, the RF (out) signal is electrically isolated using inductor and capacitor elements. Figure 8.12 contains a circuit schematic of a single-ended common-emitter circuit with a balun output. The network also contains other functions, such as inductor emitter degeneration elements and d.c. biasing elements. Note that in this circuit, d.c. blocking capacitors exist on the single input pad as well as the output pads.

From an ESD perspective, the balun electrically isolates the output pads from the electrical circuit except through the coils of the balun element. Additionally, from an ESD testing perspective, pin-to-pin stress can be applied between the two output signals. In that case, the elements between the two output pads consist of the two d.c. blocking capacitor and balun output inductor, forming an LC network. Hence, as a key point, the ESD robustness of the output stage will be a function of the ESD robustness of the passive inductor and capacitor elements. Moreover, ESD failure can occur across the balun dielectric layers between the two inductive coil regions.

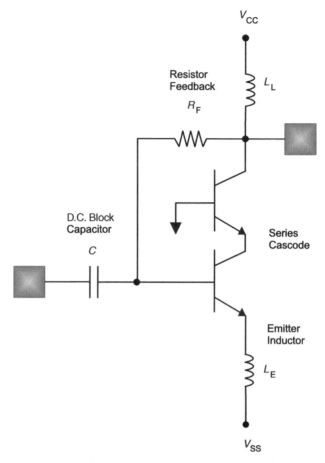

Figure 8.13 Bipolar single-ended cascode common-emitter receiver with emitter inductive degeneration element, resistor feedback element, and d.c. blocking capacitor

8.2.6 Bipolar Single-Ended Series Cascode Receiver Circuits

Series cascode networks, in a common-emitter configuration, are used in bipolar receiver networks to allow for a higher power supply voltage. Series cascode networks are common in both CMOS and bipolar receiver and transmitter architectures.

Figure 8.13 shows an example of a single-ended series cascode common-emitter bipolar receiver with inductive emitter degeneration and series shunt feedback element. In this configuration, with two bipolar transistors in series, and inductive elements, the current flow is limited between the V_{CC} and V_{SS} power supply until both transistors undergo bipolar breakdown. Additionally, with the introduction of the second transistor element, the collector of the *npn* receiver input is electrically separated from the output pad. As stated in the single common-emitter network, the feedback element and the series input resistor are vulnerable during pin-to-pin testing of RF (out) to RF (in).

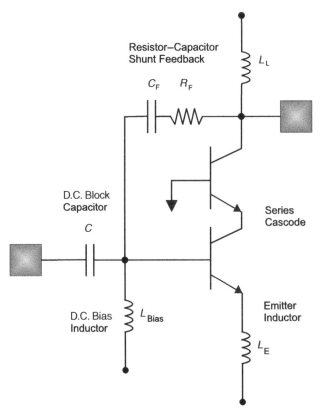

Figure 8.14 Bipolar single-ended cascode common-emitter receiver with emitter degeneration inductor element, RC-feedback, d.c. blocking capacitor, and d.c. bias inductor

Figure 8.14 shows another example of a series cascode receiver network with inductive degeneration, an RC-feedback element, d.c. blocking capacitor, and d.c. biasing inductor. In this implementation, as in the last implementation, the feedback network formed by the resistor and capacitor passive elements can be vulnerable during ESD pin-to-pin testing.

In another bipolar receiver network, the base inductor is used in series with the bipolar receiver base element (Figure 8.15). Additionally, the network contains a bias network. In this network, the base inductor provides a constraint on current excursions into the base element, providing a $L(di/dt)$ voltage drop from transient phenomenon. From an ESD perspective, this reduces the vulnerability of the emitter–base junction from current transients.

Lastly, series cascode receivers can also introduce baluns on the RF (out) pads (Figure 8.16). Between the output balun and the cascode transistors, the input bipolar transistor is electrically isolated from the output pads. As in our prior discussion, the output pin ESD robustness will be a function of the ESD robustness of the passive capacitor and balun elements.

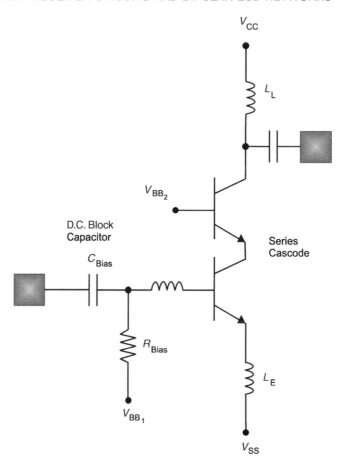

Figure 8.15 Bipolar single-ended cascode common-emitter receiver with base inductor and bias networks

8.3 BIPOLAR DIFFERENTIAL RECEIVER CIRCUITS

Differential receiver networks are used to improve the signal-to-noise ratio in bipolar networks. Differential receiver networks use a differential pair of identical *npn* bipolar transistors in a common-emitter mode. For differential bipolar receivers, two input pads are electrically connected to the base contacts of the identical *npn* transistors and the two emitters are connected together. Below the emitter connection, additional circuitry, a current source or a resistor element is commonly used (Figure 8.17).

One of the unique problems with differential receiver networks is the pin-to-pin ESD failure mechanism. In ESD testing, we can apply an ESD pulse event to one of the two differential signal pads, and use the second differential signal pad as the ground reference. In differential pair bipolar receiver networks, for positive polarity HBM ESD events, as the base voltage increases the base-to-emitter voltage of the first transistor increases leading to forward bias of the base–emitter junction. The base–emitter junction

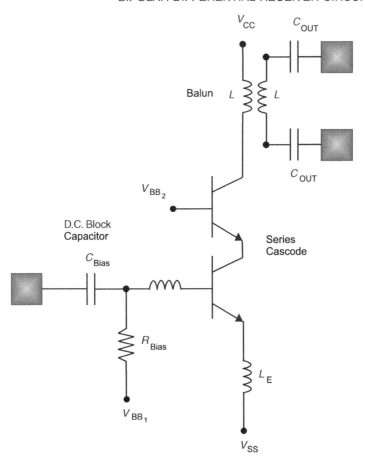

Figure 8.16   Bipolar single-ended cascode common-emitter receiver with output balun

becomes forward active, leading to current flow from the base to the emitter region. For the second *npn* bipolar transistor, the base–emitter region is reverse biased. As the voltage on the first signal pad increases, the base–emitter reverse-bias voltage across the second transistor base–emitter metallurgical junction increases. Avalanche breakdown occurs in the emitter–base metallurgical junction, leading to an increase in the current flow through the emitter and base regions; this leads to a thermal runaway and a bipolar second breakdown in the grounded second bipolar transistor of the differential pair. Note that the degradation of the second transistor prior to the first transistor can also lead to a differential offset hampering the matching of the two sides of the differential pair. It is possible that the failure criteria are associated with a *npn* mismatch prior to the ESD failure of either *npn* device.

An ESD design solution used to provide improved ESD results in a differential pair bipolar receiver network is to place a *p–n* diode element in parallel with the *npn* bipolar transistor emitter–base junction (Figure 8.18). Using a parallel element, the *p–n* junction is placed such that the anode is electrically connected to the *npn* emitter and the cathode is electrically to the base region. In this manner, an alternate forward-bias current path is established between both sides of the differential pair. Secondly, another method is to introduce a back-to-back diode

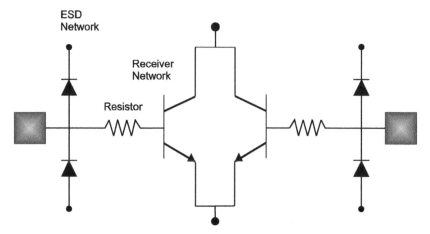

Figure 8.17 Bipolar differential pair receiver network

string between both sides of the differential pair. This has the advantage of allowing a higher current between both sides of the differential pair, but at the same time the disadvantage of asymmetry matching and capacitance loading performance degradation.

8.3.1 Bipolar Differential Cascode Common-Emitter Receiver Circuits

Differential receiver networks can also be placed in a series cascode configuration. Figure 8.19 contains a circuit schematic of differential cascode receiver. In a differential cascode circuit, with the cascoded bipolar transistor isolates, the RF (out) and RF (in) signal pads through the second cascoded transistor element. In this network, the ESD pin-to-pin

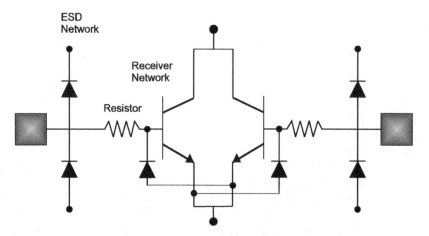

Figure 8.18 Bipolar differential pair with ESD protection to avoid pin-to-pin ESD failures

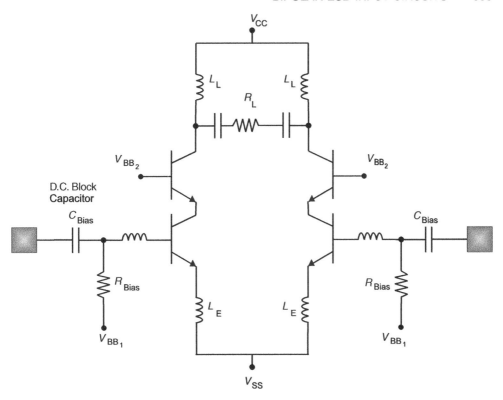

Figure 8.19   Differential cascode receiver

concern exists across the differential pair. In this implementation, with the cascode network, the breakdown voltage between the output signals to the ground is the sum of the two collector-to-emitter breakdown voltages of the transistors.

8.4 BIPOLAR ESD INPUT CIRCUITS

The ESD robustness of silicon homojunction bipolar transistors has been of interest since the 1960s. In the early development years, the focus was primarily on the understanding of second breakdown phenomena; this was followed by the focus on the ESD sensitivity of bipolar transistors, the statistical variation of the ESD robustness, and modeling.

In recent times, with the focus on silicon germanium (SiGe), silicon germanium carbon (SiGeC), gallium arsenide (GaAs), and indium phosphide (InP) technologies, the interest and focus on the ESD protection of bipolar technology has shifted toward the new technologies. The interest in ESD protection of heterojunction bipolar transistors (HBT) has only been of recent interest with the introduction of these new technologies.

In bipolar technology, digital, analog, and RF circuits are constructed of bipolar transistors and bipolar transistor derivatives, as well as active and passive elements. In a BiCMOS technology, digital circuits are constructed of CMOS elements, whereas analog and RF circuits are CMOS, bipolar, or a hybrid mixture of CMOS and bipolar elements. Bipolar elements can be utilized either as three- or two-terminal devices. As a three-terminal ESD device, the physics of the bipolar transistor and the breakdown mechanisms come into play. As a two-terminal ESD device, the transistor is configured as a diode element, and the various regions of the bipolar transistor can be utilized. In the three-terminal configuration, the bipolar circuits can play the role as an ESD input circuit or an ESD power clamp; the distinction is the physical size and application.

There are advantages for the usage of bipolar transistors and their derivatives for ESD input protection. They are as follows:

- The processes and structures of the bipolar transistors are inherently current-carrying devices.

- The processes and structures of the bipolar transistors are optimized as both a performance and power design.

- Bipolar transistors are optimized as low noise elements.

- Bipolar transistors have low resistance collector regions from subcollector implants and reach-through implants serving as a low resistance cathode for bipolar transistors and corresponding bipolar derivative elements.

- Bipolar transistors are designed to minimize substrate injection minimizing latchup risks.

- Bipolar transistors contain low capacitance collector-to-substrate junctions using steep subcollector-to-substrate doping profiles.

- Bipolar transistors provide deep trench (DT) and trench isolation (TI) structures that provide low capacitance collector-to-substrate junctions.

- Bipolar transistors provide DT and TI structures that eliminate parasitic lateral devices, lower junction capacitance, minimize substrate injection and latchup concerns.

- Bipolar transistor junction breakdown voltages (e.g., breakdown voltage collector-to-emitter BV_{CEO}) scales with technology generation and performance allowing for scalable ESD solutions.

- Bipolar transistor technologies typically provide one, two, or three different transistor types with different corresponding breakdown voltages that can be utilized for ESD input protection.

- Bipolar transistors are also offered in different design configurations in both single and multiemitter designs.

- Bipolar transistor technology has many transistor derivative devices (e.g., base–collector varactors, hyperabrupt [HA] varactors, Schottky diodes, silicon controlled rectifiers [SCR], resistors, and capacitor elements).

- BiCMOS technology has many implants for hybrid bipolar–CMOS elements.

- The operation of the bipolar transistors is 'isolated' from the substrate potential, allowing for negative substrate potentials.

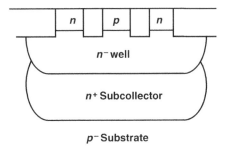

Figure 8.20   ESD shallow trench isolation (STI) p^+/n-well diode element with subcollector

- BiCMOS technology has both 'isolated' and 'nonisolated' MOSFET structures.

On the first point, ESD events are associated with high current conditions. The processes and structures of the bipolar transistors are inherently current-carrying devices and are designed to support high currents. In contrast, MOSFETs are voltage-controlled devices. Bipolar homojunction transistors (BJT) and HBT devices are optimized to carry a high collector-to-emitter current for a given unity current gain cutoff frequency (f_T). Hence, a bipolar transistor is well suited for an ESD input protection between input pads and ground when used in a collector-to-emitter configuration.

The processes and structures of the bipolar transistors are also optimized as both a performance and power design point. In most bipolar and BiCMOS technologies, the design point objective is both for a high unity current gain cutoff frequency (f_T) and a high unity power gain cutoff frequency (f_{MAX}). Technology design point objective is a dual design point that is intended to achieve both goals. CMOS technology design point is focused on MOSFET performance, which is a frequency design point. For optimization of a technology for ESD performance, a power design point is more desirable.

Bipolar transistors are optimized as low noise elements. Bipolar transistors are inherently lower noise sources compared to MOSFET transistors. MOSFET transistor elements are a source of $1/f$ noise. Hence in many bipolar or BiCMOS applications, large MOSFET ESD input circuits are sources of noise in a bipolar or BiCMOS receiver.

Bipolar transistors have heavily doped subcollector regions that can serve as low resistance shunt for ESD events. Bipolar transistors have low resistance collector regions from subcollector implants and reach-through implants serving as a low resistance cathode for bipolar transistors and corresponding derivative elements. The low resistance cathode can be utilized in bipolar transistors for ESD events in bipolar transistors, base–collector diode varactors, HA base–collector varactors, Schottky diodes, and collector-to-substrate diodes. Figure 8.20 shows an example of utilization of subcollectors in CMOS-based ESD elements.

Bipolar transistors are designed to minimize substrate carrier injection. Bipolar transistors have heavily doped subcollector regions. Auger recombination occurs in the collector region leading to a high recombination of minority carriers prior to injection into the substrate region.

Bipolar transistors contain low capacitance collector-to-substrate metallurgical junctions using steep subcollector-to-substrate doping profiles. Bipolar transistors provide DT and TI structures which provide low capacitance collector-to-substrate junctions [1–3]. The low capacitance junction elements can reduce the capacitance loading when this element is used

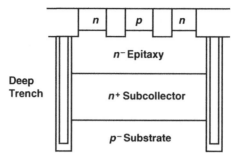

Figure 8.21   Deep trench (DT) defined ESD p^+/n-well diode structure

on input nodes in double-diode ESD networks. Deep trench and TI structures can be added to CMOS-base elements to provide improved isolation, lower noise, and lower capacitance (Figures 8.21 and 8.22) [3].

Deep trench structures for the high-performance bipolar transistors are also suitable to reduce parasitic elements which lead to undesirable ESD failure, or latchup. Bipolar transistors provide DT and TI structures that eliminate parasitic lateral devices and minimize substrate injection and latchup concerns [3].

In ESD design, it is important to have a technology that has a low-voltage, and scalable voltage trigger element, tracking, and scalable voltage trigger elements to be used as a technology is scaled to future technologies. Bipolar transistor junction breakdown voltages (e.g., breakdown voltage collector-to-emitter BV_{CEO}) scale with technology generation and performance allowing for scalable ESD solutions. As a result, scalable voltage triggering ESD input elements are valuable for scaling with receiver networks.

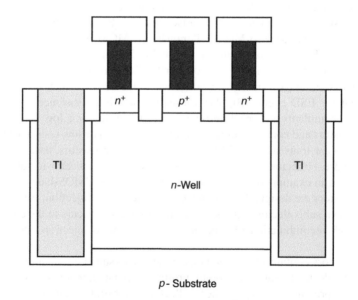

Figure 8.22   Trench isolation (TI)-defined p^+/n-well diode structure

Additionally, bipolar and BiCMOS transistor technologies typically provide one, two, or three different performance transistors with different corresponding breakdown voltages that can be utilized for ESD input protection networks. The bipolar transistor uses 'pedestal' implants in the intrinsic section of the transistor for improving performance at higher collector currents; these collector implants allow for different base-to-collector breakdown voltages. The utilization of the breakdown condition as to initiate ESD input circuits allows for different initiation or triggering conditions.

Additionally, bipolar transistors are formed in different design configurations in single-emitter and multiple-emitter designs. With three electrodes, the layout order of the designs allows for different permutations and combinations. For example, for a single emitter design, the layout order can be base–emitter–collector (BEC) or emitter–base–collector (EBC). This family can be extended with two base regions and with two collector regions forming CBEBC or BCECB configurations. With multiple emitter, base, and collector regions, a large family of different transistors can be constructed for ESD protection networks. This is a wider flexibility than is obtainable in a CMOS-based technology.

Bipolar and BiCMOS technologies also provide a wide variety of device elements and transistor derivatives that are suitable within ESD networks. Bipolar transistor technology have many transistor derivative devices (e.g., base–collector varactors, HA varactors, Schottky diodes, SCR, resistors, and capacitor elements). With these elements, a wide variety of possible frequency-triggered or voltage-triggered ESD input circuits can be constructed.

8.4.1 Diode-Configured Bipolar ESD Input Circuits

In bipolar and BiCMOS technologies, transistors can be configured to utilize the metallurgical junctions in a forward-bias or reverse-bias mode of operation. Bipolar transistors can be used in diodic operation for ESD protection. A common ESD circuit for applications is the use of a double-diode ESD network (Figure 8.23). A bipolar transistor can be configured to utilize the base–emitter or the base–collector junction in a forward-bias mode for ESD protection.

For ESD protection input networks, the base–collector junction can be used as an ESD diode element in silicon bipolar transistors, silicon germanium (SiGe), or silicon germanium carbon (SiGeC) devices for a forward bias mode of operation; this can be used instead of a p^+/n-well diode element used in a CMOS-based ESD double diode network. For an ESD double-diode network, the base region is electrically connected to the input pad and the collector region is electrically connected to the V_{CC} power supply; this will serve to provide protection for positive polarity ESD events (e.g., HBM). For negative polarity ESD events, the base region is electrically connected to the V_{SS} ground rail and the collector region is electrically connected to the input signal pad. In this application, the emitter–base are electrically shorted together to avoid an overstress of the emitter–base junction. In this implementation, the same physical element provides protection for both the negative and the positive polarity events. This implementation also allows for injection to a secondary ground rail for discharging (e.g., analog V_{SS}). In a second implementation, the collector-to-substrate junction can also be used for negative polarity ESD events. For this implementation, the emitter, base, and the collector are electrically

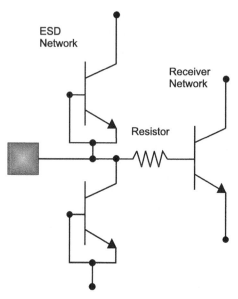

Figure 8.23 Diode-configured bipolar transistor for usage as a double-diode ESD network

shorted together. One of the advantages of using the bipolar transistor elements is the ability to use the DT or the TI structure for providing a low-capacitance collector diode element. Additionally, the capacitance of this region is optimized for the bipolar transistor to provide high performance (which is not the case for the equivalent CMOS n-well structure).

8.4.2 Bipolar ESD Input: Resistor Grounded Base Bipolar ESD Input

Bipolar ESD input devices can be established by the placement of a bipolar transistor in a common-emitter configuration between a pad and ground rail. For example, a bipolar ESD device can be established where the collector is connected to an input pad, the emitter is connected to a V_{SS} power rail. A resistor is placed between the bipolar transistor base electrical connection and the V_{SS} power rail (Figure 8.24).

The collector current can be expressed as a function of an intercept current and a generation current from avalanche generation [4–6],

$$I_C = I_S \exp\left(\frac{qV_{BE}}{kT}\right) + I_{gen}$$

The collector current can also be represented as a function of the product of the collector-to-emitter transport factor and the generation current entering the base electrode,

$$I_C = \alpha I_E + I_{gen}$$

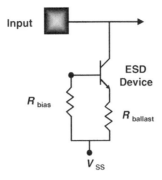

Figure 8.24   Bipolar transistor ESD protection element

The current flowing through the grounded resistor element as the current I_1 that is related to the base resistance term, and the additional resistor bias element R_1 can be represented as [6],

$$I_1 = \frac{V_{BE}}{r_b + R_1}$$

Using the expression for the relationship between the forward-bias voltage and the collector current, the current can be shown as

$$I_1 = \frac{1}{r_b + R_1} \left(\frac{kT}{q}\right) \ln\left(\frac{\alpha I_E}{I_S}\right)$$

From this expression, the partial derivative with respect to the voltage is,

$$\frac{\partial I_1}{\partial V} = \frac{1}{r_b + R_1} \left(\frac{kT}{q}\right) \frac{\partial}{\partial V} \ln\left(\frac{\alpha I_E}{I_S}\right)$$

Taking the derivative of the logarithm term, the relationship can be placed in the following form:

$$\frac{\partial I_1}{\partial V} = \frac{1}{r_b + R_1} \left(\frac{kT}{q}\right) \frac{I_S}{\alpha I_E} \frac{\partial}{\partial V}\left(\frac{\alpha I_E}{I_S}\right)$$

The total current through the bipolar ESD input circuit can be expressed as the product of the multiplication factor and collector current. This can also be expressed as the parallel current through the emitter and the base region,

$$I = MI_C = I_1 + I_E$$

From this expression, the partial derivative of current with respect to voltage, can be expressed as

$$\frac{\partial I}{\partial V} = \frac{\partial I_1}{\partial V} + \frac{\partial I_E}{\partial V}$$

or alternatively, the partial derivative of the emitter current can be expressed as

$$\frac{\partial I_E}{\partial V} = \frac{\partial I}{\partial V} - \frac{\partial I_1}{\partial V}$$

To evaluate the current as a function of voltage, the chain rule of partial differential equations is applied,

$$\frac{\partial I}{\partial V} = \frac{\partial M}{\partial V} I_C + M \frac{\partial I_C}{\partial V}$$

Substituting for current in the expression

$$\frac{\partial I}{\partial V} = \frac{\partial M}{\partial V} (\alpha I_E + I_{gen}) + M \frac{\partial}{\partial V} (\alpha I_E + I_{gen})$$

Assuming that the generation mechanism is voltage independent, then it can be expressed as

$$\frac{\partial I}{\partial V} = \frac{\partial M}{\partial V} (\alpha I_E + I_{gen}) + M \alpha \frac{\partial}{\partial V} (I_E)$$

Hence, the partial derivative of the emitter current with respect to voltage,

$$\frac{\partial I_E}{\partial V} = \frac{\partial I}{\partial V} - \frac{\partial I_1}{\partial V}$$

Substituting in for the base current term,

$$\frac{\partial I_1}{\partial V} = \frac{1}{r_b + R_1} \left(\frac{kT}{q}\right) \frac{I_S}{\alpha I_E} \frac{\partial}{\partial V} \left(\frac{\alpha I_E}{I_S}\right)$$

or

$$\frac{\partial I_1}{\partial V} = \frac{1}{r_b + R_1} \left(\frac{kT}{q}\right) \frac{1}{I_E} \frac{\partial I_E}{\partial V}$$

The partial derivative of the emitter term can be factored and expressed as a function of the partial derivative of the current I [6–8],

$$\frac{1}{r_b + R_1} \left(\frac{kT}{q}\right) \frac{1}{I_E} \frac{\partial I_E}{\partial V} + \frac{\partial I_E}{\partial V} = \frac{\partial I}{\partial V}$$

or

$$\left[1 + \frac{1}{r_b + R_1} \left(\frac{kT}{q}\right) \frac{1}{I_E}\right] \frac{\partial I_E}{\partial V} = \frac{\partial I}{\partial V}$$

This can be expressed as

$$\frac{\partial I_E}{\partial V} = \frac{\dfrac{\partial I}{\partial V}}{\left[1 + \dfrac{1}{r_b + R_1} \left(\dfrac{kT}{q}\right) \dfrac{1}{I_E}\right]}$$

The other form of the partial derivative of the emitter current with respect to the voltage, is

$$\frac{\partial I_E}{\partial V} = \frac{1}{M\alpha} \left[\frac{\partial I}{\partial V} - \frac{\partial M}{\partial V}(\alpha I_E + I_{gen}) \right]$$

Equating the two expressions, the partial derivative of current can be obtained,

$$\frac{\partial I}{\partial V} = \frac{\dfrac{\partial M}{\partial V}(\alpha I_E + I_{gen})}{1 - \dfrac{\alpha M}{\left[1 + \dfrac{1}{r_b + R_1}\left(\dfrac{kT}{q}\right)\dfrac{1}{I_E} \right]}}$$

From this form, the electrical stability of the circuit can be determined. When the partial derivative of current with respect to voltage is positive, the circuit is electrically stable. When the partial derivative of current with respect to voltage is negative, the system is electrically unstable. The condition of electrical instability is when the denominator is equal to zero,

$$1 - \frac{\alpha M}{\left[1 + \dfrac{1}{r_b + R_1}\left(\dfrac{kT}{q}\right)\dfrac{1}{I_E} \right]} = 0$$

Solving for the avalanche multiplication condition where this occurs in this electrical network can be expressed as

$$\alpha M = 1 + \frac{1}{r_b + R_1}\left(\frac{kT}{q}\right)\frac{1}{I_E}$$

Solving for the multiplication term [6],

$$M = \frac{1}{\alpha}\left[1 + \frac{1}{r_b + R_1}\left(\frac{kT}{q}\right)\frac{1}{I_E} \right]$$

The avalanche multiplication form can also be expressed as a voltage relationship, and as a function of the collector-to-base breakdown,

$$M(V) = \frac{1}{1 - \left[\dfrac{V}{BV_{CBO}}\right]^n}$$

From this form, combining the two expressions and solving for the voltage at which this occurs, the voltage of avalanche runaway is obtained as

$$V_T = BV_{CBO}\left[1 - \alpha \left\{ \frac{1}{1 + \dfrac{1}{r_b + R_1}\left(\dfrac{kT}{q}\right)\dfrac{1}{I_E}} \right\} \right]^{n/2}$$

8.5 BIPOLAR-BASED ESD POWER CLAMPS

8.5.1 Bipolar Voltage-Triggered ESD Power Clamps

Bipolar ESD power clamps can be initiated using different triggering methods [1,2]. These can include frequency triggering, capacitance-coupling triggering, and voltage triggering. Voltage-initiated triggered ESD power clamps can utilize the forward-biased networks or reverse-biased voltage breakdown networks. Voltage-triggered bipolar ESD power clamps typically contain a bipolar transistor between the first and second power rail where the first power rail is electrically connected to the bipolar transistor collector, and the second power rail is electrically connected to the bipolar transistor emitter. A bias resistor element is electrically connected to the base of the output clamp device. The bias resistor sets the base to a low potential to prevent the 'turn-on' of the output clamp. Examples of a voltage-triggered bipolar ESD power clamps can consist of the following trigger networks:

- forward-bias diode series configured voltage trigger;

- forward-bias diode Schottky diode configured voltage trigger;

- Zener-breakdown voltage trigger;

- bipolar collector-to-emitter breakdown voltage (BV_{CEO}) trigger.

In all these cases, the output clamp takes advantage of the current-carrying capability of a bipolar output clamp element that can discharge the current from the first to the second power rail. Additionally, when the voltage condition is reached, the trigger current serves as base current to the bipolar output clamp element, initiating the discharge of the ESD event.

8.5.2 Zener Breakdown Voltage-Triggered ESD Power Clamps

For a voltage-triggered bipolar ESD power clamp, where the voltage trigger is associated with a breakdown voltage of a Zener diode structure, the conditions for triggering the circuit are different from the grounded-base ESD power clamp (Figure 8.24). In this implementation, the current flowing through the Zener diode structure must be taken into account in the voltage and current equations.

Expressing the current through the structure and the Zener diode structure,

$$I = MI_C + I_D$$

The bipolar output clamp collector current can be represented by the product of the collector-to-emitter transport efficiency and the emitter current, and the generation current. The total current through the structure can be represented as

$$I = M(\alpha I_E + I_{gen}) + I_D$$

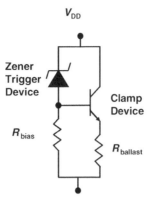

Figure 8.25 Zener-triggered ESD power clamp

Zener-triggered ESD networks have a high trigger voltage. This is an advantage for high-voltage applications, but has limited value for an advanced CMOS technology, a BiCMOS technology or an advanced bipolar technology. Given the scaling of the bipolar output transistor, the triggering of the Zener diode must be below the avalanche breakdown of the output device when utilized as an ESD power clamp. Given the Zener trigger breakdown voltage is above the avalanche condition of the ESD output clamp device, the ESD power clamp will not scale with the output clamp total perimeter. As a result, this network has value in high-voltage applications, or ESD networks that have bipolar transistors with high avalanche conditions (e.g., BV_{CER} or BV_{CEO} exceeds the Zener trigger voltage).

For a voltage-triggered bipolar ESD input device, where the voltage trigger is associated with a breakdown voltage of a Zener diode structure, the conditions for triggering the circuit are different from the grounded-base ESD input. In this implementation, in Figure 8.25, the current flowing through the Zener diode structure must be taken into account in the voltage and current equations [7,8].

Expressing the current through the structure as the current through the Zener diode structure, it was shown by Joshi *et al.* [7,8] as

$$I = MI_C + I_D$$

The bipolar collector current can be represented by the product of the collector-to-emitter transport efficiency and the emitter current, and the generation current. The total current through the structure can be represented as

$$I = M(\alpha I_E + I_{gen}) + I_D$$

To evaluate the stability of the circuit, we can solve for the partial derivative of current with respect to voltage. This can be expressed as follows,

$$\frac{\partial I}{\partial V} = \frac{\partial M}{\partial V}(\alpha I_E + I_{gen}) + M\frac{\partial}{\partial V}(\alpha I_E + I_{gen}) + \frac{\partial I_D}{\partial V}$$

Using Kirchoff's voltage law, the voltage loop about the emitter–base junction can be represented as forward voltage drop across the emitter–base junction, the base resistor, and the bias resistor.

$$V_{BE} = I_1(r_b + R_1) + I_D R_1$$

Solving for the current through the bias resistor, we can express the bias resistor current as a function of the forward voltage, and the Zener diode current,

$$I_1 = \frac{V_{BE}}{r_b + R_1} - \frac{R_1}{r_b + R_1} I_D$$

For stability analysis, it is important to evaluate the current as a function of the voltage,

$$\frac{\partial I_1}{\partial V} = \frac{1}{r_b + R_1} \frac{\partial V_{BE}}{\partial V} - \frac{R_1}{r_b + R_1} \frac{\partial I_D}{\partial V}$$

From the forward bias relationship, we can solve for the change in the forward-bias voltage as a function of the voltage across the ESD bipolar input device,

$$\frac{\partial V_{BE}}{\partial V} \frac{\partial}{\partial V} \left[\frac{kT}{q} \ln \left\{ \frac{\alpha I_E}{I_S} \right\} \right] = \frac{kT}{q} \frac{1}{I_E} \frac{\partial I_E}{\partial V}$$

Substituting in for the derivative of the forward-bias voltage term,

$$\frac{\partial I_1}{\partial V} = \frac{1}{r_b + R_1} \left(\frac{kT}{q} \right) \frac{1}{I_E} \frac{\partial I_E}{\partial V} - \frac{R_1}{r_b + R_1} \frac{\partial I_D}{\partial V}$$

Let us define a resistance value associated with the thermal voltage and the emitter current as,

$$R_E^* = \frac{(kT/q)}{I_E}$$

Then the current through the base can be defined as resistance ratios [7,8]

$$\frac{\partial I_1}{\partial V} = \frac{R_E^*}{r_b + R_1} \frac{\partial I_E}{\partial V} - \frac{R_1}{r_b + R_1} \frac{\partial I_D}{\partial V}$$

Rearranging the terms as a function of the emitter current

$$\frac{R_E^*}{r_b + R_1} \frac{\partial I_E}{\partial V} = \frac{\partial I_1}{\partial V} + \frac{R_1}{r_b + R_1} \frac{\partial I_D}{\partial V}$$

From Kirchoff's current law, the partial derivatives of the current terms can be expressed as

$$\frac{\partial I_E}{\partial V} = \frac{\partial I}{\partial V} - \frac{\partial I_D}{\partial V} - \frac{\partial I_1}{\partial V}$$

Adding the above two equations,

$$\left[1 + \frac{R_E^*}{r_b + R_1}\right]\frac{\partial I_E}{\partial V} = \frac{\partial I}{\partial V} + \left[\frac{R_1}{r_b + R_1} - 1\right]\frac{\partial I_D}{\partial V}$$

Rearranging the terms, this can be expressed as

$$\frac{\partial I_E}{\partial V} = \frac{\dfrac{\partial I}{\partial V} - \left[\dfrac{r_b}{r_b + R_1}\right]\dfrac{\partial I_D}{\partial V}}{\left[1 + \dfrac{R_E^*}{r_b + R_1}\right]}$$

Evaluating the voltage drops in the network, letting V_D be the voltage across the Zener diode element,

$$V_D = V - I_D(r_D + R_1) - I_1 R_1$$

Taking the derivative,

$$\frac{\partial V_D}{\partial V} = \frac{\partial V}{\partial V} - (r_D + R_1)\frac{\partial I_D}{\partial V} - R_1\frac{\partial I_1}{\partial V}$$

where substitutions are made for the derivatives

$$\frac{\partial I_1}{\partial V} = \frac{R_E^*}{r_b + R_1}\frac{\partial I_E}{\partial V} - \frac{R_1}{r_b + R_1}\frac{\partial I_D}{\partial V}$$

and

$$\frac{\partial I_E}{\partial V} = \frac{\dfrac{\partial I}{\partial V} - \left[\dfrac{r_b}{r_b + R_1}\right]\dfrac{\partial I_D}{\partial V}}{\left[1 + \dfrac{R_E^*}{r_b + R_1}\right]}$$

$$\frac{\partial V_D}{\partial V} = 1 - (r_D + R_1)\frac{\partial I_D}{\partial V} - R_1\left[\frac{R_E^*}{r_b + R_1}\frac{\partial I_E}{\partial V} - \frac{R_1}{r_b + R_1}\frac{\partial I_D}{\partial V}\right]$$

then

$$\frac{\partial V_D}{\partial V} = 1 - \left[(r_D + R_1) - \frac{R_1^2}{r_b + R_1}\right]\frac{\partial I_D}{\partial V} - R_1\left[\frac{R_E^*}{r_b + R_1}\frac{\partial I_E}{\partial V}\right]$$

This can be expressed as

$$\frac{\partial V_D}{\partial V} = 1 - \left[(r_D + R_1) - \frac{R_1^2}{r_b + R_1}\right]\frac{\partial I_D}{\partial V} - R_1\left[\frac{R_E^*}{r_b + R_1}\right]\frac{\frac{\partial I}{\partial V} - \left[\frac{r_b}{r_b + R_1}\right]\frac{\partial I_D}{\partial V}}{\left[1 + \frac{R_E^*}{r_b + R_1}\right]}$$

From the chain rule of differentiation,

$$\frac{\partial I_D}{\partial V} = \frac{\partial I_D}{\partial V_D}\frac{\partial V_D}{\partial V}$$

Substituting in the Zener diode derivative expression [7,8],

$$\frac{\partial V_D}{\partial V} = 1 - \left[(r_D + R_1) - \frac{R_1^2}{r_b + R_1}\right]\left(\frac{\partial I_D}{\partial V_D}\frac{\partial V_D}{\partial V}\right) - R_1\left[\frac{R_E^*}{r_b + R_1}\right]\frac{\frac{\partial I}{\partial V} - \left[\frac{r_b}{r_b + R_1}\right]\left(\frac{\partial I_D}{\partial V_D}\frac{\partial V_D}{\partial V}\right)}{\left[1 + \frac{R_E^*}{r_b + R_1}\right]}$$

Factoring the terms,

$$\frac{\partial V_D}{\partial V}\left[1 + \left((r_D + R_1) - \frac{R_1^2}{r_b + R_1}\right)\frac{\partial I_D}{\partial V_D} - R_1\frac{\left(\frac{R_E^*}{r_b + R_1}\right)\left(\frac{r_b}{r_b + R_1}\right)}{\left[1 + \frac{R_E^*}{r_b + R_1}\right]}\frac{\partial I_D}{\partial V_D}\right]$$

$$= 1 - R_1\left[\frac{R_E^*}{r_b + R_1}\right]\frac{\frac{\partial I}{\partial V}}{\left[1 + \frac{R_E^*}{r_b + R_1}\right]}$$

$$\frac{\partial V_D}{\partial V}\left[1 + \left\{\left((r_D + R_1) - \frac{R_1^2}{r_b + R_1}\right) + R_1\frac{\Gamma\left(\frac{r_b}{r_b + R_1}\right)}{[1 + \Gamma]}\right\}\frac{\partial I_D}{\partial V_D}\right] = 1 - R_1\left(\frac{\Gamma}{1 + \Gamma}\right)\frac{\partial I}{\partial V}$$

Dividing both sides by the bracketed expression, solving for the derivative of the Zener diode voltage,

$$\frac{\partial V_D}{\partial V} = \frac{\left[1 - R_1\left(\frac{\Gamma}{1 + \Gamma}\right)\frac{\partial I}{\partial V}\right]}{1 + \frac{\partial I_D}{\partial V_D}\left[r_D + \left(\frac{\Gamma}{1 + \Gamma}\right)(r_b \parallel R_1)\right]}$$

From the chain rule,

$$\frac{\partial I_D}{\partial V} = \frac{\partial I_D}{\partial V_D}\frac{\partial V_D}{\partial V}$$

Then we can express the current as,

$$\frac{\partial I_D}{\partial V} = \frac{\left(\dfrac{\partial I_D}{\partial V_D}\right)\left[1 - R_1\left(\dfrac{\Gamma}{1+\Gamma}\right)\dfrac{\partial I}{\partial V}\right]}{1 + \dfrac{\partial I_D}{\partial V_D}\left[r_D + \left(\dfrac{\Gamma}{1+\Gamma}\right)(r_b \parallel R_1)\right]}$$

Solving for the current as a function of the voltage [7,8],

$$\frac{\partial I}{\partial V} = \frac{\left(\dfrac{\partial M}{\partial V}\right)[\alpha I_E + I_{gen}]\Theta}{1 - \dfrac{\alpha M}{1+\Gamma} + R_1\Theta\dfrac{\Gamma}{1+\Gamma}}$$

$$\Theta = \frac{\left(\dfrac{\partial I_D}{\partial V_D}\right)\left[1 - \alpha M\left(\dfrac{r_b}{r_b + R_1}\right)\left(\dfrac{1}{1+\Gamma}\right)\right]}{1 + \dfrac{\partial I_D}{\partial V_D}\left[r_D + \left(\dfrac{1}{1+\Gamma}\right)(r_b \parallel R_1)\right]}$$

From this form, the electrical stability of the circuit can be determined. When the partial derivative of current with respect to voltage is positive, the circuit is electrically stable. When the partial derivative of current with respect to voltage is negative, the system is electrically unstable. The condition of electrical instability is when the denominator is equal to zero, or

$$\frac{\partial I}{\partial V} = \frac{\left(\dfrac{\partial M}{\partial V}\right)[\alpha I_E + I_{gen}]\Theta}{1 - \dfrac{\alpha M}{1+\Gamma} + R_1\Theta\dfrac{\Gamma}{1+\Gamma}} = \infty$$

From this expression, when the denominator is equal to zero, the system transitions from a stable network to an unstable network. Hence,

$$1 - \frac{\alpha M}{1+\Gamma} + R_1\Theta\frac{\Gamma}{1+\Gamma} = 0$$

Solving for the avalanche multiplication value that leads to instability, the following expression is obtained,

$$\alpha M = (1+\Gamma)\left[1 + R_1\Theta\frac{\Gamma}{1+\Gamma}\right]$$

8.5.3 BV$_{CEO}$ Voltage-Triggered ESD Power Clamps

A bipolar-based ESD power clamp that utilizes the breakdown of a bipolar transistor in a collector-to-emitter configuration can be synthesized using the first transistor for the trigger element and the second transistor as the output clamp device [9–12]. A BV$_{CEO}$ voltage-triggered bipolar ESD power clamps contains an output bipolar transistor between the first and second power rail where the first power rail is electrically connected to the bipolar transistor collector, and the second power rail is electrically connected to the bipolar transistor emitter. A bias resistor element is electrically connected to the base of the output clamp device. The bias resistor sets the base to a low potential to prevent the 'turn-on' of the output clamp.

Bipolar BV$_{CEO}$ breakdown voltage-triggered ESD power clamp in bipolar and BiCMOS technologies have ESD advantages as follows:

- low trigger voltages conditions;

- scalable;

- compatibility and design integration with analog and RF circuits;

- compatibility with bipolar transistors;

- use of supported bipolar transistor (e.g., nonuse of parasitic devices);

- circuit simulation;

- utilization of multiple transistors;

- low noise source;

- use for positive or negative polarity power supplies.

The BV$_{CEO}$ breakdown voltage-triggered ESD power clamp can utilize a transistor in a common-emitter mode and initiate the output clamp at this voltage condition. A unique aspect of this implementation is by using the BV$_{CEO}$ condition; there is an inherent interrelation with the unity current cutoff frequency of the transistor. From the Johnson Limit relationship, its power formulation is given as

$$(P_m X_C)^{1/2} f_T = \frac{E_m v_S}{2\pi}$$

where P_m is the maximum power, X_C the reactance $X_C = \frac{1}{2}(\prod f_T C_{bc})$, f_T the unity current gain cutoff frequency, E_m the maximum electric field, and v_S the electron saturation velocity. The product of the maximum voltage, V_m, and the cutoff frequency is expressed as

$$V_m f_T = \frac{E_m v_S}{2\pi}$$

Hence from the Johnson Limit equation [2,9–12],

$$V_m^* f_T^* = V_m f_T = \frac{E_m v_S}{2\pi}$$

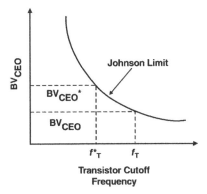

Figure 8.26   Johnson Limit relationship of BV_{CEO} versus f_T relationship

where $V_m^* f_T^*$ is associated with the first transistor and $V_m f_T$ is associated with the second transistor. The ratio of breakdown voltages can be determined as

$$\frac{V_m^*}{V_m} = \frac{f_T}{f_T^*}$$

Using this Johnson relationship, an ESD power clamp can be synthesized where a trigger device with the lowest breakdown voltage can be created by using the highest cutoff frequency (f_T) transistor and a clamp device with the highest breakdown device will have the lowest cutoff frequency (f_T) [9–12].

A BV_{CEO} breakdown voltage-triggered bipolar power clamp can be synthesized from this relationship between the power supplies (Figure 8.26). In this configuration, ESD power clamp is in a common-collector configuration. For this configuration to be suitable as an ESD power clamp, we can take advantage of the inverse relationship between the BV_{CEO} breakdown voltage and the unity current gain cutoff frequency, f_T, of the device. For an ESD power clamp, the ESD output clamp device must have a high breakdown voltage in order to address the functional potential between the V_{CC} power supply and ground potential. This ESD power clamp requires an f_T value above the ESD pulse frequency to discharge the current effectively. For the bipolar trigger device, a low-BV_{CEO} breakdown voltage device is needed in order to initiate base current into the clamp device at an early enough voltage (Figure 8.27).

This circuit can be constructed in a homojunction silicon bipolar junction transistor (BJT), or a silicon germanium, silicon germanium carbon, or gallium arsenide HBT. The bipolar-based BV_{CEO}-triggered ESD power clamp trigger network consists of a high f_T SiGe HBT with a bias resistor. When the transistor collector-to-emitter voltage is below the breakdown voltage, no current flows through the trigger transistor. The bias resistor holds the base of the SiGe HBT clamp transistor to a ground potential. With no current flowing, the output clamp can be visualized as a 'grounded base' *npn* device between the power supplies. When the voltage on V_{CC} exceeds the collector-to-emitter breakdown voltage, BV_{CEO}, in the high f_T SiGe HBT, current flows into the base of the SiGe HBT high breakdown device. This leads to the current discharge from the V_{CC} (or V_{DD}) power rail to

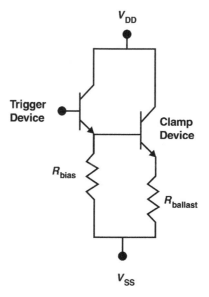

Figure 8.27 SiGe HBT ESD power clamp

the V_{SS} ground power rail. Without the initiation of the trigger element, the circuit will trigger at the BV_{CER} condition,

$$BV_{CER} = BV_{CBO} \sqrt[n]{1 - \frac{\alpha}{1 + \frac{kT}{q} \frac{1}{R_{bias}} \frac{1}{I_E}}}$$

As the bias resistor value increases, eventually, the clamp voltage will begin to appear as an 'open-base' type condition. Applying the development of Reisch [6] for open-base common-emitter configuration, for our open-base trigger element, the collector current equals the emitter current with the condition of

$$I_{C_{Trigger}} = I_{E_{Trigger}} = \frac{MI_{co}(1 + \beta)}{1 - \beta(M - 1)}$$

where the current gain is the current gain of the trigger device. This current serves as the base current to the bipolar power clamp output transistor. In the condition that the clamp is not in an avalanche state, the trigger device is the current through the clamp,

$$I_{C_{Clamp}} = \beta_{Clamp} \frac{MI_{co}(1 + \beta_{TR})}{1 - \beta_{TR}(M - 1)}$$

Figure 8.28 and Table 8.1 show HBM experimental of this network implemented in a BiCMOS SiGe technology as a function of structure size.

Resistor ballasting is introduced into this breakdown trigger by adding a resistor element between the bipolar output clamp emitter and the lower power rail. The introduction

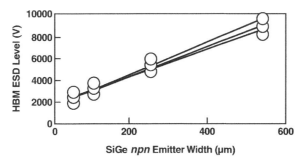

Figure 8.28 ESD results of a BV_{CEO} breakdown-triggered SiGe HBT Darlington clamp with a low-voltage BV_{CEO} trigger and high clamp element as a function of the structure size

of ballasting resistors improves the thermal stability and improves the current distributions between the segments of the ESD power clamp. Table 8.2 compares the 216 µm emitter SiGe HBT device with and without ballasting resistors. These results show that although the ballast resistors add more resistance in series with the SiGe HBT clamp device, the resistors improve the ESD stability from HBM pulses.

TLP measurements do show that the added ballast resistance does change the on-resistance slope.

Alternative ESD power clamps can be formed using the different bipolar transistors in a BiCMOS technology. For example, it is possible to use a higher BV_{CEO} trigger element in the bipolar BV_{CEO}-triggered network (Figure 8.29). Using a high-BV_{CEO} trigger and a high-BV_{CEO} clamp device, the trigger circuit can be delayed for peripheral voltage conditions below BV_{CEO} of the SiGe HBT device. As the trigger device breakdown voltage approaches the

Table 8.1 HBM test results of two-stage Darlington circuit with low-breakdown trigger and high-breakdown clamp device

Trigger (GHz)	Clamp (GHz)	Size (µm)	HBM (kV)
47	27	53.9	1.7
		108	3.1
		216	5.3
		532	8.5

Table 8.2 SiGe HBT high f_T-trigger ESD power clamp with and without emitter ballasting resistor elements

		HBM (kV)	
Trigger (GHz)	Clamp (GHz)	Ballast resistor	No ballast resistor
47	27	5.3	4.0
		5.4	4.2
		5.3	4.1

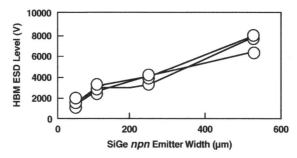

Figure 8.29   HBM ESD results of a SiGe HBT Darlington clamp with a high-voltage trigger and high-BV$_{CEO}$ clamp element with the emitter–base trigger open

output device breakdown voltage, the trigger element may be less effective and more inconsistent in initiating a base current in the bipolar output clamp element.

These experimental results show that as the trigger voltage approaches the clamp breakdown voltage BV$_{CER}$, the site-to-site variation increases with less assurance of the response of the trigger network. Hence, an ESD design practice in this implementation to insure ESD power clamp operation consistency is to increase the voltage margin between the trigger element and the breakdown of the clamp as large as possible in the given application.

Table 8.3 shows the MM ESD results for the BV$_{CEO}$-triggered SiGe HBT ESD power clamp with the high-frequency/low-breakdown trigger and high-breakdown/low-frequency clamp network with the base floating.

Figure 8.30 shows the TLP measurement of the 532 μm SiGe HBT ESD power clamp. From the TLP characteristics, the ESD power clamp trigger voltage is dependent on the BV$_{CEO}$ of the 47 GHz SiGe HBT at approximately 4 V. In this structure, the leakage increased from 1.7 to 27 pA prior to the significant increase in the leakage current. TLP failure current increased with the size of the SiGe HBT device. The largest SiGe HBT multifinger power clamp achieved 4.4 A.

Low-breakdown BV$_{CEO}$-triggered ESD power clamps can be used for internal core power grids or low-voltage applications where a low-breakdown power clamp may provide ESD advantage. As the core power supply voltage may be significantly less than the peripheral circuitry, it is possible to lower the trigger condition and the clamp voltage prior to the avalanche breakdown.

Using a high f_T SiGe HBT device for both the trigger element and the clamp element in a bipolar BV$_{CEO}$ triggered Darlington configuration, a low-voltage ESD power clamp can be

Table 8.3   MM test results of the BV$_{CEO}$-triggered ESD power clamp with low-breakdown trigger and high-breakdown clamp device (base trigger floating)

Trigger (GHz)	Clamp (GHz)	Size (μm)	MM (kV)
47	27	53.9	0.2
		108	0.35
		216	0.60
		532	1.20

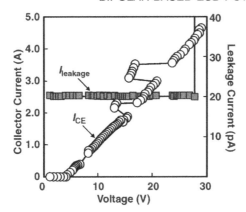

Figure 8.30   TLP I–V and leakage measurements of 532 μm BV_{CEO}-triggered SiGe HBT Darlington ESD power clamp

constructed. Figure 8.31 shows the SiGe HBT power clamp with the two different SiGe HBT clamp device and identical trigger elements.

8.5.4 Mixed Voltage Interface Forward-Bias Voltage and BV_{CEO}-Breakdown Synthesized Bipolar ESD Power Clamps

Bipolar ESD power clamps can consist of voltage-initiated trigger networks to initiate the ESD power clamp network. Voltage-trigger network can be a forward-bias voltage-initiated network or a breakdown-initiated network. Given that the application voltage exceeds the breakdown voltage of a trigger element, new voltage-trigger networks can be established that synthesize both the forward-bias trigger elements and the breakdown trigger elements [11]. Examples of ESD networks that integrate both forward-bias trigger elements and breakdown-voltage trigger elements can be as follows:

- forward bias: Schottky diodes;
- forward bias: p^+/n-well diodes;

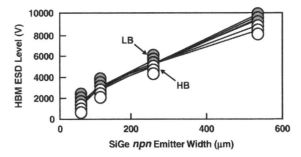

Figure 8.31   HBM results of a BV_{CEO} voltage-triggered SiGe HBT power clamp comparing a low-BV_{CEO} and high-BV_{CEO} clamp element with a low BV_{CE_o} trigger element

- forward bias: polysilicon-gated diodes;

- forward bias: bipolar base–collector varactor;

- reverse bias: bipolar in collector-to-emitter configuration with base-floating (BV_{CEO});

- reverse bias: bipolar in collector-to-emitter configuration with a base resistor (BV_{CER});

- reverse bias Zener diode.

When a bipolar ESD power clamp is BV_{CEO}-initiated, the network is constrained to the Johnson Limit. The limitation of this network is that the trigger condition is constrained to the unity current gain cutoff frequency, f_T, and not suitable for I/O, mixed power supply applications, or nonnative implementations.

For mixed voltage applications, the peripheral voltage is typically higher than the native voltage power supply. A new variable trigger implementation is developed where additional elements are placed in series with the trigger element. Placing diodes or varactors in series, the trigger condition can be level-shifted to a higher breakdown condition. In a diode string implementation where each element is of equal area, the turn-on condition is

$$V_T = NV_f - (kT/q)(N - 1)(N/2)\ln(\beta + 1)$$

where N is the number of pnp elements, V_f the forward diode voltage, and β the pnp current gain of the 'diode' element. Various implementations can be used for the forward-bias diode element are as follows:

- CMOS-based p^+/n-well diode;

- bipolar-based bipolar transistor in base–collector mode;

- silicon–germanium-based base–collector varactor.

For example, a SiGeC varactor structure is used in a forward-bias mode of operation. The varactor structure consists of a SiGe selective epitaxial p^+ anode and collector/subcollector n^{++} cathode. By placing the SiGe HBT device (used in a BV_{CEO}-breakdown mode) in series with a SiGeC varactor diode string (used in a forward-bias mode), a new trigger condition is established for the circuit [11],

$$V_T = E_m v_S/2\Pi f_T + NV_f - (kT/q)(N - 1)(N/2)\ln(\beta + 1)$$

This trigger condition provides a set of design contours of trigger values where the number of elements and the cutoff frequency are the trigger parameters (Figure 8.31). In the case of no extra series diode elements ($N = 0$), the turn-on voltage is the Johnson Limit characteristic of the relationship of BV_{CEO} and the unity current gain cutoff frequency, f_T. As the number of series diode elements increases, the ESD trigger network turn-on voltage shifts the Johnson Limit curve on the y-axis. In the case of an ideal diode string, the turn-on voltage would shift along the y-axis in equal increments associated with the ideal diode forward-bias condition. But, as a result of the vertical parasitic pnp nonlinearity factor, the net increase with each successive element is less than the ideal diode forward voltage value (e.g., $V_{BE} = 0.7$ V at ambient temperature).

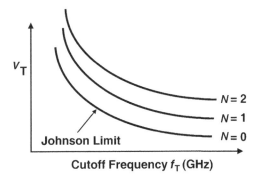

Figure 8.32 Trigger design contours as a function of series varactors and cutoff frequency

To evaluate the nonideality factor, we can evaluate the trigger condition as a function of the number of diode elements

$$\frac{\partial V_T}{\partial N} = \frac{\partial}{\partial N}\{E_m v_S/2\Pi f_T + NV_f - (kT/q)(N-1)(N/2)\ln(\beta+1)\}$$

where on differentiating with respect to the number of series elements, the change in the turn-on voltage as a function of element number can be evaluated as

$$\frac{\partial V_T}{\partial N} = \left\{ V_f - \left(\frac{kT}{q}\right)\left(\frac{2N-1}{2}\right)\ln(\beta+1)\right\}$$

This expression shows that as the number of diode elements increases, there is a correction factor from the ideal, which is a function of the number of elements. To evaluate the trigger voltage sensitivity as a function of the bipolar gain characteristic,

$$\frac{\partial V_T}{\partial \beta} = \frac{\partial}{\partial \beta}\{E_m v_S/2\Pi f_T + NV_f - (kT/q)(N-1)(N/2)\ln(\beta+1)\}$$

Differentiating with respect to the bipolar current gain,

$$\frac{\partial V_T}{\partial \beta} = -\left(\frac{kT}{q}\right)\left(\frac{1}{\beta+1}\right)\frac{(N-1)N}{2}$$

As the frequency of the transistor increases, the turn-on voltage sensitivity is

$$\frac{\partial V_T}{\partial f_T} = \frac{\partial}{\partial f_T}\{E_m v_S/2\Pi f_T + NV_f - (kT/q)(N-1)(N/2)\ln(\beta+1)\}$$

or the sensitivity of the trigger condition as a function of the frequency is

$$\frac{\partial V_T}{\partial f_T} = -\frac{E_m v_S}{2\pi f_T^2}$$

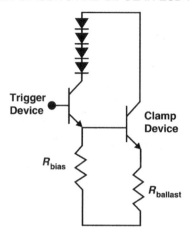

Figure 8.33 BiCMOS forward-bias and BV_{CEO}-initiated trigger network bipolar ESD power clamp

Figure 8.33 shows the bipolar ESD clamp, where the trigger network is represented as a series of diode elements in parallel with the bipolar trigger transistor in a collector-to-emitter configuration with the base electrode floating. For the network to be operable, the bipolar ESD clamp network must have a breakdown voltage that exceeds the trigger voltage condition. With the presence of the trigger element, the output clamp element will breakdown at the BV_{CER} voltage condition, where the bias resistor serves as the base resistance. The ballast resistance also establishes a base–emitter debiasing voltage state as current flows through the ballast elements.

For the analysis of the operation of this mixed voltage BiCMOS SiGe ESD power clamp, a matrix of studies varied the SiGe varactor number, the size of the output clamp, and the role of emitter ballasting resistors. To understand the operation of the circuit, HBM, MM, and TLP measurements are first taken for the case of no additional varactors. The varactor number was modified to study the variation of the ESD results with varactor number. Table 8.4 shows the HBM and MM ESD results as a function of the clamp size (emitter width). As the size of the structure increases, both HBM and MM ESD results increase. In the first observation, through the comparison of a 120 GHz f_T SiGeC trigger circuit to a prior generation technology 47 GHz f_T trigger network, we find that there

Table 8.4 HBM and MM ESD results with a 120 GHz SiGeC transistor and 90 GHz SiGeC clamp device

Trigger (GHz)	Clamp (GHz)	Clamp length (μm)	HBM (V)	MM (V)
120	90	50	2500	240
		100	3100	390
		150	4700	480
		200	5000	600
		250	5900	630

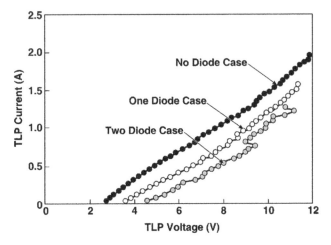

Figure 8.34 TLP *I–V* characteristics as a function of the trigger condition

is no scaling impact with successive technology generation of this clamp design (Figure 8.34). Various trigger element sizes were used to evaluate the ESD robustness of the trigger networks. Table 8.5 contains the HBM and MM ESD results as a function of cathode finger number for SiGeC varactor structures. Evaluation of the HBM and MM results for the bipolar ESD power clamp network trigger provides an insight into the operation of the circuit. The first key discovery in the HBM, MM, and TLP experimental work of the network is that as the trigger voltage value is increased (utilizing additional 'diode' elements), HBM and MM ESD results decrease for a fixed output clamp size. With the increased voltage turn-on of the trigger network, the ESD robustness of the network decreases.

Table 8.5 shows the TLP measurement of the failure current as a function of structure size and the number of SiGeC varactors in series. From the table, it can be seen that the bipolar power clamp current-to-failure decreases with the increase in the trigger voltage condition.

Figure 8.34 shows the TLP *I–V* characteristics of the bipolar ESD power clamp. Figure 8.34 shows the change in the TLP *I–V* characteristics as the trigger voltage condition is varied. The data show that as the number of successive trigger elements increases, the TLP

Table 8.5 TLP current-to-failure of the SiGe HBT ESD power clamp as a function of the SiGe *npn* clamp size and number of varactors

Clamp length (μm)	Failure current (0 Var) (A)	Failure current (1 Var) (A)	Failure current (2 Var) (A)
50	0.7	0.72	0.58
100	1.25	1.05	1.0
150	1.7	1.5	1.3
200	1.8	1.6	1.3
250	2.1	1.6	

I–V characteristics shift along the voltage axis (e.g., *x*-axis). From the data, it can also be observed that the last measurement decreases with the increasing number of series diode elements [14].

As the bipolar ESD power clamp trigger voltage increases, the margin between the breakdown voltage of the output clamp and the trigger network decreases. Hence an ESD metric of interest in this bipolar ESD power clamp circuit is the following equation

$$BV_{CER} - V_T = BV_{CER} - \left\{ \frac{E_m v_S}{2\pi f_T} + NV_f - \left(\frac{kT}{q}\right)\left(\frac{N(N-1)}{2}\right)\ln(\beta+1) \right\}$$

where BV_{CER} is the output clamp breakdown voltage given as

$$BV_{CER} \cong BV_{CBO}\left(1 - \frac{I_{co}R_B}{V_{BE}}\right)^{1/n}$$

8.5.5 Ultra-Low Voltage Forward-Biased Voltage-Trigger BiCMOS ESD Power Clamps

As the faster transistors are produced in bipolar, BiCMOS SiGe and GaAs technologies, low-voltage trigger ESD networks will be required to achieve good ESD protection. It is only recently that SiGe HBT devices achieved unity current gain cutoff frequency (f_T) levels of 120 GHz. ESD solutions for the RF input nodes and ESD power clamps are key for the success in RF applications. As the BiCMOS SiGe transistor is scaled, the power supply voltage is also scaled, allowing for the scaling of the ESD power clamp trigger condition. With the rapid scaling of the BJT and HBT devices to higher cutoff frequencies, low trigger voltage devices may be required, whose trigger condition is not limited to the Johnson Limit and can be used for power supply V_{CC} voltage. In this section, the usage of bipolar-based HBT ESD power clamps whose trigger condition are not limited to, and lower than, the Johnson Limit is explored [13–16].

To avoid the Johnson Limit bottleneck, a forward-bias diode-trigger network instead of a reverse-bias BV_{CEO} breakdown trigger network allows for the lowering of the trigger condition. Eliminating the open-base bipolar HBT BV_{CEO}-configured device trigger and replacing the trigger circuit with a forward-bias diode voltage-triggered network, the ESD design box is increased, allowing lower trigger conditions. Figure 8.35 shows a forward-bias diode voltage-triggered bipolar ESD power clamp. In this implementation, the trigger condition can be raised by adding more diode elements.

In a diode string implementation where each element is of equal area, the turn-on condition reduces to

$$V_T = NV_f - (kT/q)(N-1)(N/2)\ln(\beta+1)$$

where N is the number of *pnp* elements, V_f the forward diode voltage, and β the *pnp* current gain of the 'diode' element. For example, in a BiCMOS SiGeC technology, a SiGeC varactor

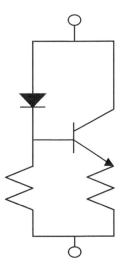

Figure 8.35   Ultra-low forward-bias diode voltage-triggered ESD power clamp

structure can be used in a forward bias. The varactor structure can consist of a SiGe selective epitaxial p^+ anode and collector/subcollector n^{++} cathode. By replacing the breakdown BV_{CEO} HBT trigger device with N elements of a SiGeC varactor, a new trigger condition can be established for the circuit, where the diode string voltage is less than the breakdown voltage-initiated trigger network [13],

$$NV_f - \left(\frac{kT}{q}\right)\frac{N(N-1)}{2}\ln(\beta + 1) \leq \frac{E_m v_S}{2\pi f_T}$$

This expression can be stated according to the following form,

$$NV_f \leq \left(\frac{kT}{q}\right)\frac{N(N-1)}{2}\ln(\beta_{pnp} + 1) + \frac{E_m v_S}{2\pi f_T}$$

where the desired condition is a trigger circuit in which the trigger is lower than the Johnson Limit of the bipolar BV_{CEO} reverse breakdown trigger element. In the case of a heavily doped subcollector, the vertical parasitic *pnp* current gain is low. In this case, the *pnp* term of the expression is small, and the inequality can be expressed as

$$V_T \approx NV_f \leq \frac{E_m v_S}{2\pi f_T}$$

Figure 8.36 shows a set of design curves in which the new trigger condition is plotted against the Johnson condition. This trigger condition provides a set of design contours of trigger values, where the number of elements and the cutoff frequency are the trigger parameters (Figure 8.36).

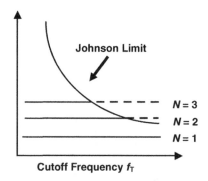

Figure 8.36 ESD power clamp design curve: trigger voltage versus unity current gain cutoff frequency (f_T) plot highlighting frequency independent design contours diode-configured implementation to the Johnson Limit BV_{CEO} contour condition

In the design plot, each horizontal line represents an additional diode overlaid on the Johnson Limit curve. At some number of diode elements, the diode-trigger network will exceed the Johnson Limit of the SiGe HBT device. Hence, the desired design space on the V_T–f_T plot is the space below the Johnson Limit characteristic. The number of diodes that fulfills this relationship is the integers from zero to N where

$$N \le \frac{E_m v_S}{2\pi f_T V_f}$$

For the forward-bias voltage-trigger element to initiate before the output clamp element, the trigger voltage must be less than the breakdown voltage of the output clamp. The output clamp device is in a common-emitter configuration. For bipolar transistors, the ordering of the breakdown voltages can be expressed as

$$BV_{CEO} \le BV_{CER} \le BV_{CBO}$$

where we can express BV_{CEO} from BV_{CBO} as

$$BV_{CEO} = BV_{CBO}(1 - \alpha_{npn})^{1/n}$$

where α_{npn} is the collector-to-emitter current transport factor for the *npn* output transistor. This can be also expressed as a function of the vertical bipolar gain, β_{npn},

$$BV_{CEO} = BV_{CBO}\left(\frac{1}{\beta_{npn}}\right)^{1/n}$$

In our network, a base resistance exists that decreases the trigger condition below the BV_{CBO} voltage condition, known as BV_{CER}

$$BV_{CER} \cong BV_{CBO}\left(1 - \frac{I_{co}R_B}{V_{BE}}\right)^{1/n}$$

where I_{co} is the reverse collector-to-base current and R_B the effective base resistance. Hence we can define the condition for a forward-bias diode-triggered bipolar ESD power clamp as

$$V_T \leq \frac{E_m v_S}{2\pi f_T} \leq BV'_{CEO} \leq \frac{E_m v_S}{2\pi (f'_T)} \leq BV'_{CER}$$

where the cutoff frequency f'_T is the clamp cutoff frequency, and the clamp breakdown voltage with a resistor element is

$$BV'_{CER} \cong BV'_{CBO} \left(1 - \frac{I_{co} R_B}{V'_{BE}}\right)^{1/n}$$

and the trigger voltage is defined as

$$V_T = NV_f - \left(\frac{kT}{q}\right) \frac{N(N-1)}{2} \ln(\beta_{npn} + 1)$$

As the transistor is scaled, the unity current gain cutoff frequency increases, leading to lower breakdown voltages of the SiGe npn transistor. From the expression,

$$V_T \approx NV_f \leq \frac{E_m v_S}{2\pi f_T}$$

we can anticipate that the number of series diode elements, N, must be scaled in future generations. Hence the cutoff frequency scaling, $f'_T = f_T \alpha$, will drive the value for the scaled number of elements, N'.

As an example implementation, the forward-bias trigger ESD network was constructed in a BiCMOS SiGe technology. The structures tested in this technology contain a 200/280 GHz (f_T/f_{MAX}) SiGe HBT device with carbon (C) incorporated in the raised extrinsic base region. Diode-configured SiGe HBT trigger elements are used in a SiGeC HBT power clamp network in a 200/285 GHz f_T/f_{MAX} SiGe HBT technology in a 0.13-μm CMOS technology base. Voldman and Gebreselasie demonstrated the operation of the forward-bias voltage-trigger diode ESD power clamp network using a 200/285 GHz (f_T/f_{MAX}) clamp element [13]. Figure 8.37 is a TLP I–V characteristics for the forward-bias diode-configured trigger SiGe ESD power clamp with two different size output clamp elements. The turn-on voltage of the ESD power clamp does not change with the size of the output clamp, but is determined by the trigger network. The ESD structure demonstrates a low-voltage turn-on at approximately 1.8 V. The turn-on voltage will be the sum of the forward-bias base–collector voltage and on-resistance of the trigger element, the forward-bias emitter–base voltage and base resistance of the clamp element, as well as the voltage drop across the emitter ballasting resistor elements. The TLP on-resistance decreased with the larger ESD power clamp. The TLP current-to-failure increased from approximately 1.0 to 1.7 A as the size of the output device doubled in size, using the same size trigger element.

Figure 8.38 is a TLP I–V characteristics for the forward-biased voltage-triggered SiGe power clamp with and without the emitter ballast resistors; this demonstrates that the external emitter ballast resistor leads to a high on-resistance and requires a higher voltage for

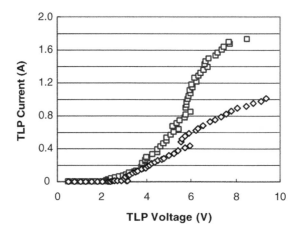

Figure 8.37   TLP *I–V* characteristics for forward-biased voltage-triggered SiGe ESD power clamp with emitter ballast resistor for the first and second structure size

discharging an equivalent source current from the unballasted elements. At low currents, the voltage drop across the ballast resistor is small, hence the turn-on voltage is not significantly influenced. When the bipolar ESD power clamp turn-on voltage is exceeded, the on-resistance is significantly low [13].

8.5.6 Capacitively-Triggered BiCMOS ESD Power Clamps

In advanced RF technologies, additional ESD power clamp design considerations are required to prevent the ESD power clamps from impacting circuit characteristics. ESD power clamps that do not impact d.c. characteristics such as leakage, and RF characteristics such as gain, bandwidth, and linearity, and circuit stability criteria are desired for silicon

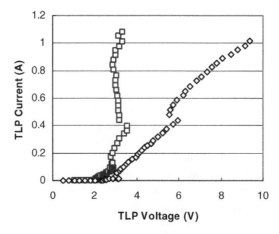

Figure 8.38   TLP *I–V* characteristics for forward-bias voltage-triggered SiGe ESD power clamp with and without emitter ballast resistor

germanium, GaAs, and InP technologies. In addition, these power clamps must be effective in providing improved ESD protection to the input circuitry by providing a low impedance path through the ESD current loop.

Bipolar ESD power clamps that establish a current path either through the trigger network or the output clamp element will impact the total leakage in a RF semiconductor chip. This will be significant if the total number of circuit elements in the chip is small. In high-density high circuit count CMOS logic chips, the percentage of leakage associated with the ESD power clamps is small compared to the entire core logic leakage. In a small BiCMOS or RF semiconductor chip, where the number of total circuit elements may be small, the ESD power clamp can be a significant part of the total dissipated power.

One technique is to utilize a capacitor in series with the trigger network to prevent a d.c. path through the trigger network. Using a reverse-breakdown element, or a capacitor, no current will flow through the trigger network until the breakdown voltage is achieved. In the case of a capacitor element, where the capacitance breakdown is significantly higher, the capacitor will serve as a capacitive coupling to activate the ESD bipolar network.

Ma and Li [17,18] compared a capacitor-triggered Darlington-configured ESD power clamp to a forward-bias diode-triggered Darlington-configured ESD power clamp. Instead of a forward-bias diode voltage trigger ESD bipolar power clamp, a capacitor element was used to replace the voltage-trigger diode string between the top node and the trigger initiation. These ESD bipolar ESD networks were constructed in a InGaP/GaAs technology.

Figure 8.39 shows a Darlington-configured ESD bipolar power clamp with a voltage-triggered forward-bias diode string network. In this ESD bipolar power clamp, when the turn-on voltage of the diode string is achieved, current will flow to the base of the first stage of the Darlington ESD power clamp. In the network, a diode is placed in series with the first- and second-stage bipolar elements.

Figure 8.40 shows a capacitor-triggered Darlington ESD power clamp. The forward-bias voltage-trigger diode string is replaced with a capacitor element. A capacitor is placed between the upper power rail and the base of the first stage of the Darlington ESD power clamp. In this case, when a pulse is applied, the capacitor increases the base potential leading to turn-on of the first stage of the Darlington-configured network. Ma and Li [17,18] showed that the d.c. leakage current remains low for high levels of RF power. As the RF power

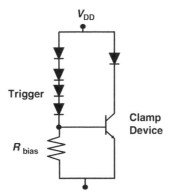

Figure 8.39 Darlington-configured ESD bipolar power clamp with a forward-bias diode-trigger network

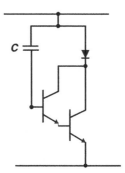

Figure 8.40 Capacitor-triggered Darlington ESD power clamp

increases, the diode-string trigger network dissipates current whereas the capacitor-trigger network does not dissipate leakage current. RF simulation shows that the Third Order Intermodulation Product (3rd OIP) was also shown to have a larger negative value for the capacitor-triggered networks compared to the forward-bias voltage diode-trigger network. From the TLP *I–V* characteristics, the capacitor-trigger network Darlington bipolar ESD power clamp was shown to have a turn-on voltage lower than 4 V, whereas the diode-triggered Darlington maintained over 10 V. Although the trigger conditions were modulated by the trigger network type, the TLP current-to-failure was approximately equal in all cases. Hence the trigger network influenced the RF and loading effects and the trigger voltage but had minor influence on the net ESD robustness of the bipolar power clamp network. In this comparison, it is noted that the capacitance loading effect of the capacitor-trigger network is higher than that of the diode-triggered network; this impacts the insertion loss RF characteristic and the loading effect. In the case of a ESD power clamp, this capacitive loading effect is not an issue. It was noted by Ma and Li [17,18] that the capacitor-triggered ESD network can be distributed within a network forming a L–C transmission line, reducing the capacitive impact to a signal pin.

As an ESD design practice, the utilization of a capacitor-triggered bipolar ESD network provides the following advantages for low leakage, good linearity, and ESD protection capability. The loading effect of the capacitance can be addressed by the distribution of the network and utilization of the 'distributed load' circuit design techniques.

8.6 BIPOLAR ESD DIODE STRING AND TRIPLE-WELL POWER CLAMPS

ESD diode strings can also be constructed in BiCMOS technology, and triple-well technologies between the power rail and ground rails. In RF bipolar networks, these can also be utilized on the output of RF power amplifier. These have been extensively discussed in other texts [1,2].

In BiCMOS technology, triple-well diode strings that eliminate the parasitic bipolar amplification can be used. Triple-well BiCMOS diode strings can utilize DT structures, subcollectors, and triple-well implants to create low-leakage diode strings [19–23]. Chen *et al.* [21–22] demonstrated the ability to eliminate the Darlington amplification effect using DT and subcollector isolated STI-bound diode structures in an isolated *p*-well region.

Additionally, Wu and Ker [23] demonstrated low-leakage diode strings that can be utilized as trigger networks in BiCMOS SiGe network.

8.7 SUMMARY AND CLOSING COMMENTS

In this chapter we have discussed the ESD protection of bipolar receiver and transmitter circuits.

In the bipolar receiver networks, the transistor element as well as the passive elements used in the d.c. blocking, a.c. blocking, matching networks, and feedback networks are potentially vulnerable to ESD events.

ESD bipolar power clamps can be initiated from reverse breakdown phenomenon or forward-biased elements; these trigger elements utilizing Zener diode breakdown, bipolar junction transistor breakdown, and forward-bias trigger networks are also discussed. The best ESD trigger solutions allow for turn-on voltage reduction as the technology progresses to future generations. Additionally, capacitive-coupled bipolar ESD power clamps are shown, which are not limited by breakdown phenomenon. These concepts are generally applicable to all bipolar homojunction or heterojunction bipolar transistor technologies.

In the next chapter, Chapter 9, ESD-RF design methods will be discussed. In order to cosynthesize ESD protection networks and RF elements, new methods of design and integration are needed. In this chapter we have focused on new methodologies for designing ESD elements in RF CMOS, RF BiCMOS, and RF BiCMOS SiGe technologies.

PROBLEMS

1. Show the voltage conditions on the input for a single-ended bipolar receiver network with the emitter connected to substrate ground potential.

2. Show the voltage conditions on the input for a single-ended bipolar receiver network with the emitter connected to a resistor in series between the emitter and substrate ground potential.

3. Show the voltage conditions on the input for a single-ended bipolar receiver network with the emitter connected to an inductor in series between the inductor and substrate ground potential.

4. Given an RF differential bipolar receiver network, show the voltage bias cases for an ESD stress between the two sides of positive and negative signal pins. What is the voltage differential during ESD stress? What is the voltage to failure?

5. Given an RF differential bipolar receiver network, a diode is placed across the emitter–base junction. Estimate the voltage differential when the network is conducting current pin-to-pin.

6. Given an RF differential voltage swing, what is the least number of diode elements that can be placed across the two differential signal pairs?

7. Given a resistor feedback between the RF input and RF output, calculate the voltage conditions when the RF output is stressed and the RF input is at ground potential. What

is the voltage stress across the resistor? How is that compared to the breakdown voltage collector-to-emitter of the RF bipolar output device?

8. Given a resistor and capacitor series feedback between the RF input and RF output, calculate the voltage conditions when the RF output is stressed, and the RF input is at ground potential. What is the voltage stress across the resistor and across the capacitor? How is that compared to the breakdown voltage collector-to-emitter of the RF bipolar output device?

9. Show the opposite cases of Problems 7 and 8 when the RF (input) is stressed and RF (output) is at ground potential.

10. Given a back-to-back diode string is placed across the differential pair, what is the minimum voltage condition for turn-on (and what is the corresponding number of diodes) and what is the maximum condition for turn-on (and what is the corresponding number of diodes)? Assume that the voltage-to-failure occurs when the emitter–base breakdown voltage is reached on the reverse-biased element. What is the capacitance of the diode string for the two cases?

11. Given a Zener breakdown triggered ESD bipolar power clamp, it was shown that the following results of the TLP I–V characteristics were obtained (Joshi *et al.* [7]). Calculate the power at the trigger point. Calculate the power at the holding point.

V_{BR} (V)	V_{on} (V)	V_{t1} (V)	I_{t1} (mA)	V_H (V)	I_H (A)
27	27.5	37.5	750	27	0.4
32	33	41	600	27	0.6
36	37	42	500	27	0.65
45	43	47	400	27	0.8

12. Derive the equations for an ESD Darlington-configured power clamp with a forward-bias diode string-trigger network in series with a resistor element. The last trigger diode element is electrically connected to the base of the first Darlington stage, and the emitter of the first stage is electrically connected to the base of the second output Darlington stage. The first and second stages have a diode in series with the collector node. The second-stage Darlington emitter is electrically connected to the lower power rail.

13. Derive the equations for an ESD Darlington-configured power clamp with a capacitor-trigger network in series with a resistor element. The capacitor element is electrically connected to the base of the first Darlington stage, and the emitter of the first stage is electrically connected to the base of the second output Darlington stage. The first and second stages have a diode in series with the collector node. The second-stage Darlington emitter is electrically connected to the lower power rail.

14. Derive the equations for an ESD Darlington-configured power clamp with a capacitor-trigger network where the capacitor is electrically connected to the base of the first

Darlington stage, and the emitter of the first stage is electrically connected to the base of the second output Darlington stage. The first and second stages have a diode in series with the collector node. The second-stage Darlington emitter is electrically connected to the lower power rail. Note that no resistor element is in series with the capacitor trigger element.

15. From the capacitor-triggered ESD bipolar Darlington power clamp, in the above problem, draw a set of ESD power clamps separated by an incremental inductor L between each successive power clamp. Represent the network as an LC transmission line.

16. From the capacitor-triggered ESD bipolar Darlington power clamp, in the above problem, draw a set of ESD power clamps separated by an incremental inductor L between each successive power clamp. Represent the network as an LGC transmission line, where the diode and Darlington network is represented as a conductance, G. Solve for G.

17. Assume a triple-well diode string with a p^+ diffusion in the p-well, an n^+ diffusion in the p-well, and a separate n-band region for each successive stage. Show all parasitic *pnp* and *npn* transistor elements. Derive a relationship, assuming the n-band is serving as a cathode element in the network.

18. Assume a triple well diode string with a p^+ diffusion in the p-well, an n^+ diffusion in the p-well, and a separate n-band region for each successive stage. Show all parasitic *pnp* and *npn* transistor elements. Derive a relationship, assuming the n-band is electrically biased to a separate potential.

19. Assume a triple well diode string, with a p^+ diffusion in the p-well, an n^+ diffusion in the p-well, and an n-band region for each successive stage is electrically connected together. Show all parasitic *pnp* and *npn* transistor elements. Derive a leakage current model for this implementation.

REFERENCES

1. Voldman S. *ESD: physics and devices*. Chichester: John Wiley and Sons, Inc.; 2004.
2. Voldman S. *ESD: circuits and devices*. Chichester: John Wiley and Sons, Inc.; 2005.
3. Voldman S. The effect of deep trench isolation, trench isolation and sub-collector doping on the electrostatic discharge (ESD) robustness of radio frequency (RF) ESD diode structures in BiCMOS silicon germanium technology. *Proceedings of the Electrical Overstress/Electrostatic Discharge (EOS/ESD) Symposium*, 2003. p. 214–23.
4. Grove AS. *Physics and technology of semiconductor devices*. New York: John Wiley and Sons, Inc.; 1967.
5. Sze SM. *Physics of semiconductor devices* (2nd ed.). New York: John Wiley and Sons, Inc.; 1981.
6. Reisch M. On bistable behavior and open-base breakdown of bipolar transistors in the avalanche regime – modeling and applications. *IEEE Transactions on Electron Devices*, June 1992. p. 1398–409.
7. Joshi S, Juliano P, Rosenbaum E, Katz G, Kang SM. ESD protection for BiCMOS circuits. *Proceedings of the Bipolar Circuit Technology Meeting (BCTM)*, 2000. p. 218–21.

8. Joshi S, Ida R, Givelin P, Rosenbaum E. An analysis of bipolar breakdown and its application to the design of ESD protection circuits. *Proceedings of the International Reliability Physics Symposium (IRPS)*, 2001. p. 240–5.

9. Voldman S, Botula A, Hui D. Silicon germanium heterojunction bipolar transistor ESD power clamps and the Johnson Limit. *Proceedings of the Electrical Overstress/Electrostatic Discharge (EOS/ESD) Symposium*, 2001. p. 326–36.

10. Botula A, Hui DT, Voldman S. Electrostatic discharge power clamp circuit. U.S. Patent No. 6,429,489 (August 6, 2002).

11. Voldman S. Variable trigger voltage ESD power clamps for mixed voltage applications using a 120 GHz/100 GHz (f_T/f_{MAX}) silicon germanium heterojunction bipolar transistor with carbon incorporation. *Proceedings of the Electrical Overstress/Electrostatic Discharge (EOS/ESD) Symposium*, 2002. p. 52–61.

12. Voldman S, Botula A, Hui DT. Electrostatic discharge power clamp circuit. U.S. Patent No. 6,549,061 (April 15, 2003).

13. Voldman S, Gebreselasie E. Low-voltage diode-configured SiGe:C HBT triggered ESD power clamps using a raised extrinsic base 200/285 GHz (f_T/f_{MAX}) SiGe:C HBT device. *Proceedings of the Electrical Overstress/Electrostatic Discharge (EOS/ESD) Symposium*, 2004. p. 57–66.

14. Voldman S. Electrostatic discharge input and power clamp circuit for high cutoff frequency technology radio frequency (RF) applications. U.S. Patent No. 6,946,707 (September 20, 2005).

15. Voldman S. A review of CMOS latchup and electrostatic discharge (ESD) in bipolar complimentary MOSFET (BiCMOS) silicon germanium technologies: Part I – ESD. *Journal of Microelectronics and Reliability* 2005;**45**:323–40.

16. Voldman S. A review of CMOS latchup and electrostatic discharge (ESD) in bipolar complimentary MOSFET (BiCMOS) silicon germanium technologies: Part II – Latchup. *Journal of Microelectronics and Reliability* 2005;**45**:437–55.

17. Ma Y, Li GP. A novel on-chip ESD protection circuit for GaAs HBT RF power amplifiers. *Proceedings of the Electrical Overstress/Electrostatic Discharge (EOS/ESD) Symposium*, 2002. p. 83–91; *Journal of Electrostatics* 2003;**59**:211–27.

18. Ma Y, Li GP. InGaP/GaAs HBT DC-20 GHz distributed amplifier with compact ESD protection circuits. *Proceedings of the Electrical Overstress/Electrostatic Discharge (EOS/ESD) Symposium*, 2004. p. 50–4.

19. Pequignot JP, Sloan JH, Stout DW, Voldman S. Electrostatic discharge protection networks for triple well semiconductor devices. U.S. Patent No. 6,891,207 (May 10, 2005).

20. Chen SS, Chen TY, Tang TH, Su JL, Shen TM, Chen JK. Low leakage diode string designs using triple well technologies for RF ESD applications. *IEEE Electron Device Letters* 2003;**ED-24**: 595–7.

21. Chen JK, Chen TY, Tang TH, Su JL, Shen TM, Chen JK. Designs of low-leakage deep trench diode string for ESD applications. *Proceedings of the Taiwan Electrostatic Discharge Conference (T-ESDC)* 2003;151–6.

22. Chen SS, Chen TY, Tang TH, Su JL, Chen JK, Chou CH. Characteristics of low leakage deep trench diode for ESD protection design in 0.18-μm SiGe BiCMOS process. *IEEE Transactions on Electron Devices* 2003;**ED-50**(7):1683–9.

23. Wu WL, Ker MD. Design on diode string to minimize leakage current for ESD protection in 0.18-um BiCMOS SiGe process. *Proceedings of the Taiwan Electrostatic Discharge Conference (T-ESDC)*, 2004. p. 72–6.

9 RF and ESD Computer-Aided Design (CAD)

In the RF ESD design, for a successful ESD design strategy, the method must conform to the RF design methodology of the RF design release process. This conformance must be in the test site strategy, tools, the d.c. and RF testing methods, to the RF product release process. In a foundry environment, a method must be established that allows freedom to the circuit design teams to complete their RF analysis and the flexibility to modify ESD devices. This must be done in an efficient manner that does not generate a large workload and must also have the natural flexibility to provide all design requirements. Hence, significant thought must be done based on the technology, the design process, the breadth of products and objectives. In this chapter, we will show an example of a method applied in a foundry environment in recent times that supports a significant amount of technologies and customers.

9.1 RF ESD DESIGN ENVIRONMENT

In the RF ESD environment, the design methodology must conform to the RF design methodology of the RF design release process. In the physical design and integration of ESD elements, computer-aided design (CAD) systems must provide an environment to allow the ability to cosynthesize ESD and radio frequency (RF) functional requirements. This is achievable by either providing a system that allows variable size ESD elements or variable size functional device elements, or both.

In this chapter, it is assumed that the RF elements are variable in physical dimensions, and the emphasis and focus will be on the generation of ESD elements that can be modified in physical size and form for ESD protection of RF applications.

9.1.1 Electrostatic Discharge and Radio Frequency (RF) Cosynthesis Design Methods

A design methodology is desirable that allows the optimization and tuning of ESD networks in an analog and mixed signal environment and additionally an RF design environment [1–3].

ESD: RF Technology and Circuits   Steven H. Voldman
© 2006 John Wiley & Sons, Ltd

In an analog and RF design environment, parameterized cells, also known as p-cells (or Pcells), allow for means to modify the size of the physical elements [1–4]. Additionally, these designs must be able to integrate with guard ring technology requirements [5–7] and under bond wire pads [8]. Semiconductor devices, both active and passive elements, can be constructed from 'primitive' p-cell. These primitive p-cell elements represent a single device element (e.g., resistor, capacitor, inductor, MOSFET, or bipolar element). These physical elements undergo full d.c. and a.c. RF characterization, from which the released models are constructed. In RF ESD cosynthesis, it is desirable to be able to provide a methodology that allows the ability to vary the ESD network characteristics, whether size or topology, to evaluate the RF functional impacts. Given that the design methodology allows for both size and topology variations, the RF characterization can be evaluated during the RF design phase.

An RF ESD design methodology that achieves this objective can be a method that forms ESD networks from these primitive p-cell device elements and converts to a higher order parameterized cell [1–3]. From the lowest order device primitive parameterized cells, hierarchical parameterized cells can be constructed to form a library of ESD circuits and networks. By utilizing the primitive p-cell structures into higher order networks, an ESD CAD strategy is developed to fulfill the objective as follows:

- design flexibility;

- d.c. characterization;

- a.c. and RF characterization;

- RF models;

- choice of ESD network type.

Developing an ESD design system of hierarchical system of parameterized cells, higher level ESD networks can be constructed without an additional RF characterization. In this methodology, the lowest order $O[1]$ device p-cells can be d.c. and RF characterized. The basic device library is constructed of the devices, which are fully quantified; both passive and active elements are placed into d.c. pad sets for d.c. measurements and within S-parameter pad sets for RF characterization. Additionally, the ESD testing can be completed on the base library of elements using wafer-level transmission line pulse (TLP) and human body model (HBM) testing. Note that the circuit characterization and ESD testing can be done on both the primitive $O[1]$ p-cell elements or the $O[n]$ hierarchical p-cell. To provide RF ESD design cosynthesis, it is desirable to have both the ability to vary physical size or circuit topology. In this method, schematics and symbology of both the primitive and the higher order $O[n]$ ESD p-cell networks is possible with the generation of design layout.

9.1.2 ESD Hierarchical Pcell Physical Layout Generation

In ESD design, it is desirable to change the physical layout of the ESD network. Hence an RF ESD design methodology is needed that can vary the physical layout of the design in physical size (e.g., area), form factor (e.g., length, width, and ratio), shape (e.g., rectangular, circular), and relative size of elements within a given circuit (e.g., be able to vary the size of the resistor, capacitor, or MOSFET independently within the common circuit topology) [1–3]. For example, in some chip architectures it is desirable to place multiple power clamps across a design instead of a single element. It is also desirable to place different size ESD circuits for a common topology in a common chip. Hence these features will be needed for integration and synthesis into a semiconductor chip architecture. In the formation of the hierarchical ESD *p*-cell elements, the generation of the ESD physical layout can be formed using the 'graphical method,' or the 'code method.' In the 'graphical method,' the *p*-cells and physical shapes are placed in the design environment manually. The physical sizes of the O[1] *p*-cells are defined by its parameters. The physical dimensions that are desired to be modified are passed up to the higher order *p*-cell design through 'inheritance.' The inherited parameters become free variables, which allow the physical size changes of the O[1] elements. The parameters that allowed adjustment in each O[1] element are contained in the final O[*n*] *p*-cell ESD network. Another methodology allows for the placement to be completed by software instead of manual placement of the physical elements and the electrical connections. Skill code generates the schematic directly and forms placement of the elements.

For this design method to be successful, either the elements must be scalable or quantified. This can be verified through an experimental testing of the ESD elements. Additionally, with the placement of a plurality of elements, new failure mechanisms must not occur. In our case it was found that there is a range where linearity of ESD robustness versus structure size is true and no new failure mechanisms were evident.

9.1.3 ESD Hierarchical Pcell Schematic Generation

In an RF ESD design, it is desirable to be able to evaluate the RF performance of an ESD network without emphasis or focus on the physical design. A method that allows full circuit evaluation and the ability to change the physical size of the element through circuit simulation is necessary [1–3]. Moreover, it is desirable to provide changes in the circuit topology within the ESD network. The circuit topology can influence the trigger conditions, capacitance loading, linearity, and RF circuit stability. Hence, it is desirable to have ESD networks that can change the physical topology within a circuit design environment. RF CMOS and RF BiCMOS networks also have a wide range of application voltages; ESD circuits are required that can be modified to address the different application voltages. Hence, an ESD design system that allows for both the change of circuit topology as well as the structure size in an automated fashion is critical for the RF ESD design cosynthesis. The circuit topology automation allows for the customer to autogenerate new ESD circuits and ESD power clamps without additional design work.

For this method to be successful, the interconnect and associated wiring outside the primitive elements must have a design environment that independently extracts the wiring and via interconnect. In our design environment, the wiring and via interconnect are autoextracted, which then eliminates the need for characterizing RF characteristics for every model element and every ESD structure (e.g., the primitives are RF characterized).

9.2 ESD DESIGN WITH HIERARCHICAL PARAMETERIZED CELLS

In semiconductor chip design, there are fundamental ESD functional blocks required in the semiconductor chip synthesis and floor planning. For an RF ESD design, the minimum set of classes of hierarchical parameterized cells needed to support an ESD design system are as follows:

- ESD input networks;

- ESD rail-to-rail networks;

- ESD power clamps.

In an RF ESD design system, the ESD design system should contain a family of these elements in each category, which satisfy signal types (e.g., digital, analog, and RF), RF parameters (e.g., noise, linearity, stability), and spatial placement. For RF input circuits, the loading effects of the ESD input and impedance influence are critical for RF performance. For ESD rail-to-rail networks, the issues are stability and noise coupling. Noise is a concern in digital networks; peripheral and core circuitry are isolated when the peripheral circuit noise is significant and the interior core logic networks are sensitive to noise disruption. Noise is a large concern in semiconductor chips with both digital and analog function on a common substrate. In mixed-signal applications, functional circuit blocks are separated to minimize noise concerns. Digital noise affects both the analog and d.c. circuitry impacting the noise figure (NF). Designers need the ability to estimate the noise and the stability of the circuit in the presence of multiple circuits and ESD networks. To eliminate noise, digital circuit blocks are separated from the analog and RF circuit blocks without a common ground or power bus. The introduction of the ESD elements between the grounds can address the ESD concerns, but increases the noise and stability implications. As a result, the cosynthesis of the ESD and noise concerns needs to be flexible to address both issues. For ESD power clamps, the ESD networks can influence noise, RF stability, and leakage.

9.2.1 Hierarchical Pcell Graphical Method

Using the graphical methodology, the O[2] p-cell can be created with two O[1] p-cells and metal bussing that expands with the physical design sizes of the device structures [1–3]. To allow the automation, 'Stretch Lines' are used to shift and grow the physical metal connections. Figure 9.1 shows an example of the design containing two growable p^+/n-well p-cells and three stretch lines. The top stretch line has an algorithm associated with the pitch and finger

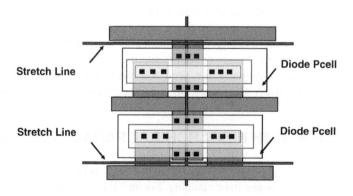

Figure 9.1   Graphical implementation of dual-diode ESD network

number to move the V_{DD} wire vertically. The lower stretch line algorithm moves the V_{SS} bus downward as cathode fingers are added to the lower p-cell element. The vertical stretch line allows the input, V_{DD}, and V_{SS} metal to grow with the length of the diode element.

For a construction of the p-cell, there are different methods of p-cell definition within the Cadence^{TM} environment. This methodology is referred to as the 'graphical' technique. The command structure for p-cell definition involves Stretch, Conditional Inclusion, Repetition, Parameterized shapes, Repeat along shape, Reference point, Inherited Parameters, Parameterized Layer, Parameterized Label, Parameterized Property, Parameters, and Compile. The Stretch function allows Stretch in X, Stretch in Y, Qualify, and Modify. The Repetition function allows for Repeat in X, Repeat in Y, and Repeat in X and Y.

Stretch commands require an algorithm to define the design 'expression for stretch.' For this p-cell, the 'expression for stretch' is defined as {{pitch*num_stripes_up}–pitch} where 'pitch' is the width of the upward diode periodicity and the 'num_stripes_up' is an inherited parameter contained in the higher order p-cell passed from the lower p-cell to address the number of fingers of the diode between the input pad and the V_{DD} power supply. Likewise for the downward diode, a second 'expression for stretch' is defined for the second p-cell diode element stretch line. The expression for stretch is defined as '{{pitch*num_stripes_down}–pitch}' for the second stretch line in the y-direction. For the first stretch line, the direction of stretch is 'up'; for the second stretch line, the direction of stretch is 'down.' For the stretch of the diode p-cells and the busses, a stretch line exists in the x-direction. For the stretch in x, an 'expression of stretch' is defined as {{a*num_segments_up} +b} where a and b are constants. The stretch direction is chosen to the right.

For the p-cell parameter summary, a typical output has the form as shown below:

Parameters defined in this parameterized cells
num_segments_up num_segments_down
num_stripes_up num_stripes_down
Stretch
Stretch Type: Vertical
Name of Expression of Stretch: 2.52*num_stripes_down–2.52
Stretch Direction: down

Stretch
Stretch Type: Vertical
Name of Expression of Stretch: 2.52*num_stripes_up–2.52
Stretch Direction: up

Stretch
Stretch Type: Horizontal
Name of Expression of Stretch:
7.6*num_segments_up + 4.2
Stretch Direction: up

Inherited Parameters
Number of instances with parameter inheritance: 2
Instance Name: I3

Inherited Parameter: Name: num_X
Inherited Parameter: Value: num_segments_up
Inherited Parameter: Type: integer
Inherited Parameter: Default: 1

Inherited Parameter: Name: num_PD
Inherited Parameter: Value: num_stripes_up
Inherited Parameter: Type: integer
Inherited Parameter: Default: 1

Inherited Parameter: Name: num_X
Inherited Parameter: Value: num_segments_down
Inherited Parameter: Type: integer
Inherited Parameter: Default: 1

Inherited Parameter: Name: num_PD
Inherited Parameter: Value: num_stripes_down
Inherited Parameter: Type: integer
Inherited Parameter: Default: 1

As part of an RF ESD CAD design system, a hierarchical O[3] parameterized cell is designed, which forms a bidirectional O[2] series diode strings that can vary the number of series diode elements and the physical width of each diode element. For example, a design may use four diodes in one direction and two in the other direction between the ground rails. The automated ESD design system has the ability to adjust the design size and the number of elements. In digital circuits, the design decision is typically decided based on the digital d.c. voltage separation required between the grounds; in high speed digital, analog, and RF circuits, the design issue is the capacitive coupling at high frequency. As more elements are added, capacitive coupling is reduced. In our ESD design system, the interconnects and wires automatically stretch and scale with the structure size. Algorithms are developed that autogenerate the interconnects on the basis of the number of diodes 'up' versus diodes 'down.' As elements are added, both the graphical layout and the physical schematics introduce the elements maintaining the electrical interconnects and pin connection.

9.2.2 Hierarchical Pcell Schematic Method

A powerful feature of our ESD design system is the ability to autogenerate ESD networks from both the 'graphical method' as well as the 'schematic method.' Typically, ESD designers start from the graphical layout of the physical design, and circuit designers start from the schematic layout to evaluate the performance objectives. To cosynthesize the performance and the ESD objectives, it is important to be able to have different modes of integration and starting points in the design methodology.

To achieve the autogeneration of ESD circuits, a design flow has been developed (Figure 9.2). The flow is based on the development of *p*-cells for both the schematic and layout cells. The *p*-cells are hierarchical; built from device O[1] primitives that have been characterized with defined models. Without the need for additional RF characterization, the design kit development cycle is compressed. Autogeneration also allows for design rule checking (DRC) correct layouts and logical-to-physical (LVS) correct circuits.

In the ESD CAD design system, the schematic *p*-cell is generated by the input variables to account for the input values of the inherited parameters. A problem with schematic autogeneration is the circuit simulation phase. The circuit may be placed as a subcircuit, however, specter simulation will only allow a single definition of a subcircuit. This prevents

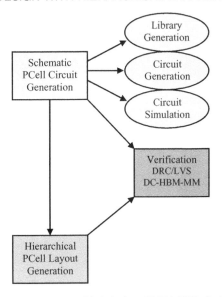

Figure 9.2   Computer-aided design (CAD) ESD design flow

the reuse of the schematic p-cell in any other configuration. To retain the ESD circuit variability, a design flow has been built around the schematic p-cell.

From the schematic methodology, four different modes of implementation were addressed:

- creation of the ESD element;

- creation and placement of an ESD element;

- placement of an existing ESD element;

- placement of an ESD schematic.

Figure 9.3 shows a representation of the different methodologies in an automated design environment. As an example of the schematic methodology, from the schematic editing screen, the user invokes AMS utils → ESD. From the ESD pull-down, four functions are defined: ESD → Create an ESD element, ESD → Create and Place an ESD element, ESD → Place an existing ESD element, and ESD → Place an ESD schematic.

In our ESD CAD design system, the schematic Pell is generated by the input variables to account for the inherited parameters input values [1–3]. A problem with schematic autogeneration is the circuit simulation phase. The circuit may be placed as a subcircuit, however, simulation will only allow a single definition of a subcircuit. This prevents the reuse of the schematic p-cell in any other configuration. To retain the ESD circuit variability, a design flow has been built around the schematic p-cell.

In one method, the designer is allowed the capability of building an ESD library with the creation of ESD cells. The designer will select the option to 'Create an ESD element.' Figure 9.4 shows an example in which the 'Create an ESD element' function initiates creation of an ESD schematic for a parameterized cell of a back-to-back diode string known as 'AntiparallelDiodeString.' To generate the electrical schematic, the ESD design system

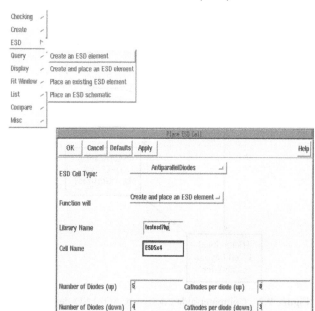

Figure 9.3 Creation of an ESD schematic of an 'antiparallel diode string' between grounds from the 'schematic method'

requests the 'number of diodes up' and the 'number of diodes down'; this determines the number of diodes in the string that are used between digital V_{SS} and analog V_{SS} (or RF V_{SS}) for grounds. For power supply rails, the 'AntiparallelDiodeString' is used between digital V_{DD} and analog V_{DD} (or RF V_{DD}). The design system also requests the number of cathode fingers in the diode structures for the 'up' string and 'down' string. The input parameters are passed into a procedure, which will build an ESD cell with the schematic p-cell built according to the input parameters and placed in the designated ESD cell. An instance of the

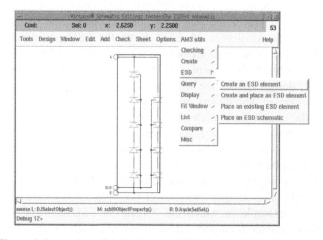

Figure 9.4 Antiparallel diode string ESD Pcell with design options

ESD layout p-cell will also be placed in the designated ESD cell. This allows for the automated building of an ESD library creating a schematic, layout, and symbol of the circuit on the basis of the input parameters. This symbol may be placed in the circuit by selecting the 'Place an ESD circuit' option.

The second method allows for the autogeneration of the schematic ESD circuit to be placed directly into the design. This procedure is available with the 'Place an ESD schematic' option and will allow the designer to autogenerate the circuit and place it in the schematic. As these cells are hierarchical, the primitive devices and autowiring are placed by creating an instance of the schematic p-cell and then flattening the element. The instance must be flattened to avoid redefinition of subcircuits.

The problem arises during the layout phase of the design. In the schematic due to the flattening, the hierarchy has been removed and only primitive elements remain. During design implementation, the primitives are placed and the hierarchy is lost. To maintain the hierarchy, an instance box is placed in the schematic retaining the input parameters, device names and characteristics as properties and the elements are recognized and the primitives are replaced with the hierarchical p-cell.

To produce multiple implementations using different inherited parameter variable input, different embodiments of the same circuit type can be created in our methodology. In this process, the schematic is renamed to be able to produce multiple implementations in a common chip or design; the renaming process allows for the design system to distinguish multiple cell views to be present in a common design.

When the inherited parameters are defined, the circuit schematic is generated according to the selected variables. Substrate, ground, and pin connections are established for the system to identify the connectivity of the circuit. The design system can also autogenerate the layout from the electrical schematic that will appear as equivalent to the previously discussed graphical implementation.

The physical layout of the ESD circuits is implemented with p-cells using existing primitives in the reference library. The circuit topology is formed within the p-cell including wiring such that all parasitics may be accounted for in preproduction test site construction.

9.3 ESD DESIGN OF RF CMOS-BASED HIERARCHICAL PARAMETERIZED CELLS

For an RF CMOS-based hierarchical parameterized cell ESD design system, the ESD design system must contain RF CMOS primitive parameterized cells in which higher order designs can be constructed.

For CMOS-based ESD input circuits, the form factor, the physical size, and the capacitance loading factor will vary for different input circuits for digital, analog, and RF signal pins. For the simple case of a CMOS double-diode network, the size of the perimeter from the diode element to V_{DD} may be different from the element to V_{SS}. In the primitive parameterized cell, there are a significant number of dimensions that can be modified for ESD optimization. The lowest number of inherited parameters for an O[2] double-diode is the four parameter case. The four parameter case allows only the width of the diode finger and number of fingers for both primitive O[1] diodes. The double-diode can be constructed of an 'up' and the 'down' diode element. The 'up' diode anode is connected to the input pad and the cathode is connected to the V_{DD} power supply rail. The 'down' diode O[1] p-cell has its cathode connected to the input pad and the anode connected to the V_{SS} rail or chip substrate. To simplify the number of design variations, the p^+/n-well O[1] p-cell is an ESD-optimized design whose ends are fixed in design style and whose metal, contact, isolation, length, and finger number are growable. Metal bussing is automatically growable with the width of the diode structures using algorithms associated with number of fingers and design pitch.

For an HBM ESD input network, a hierarchical p-cell can be formed to establish a double-diode ESD network or a grounded gate n-channel MOSFET ESD element. An ESD double-diode network can consist of two p^+/n-well defined diodes, a single p^+/n-well diode and an n-well-to-substrate diode, or a single p^+/n-well diode and an n-diffusion-to-substrate diode element. These can be constructed from three types of O[1] p-cells: the p^+/n-well diode, the n-well-to-substrate diode, and the n^+ to substrate diode O[1] p-cells. For the case of a grounded gate n-channel MOSFET (GGNMOS) ESD element, this can be constructed from an O[1] n-channel MOSFET p-cell and an O[1] resistor p-cell element.

For a CDM ESD input network, a hierarchical p-cell can be formed to establish a CDM network near a CMOS receiver input. The CDM network can consist of an O[1] resistor element and an O[1] grounded gate n-channel MOSFET (GGNMOS) ESD element. Moreover, other HBM/CDM network topologies can be constructed such as the double-diode/resistor/double-diode topology. In this case, the physical layout will consist of four O[1] diode p-cells and an O[1] resistor p-cell.

For CMOS rail-to-rail ESD networks, a hierarchical O[3] p-cell can be formed using two O[2] diode strings, as discussed previously (Figures 9.3 and 9.4). The O[2] series diode string p-cells will consist of O[1] p-cells. From the O[2] series diode string p-cells, metal bussing is used to interconnect the two series diode strings.

For ESD power clamps, an RC-triggered MOSFET power clamp can be constructed into a hierarchical O[3] p-cell [1–3]. In its construction, the RC-discriminator network can be converted into a O[2] RC p-cell whose R and C values can be modified. A series of inverters can be formed from the O[1] n-channel and O[1] p-channel MOSFET p-cells. The n-channel MOSFET output clamp element is a single O[1] n-channel MOSFET p-cell whose parameters are 'inherited' to allow variation in the size of the clamp element. For an ESD power clamp, RC-triggered MOSFET-based ESD power clamps can be a O[2] or O[3] hierarchical parameterized cell (Figure 9.5). For example, this automated hierarchical RC-triggered clamp consists of O[1] n-channel MOSFET, p-channel MOSFETs, and MIM capacitor p-cells (Figure 9.6). For digital blocks and design form factors of different size, the size of the ESD power clamp can be physically varied.

One design style can be constructed in such a way that the CMOS inverter drive network is fixed. For customer redefinition, the O[2] RC-trigger network can allow for variation in the RC time by allowing either a fixed R or fixed C, or allowing both to be free variables. In one embodiment, the O[2] RC-trigger is the second-order parameterized cell where the resistor is fixed and the capacitor is variable in size. The capacitor element grows to the left. In this

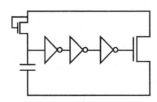

Figure 9.5   RC-triggered MOSFET ESD power clamp hierarchical parameterized cell

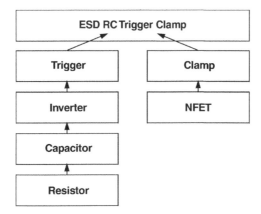

Figure 9.6   RC-triggered ESD power clamp hierarchical structure

manner, the RC can be tuned to a customer's chip design for optimization. The RC trigger is then integrated with the fixed inverter network forming the third-order p-cell triggering network. In most cases, depending on the ESD design objective, power clamps of different size are desired. The output clamp segment is automated to change in physical size and grows to the right. The customer has two inherited parameters that are passed up to the highest order circuit; the first is the capacitor size that provides RC tuning and the second is the size of the output clamp that provides the ESD robustness of the circuit.

9.4 RF BiCMOS ESD HIERARCHICAL PARAMETERIZED CELL

In an RF BiCMOS technology, ESD networks can be constructed from these primitive p-cell elements of both CMOS and bipolar elements. Hierarchical parameterized cells can be formed from bipolar-only primitive p-cells, CMOS-only primitive p-cells, or a combination of both bipolar and CMOS primitive elements. Moreover, new primitive elements using hybrid levels can be used for the primitive cell design, which can provide advantage in the RF BiCMOS implementations.

From the lowest order O[1] primitive parameterized cells, hierarchical parameterized cells can be constructed to form a library of RF BiCMOS ESD networks. The RF BiCMOS ESD library can be a 'super-set' of hierarchical parameterized cell networks that can consist of four types of O[n] hierarchical p-cells:

- RF bipolar-only ESD network consisting of bipolar primitive O[1] p-cells [9–24];

- RF CMOS-only ESD network consisting of CMOS primitive O[1] p-cells [26];

- RF BiCMOS ESD network consisting of bipolar primitive O[1] p-cells and CMOS O[1] primitive p-cells [27–32];

- RF CMOS-like ESD network consisting of hybrid bipolar/CMOS O[1] p-cells and RF CMOS primitive O[1] p-cells [27–32];

- RF BiCMOS ESD network consisting of hybrid bipolar/CMOS O[1] p-cells, bipolar O[1] p-cells, and RF CMOS primitive O[1] p-cells [27–32].

In an RF BiCMOS technology, the bipolar-only primitive p-cells can consist of high performance, medium performance, and low performance *npn* bipolar junction transistor (BJT). The RF BiCMOS technology also consists of lateral or vertical *pnp* bipolar junction transistor (BJT) primitive p-cell elements. RF BiCMOS technology allows for the formation of active and passive derivative elements off the base of the bipolar transistor. From the bipolar transistor, Schottky diodes, PIN diodes, variable capacitor (varactors), and hyper-abrupt variable capacitor (HA varactor) element primitive p-cells are generated derivative elements. These derivative elements can be used for ESD networks. In an RF BiCMOS technology, because of the wide power supply application range, dual-gate and single-gate MOSFET primitive cells exist that can be used for both external peripheral I/O and internal core RF BiCMOS circuitry. Additionally, primitive elements can consist of inductors, silicon resistors, poly-silicon resistors, and nonsilicon resistor elements. All of these physical elements can be utilized to form various ESD $O[n]$ hierarchical parameterized cell networks.

For an RF BiCMOS technology, as in RF CMOS technology, the hierarchical parameterized cells needed to support an ESD design system involve ESD networks of different types:

- HBM ESD input networks;

- CDM ESD input networks;

- mixed voltage interface ESD input networks;

- rail-to-rail ESD networks;

- ESD power clamps.

9.4.1 BiCMOS ESD Input Networks

RF BiCMOS HBM ESD input networks can consist of double-diode and diode string implementations. In an RF BiCMOS technology, the double-diode ESD input network can utilize a number of different primitive $O[1]$ p-cell elements for the diode elements between the input node and the V_{DD} power supply:

- CMOS STI-defined p^+/n-well diode;

- bipolar base–collector junction diode (e.g., base–emitter shorted);

- bipolar base–collector variable capacitor (varactor);

- bipolar p-well-to-subcollector capacitor (varactor).

In an RF BiCMOS technology, the double-diode ESD input network can utilize a number of different primitive $O[1]$ elements for the diode elements between the input node and V_{SS} power supply or p^- substrate:

- CMOS STI-defined p^+/n-well diode structure;

- CMOS n-well to substrate diode;

- CMOS STI-defined n^+ source/drain to substrate diode;

- bipolar base–collector junction diode (e.g., base–emitter shorted);

- bipolar base–collector variable capacitor (varactor);

- bipolar p-well-to-sub-collector capacitor (varactor);

- bipolar subcollector to p^- substrate diode.

Using design levels from CMOS and bipolar devices, new hybrid elements and new ESD diode elements can be formed to produce ESD input node structures. For ESD diode elements, subcollector implants, trench isolation (TI), and deep trench (DT) isolation structures can be used to enhance the performance of the diode structures [27]. Some examples of hybrid dual-well primitive O[1] p-cell diode elements are as follows:

- STI-defined p^+/n-well diode with subcollector implant;

- STI-defined p^+/n-well diode with TI bordered n-well;

- STI-defined p^+/n-well diode with DT isolation bordered n-well;

- STI-defined p^+/n-well diode with TI bordered subcollector;

- STI-defined p^+/n-well diode with DT isolation bordered subcollector;

- n-well-to-substrate diode element with TI border;

- n-well-to-substrate diode element with DT border;

- STI-defined p^+ anode in a p-well on a n^+ subcollector with a DT border.

In the above list of primitive p-cell elements, the hybrid primitive elements can provide advantages in ideality, capacitance, substrate injection, noise isolation, latchup robustness, and ESD robustness. Hence, in the integration of a design, the decision of the element to be used can influence the functional tradeoffs, latchup margin, and ESD robustness [27].

For an RF BiCMOS hierarchical O[2] p-cell and for a double-diode design p-cell, the graphical methodology can enable 'switches' in the parameterized cell that can enable or disable hybrid design levels of the bipolar element.

The double-diode design is a p-cell that allows for the inheritance of six parameters— number and the width of the diode fingers (e.g., stripes) for the 'up' and the 'down' diode element as well as two additional parameters. The fifth parameter adds the DT structure to the diode element and the sixth parameter adds the subcollector (NS). In this implementation, four different ESD devices can be constructed with a single O[2] parameterized cell. When the DT switch is enabled, the design must adjust its ground rule parameter and spacings to satisfy the correct design rule dimensions. In the case of the subcollector, the ground rule spacings must also adjust for well-to-well rules in the presence of subcollectors. Using the graphical methodology, stretch lines adjust the electrical connections to the V_{DD} and V_{SS}, and the metal width along the length of the diode elements is then increased in physical width.

For the six-parameter hybrid O[2] p-cell element, the p-cell parameter summary, a typical output has the form as shown below. In this implementation, two p^+/n-well diode O[1] p-cells that contain switches for the DT and the subcollector design levels are used:

Parameters defined in this parameterized cells
num_segments_up num_segments_down

num_stripes_up num_stripes_down
DT NS
Stretch
Stretch Type: Vertical
Name of Expression of Stretch: $2.52^*num\_stripes\_down-2.52$
Stretch Direction: down

Stretch
Stretch Type: Vertical
Name of Expression of Stretch: $2.52^*num\_stripes\_up-2.52$
Stretch Direction: up

Stretch
Stretch Type: Horizontal
Name of Expression of Stretch:
$7.6^*num\_segments\_up + 4.2$
Stretch Direction: up

Inherited Parameters
Number of instances with parameter inheritance: 2
Instance Name: 13
Inherited Parameter: Name:NS
Inherited Parameter: Value:NS
Inherited Parameter: Type:boolean
Inherited Parameter: Default:"TRUE"

Inherited Parameter: Name:DT
Inherited Parameter: Value:DT
Inherited Parameter: Type:boolean
Inherited Parameter: Default: "FALSE"

Inherited Parameter: Name: num_X
Inherited Parameter: Value: num_segments_up
Inherited Parameter: Type: integer
Inherited Parameter: Default: 1

Inherited Parameter: Name: num_PD
Inherited Parameter: Value: num_stripes_up
Inherited Parameter: Type: integer
Inherited Parameter: Default: 1

Instance Name: I1
Inherited Parameter: Name:NS
Inherited Parameter: Value:NS
Inherited Parameter: Type:boolean
Inherited Parameter: Default: "TRUE"

Inherited Parameter: Name:DT
Inherited Parameter: Value:DT
Inherited Parameter: Type:boolean
Inherited Parameter: Default: "FALSE"

Inherited Parameter: Name: num_X
Inherited Parameter: Value: num_segments_down
Inherited Parameter: Type: integer
Inherited Parameter: Default: 1

Inherited Parameter: Name: num_PD
Inherited Parameter: Value: num_stripes_down
Inherited Parameter: Type: integer
Inherited Parameter: Default: 1

In this implementation, the number of 'stripes' (e.g., fingers) is an integer number. The width of the design can be formed in a 'quantized' or 'nonquantized' implementation. In this implementation, to maintain a constant metal pitch, a 'quantized' unit cell is replicated. In a methodology that allows for any physical dimension or a 'nonquantized' width unit cell, the metal line width and the metal-to-metal space can be altered. In this implementation, the unit cell is fixed, maintaining metal pitch, metal line width, and the spacing of the physical contacts on the p^+ anode and the n^+ cathode regions.

9.4.2 BiCMOS ESD Rail-to-Rail

For a BiCMOS rail-to-rail implementation, a hierarchical O[2] or O[3] antiparallel 'diode string' p-cell can be used between the common-potential power rails (e.g., analog V_{SS} to digital V_{SS}). For example, as part of the ESD CAD design system, a hierarchical O[3] parameterized cell is designed forming a bidirectional SiGe varactor strings that can vary the number of varactors and the physical width of each varactor (Figure 9.7). This implementation can be symmetric or asymmetric in the number of varactors in either direction. An automated ESD design system has the ability to adjust the design size and the number of primitive O[1] elements. The distinction of the BiCMOS implementation of the ESD rail-to-rail hierarchical p-cell compared to the CMOS implementations is that the BiCMOS implementation lends itself to a significant variety of possible primitive O[1] p-cells that can be utilized as discussed above. The RF BiCMOS implementation lends itself to a higher flexibility in the tradeoff between the capacitive coupling, the voltage differential, the voltage turn-on, and the ESD robustness of this hierarchical element. From the list of the primitive p-cell elements, the hybrid primitive elements can provide advantages in ideality, capacitance, substrate injection, noise isolation, latchup robustness, and ESD robustness as well as interaction of adjacent primitive elements. In diode string elements, nonideal turn-on voltages can occur when the vertical bipolar current gain is significant. As a result, the understanding of the physical response of these elements and the stage-to-stage interactions are the key factors in the design integration.

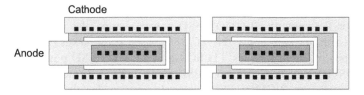

Figure 9.7   RF BiCMOS ESD network

Table 9.1 Design system asymmetric diode string parameterized cell HBM and MM ESD results

SiGeC asymmetric diode string 5:1 number of cathodes	HBM failure voltage (V)	MM failure voltage (V)
2	2300	240
4	3500	390
6	4800	510
8	6000	720
10	7200	750

In Table 9.1, the HBM and MM experimental results of an example of a BiCMOS O[3] asymmetric diode string p-cell are shown. In this implementation, a BiCMOS SiGeC-based varactor structure is used for the diode string implementation. Five forward-biased varactors are constructed in one direction and for the reverse direction, a single reverse direction varactor is used. In the design study, the number of cathode fingers (e.g., cathode stripes) is varied as part of the design study.

9.4.3 BiCMOS ESD Power Clamps

BiCMOS power clamps in the power grid are needed for digital, analog, and RF circuitry to provide a low impedance to the ground rail in the ESD current loop. Although a low impedance path is required, not all ESD power clamps are suitable for all circuit implementations. Some of the design issues for BiCMOS power clamps are influenced by the following issues:

- noise generation;
- V_{DD}-to-V_{SS} latchup;
- impact on RF circuit stability;
- voltage margin;
- temperature;
- leakage;
- negative power supply conditions.

RF CMOS-based ESD power clamps are a source of MOSFET $1/f$ noise. In analog and RF applications, this is undesirable in some circuit applications. In BiCMOS ESD power clamps, it is valuable to provide effective ESD solutions that do not impact the noise floor of an analog or RF network. BiCMOS ESD power clamps also cannot be a source of latchup between the power supply rails. CMOS-based ESD power clamps have demonstrated two types of latchup issues: (1) SCR-based CMOS ESD power clamps, and (2) p-channel MOSFET-based RC-triggered ESD power clamps forming parasitic $pnpn$ with the RC-trigger elements. BiCMOS ESD power clamps can also change the RF stability of networks. Hence, the analysis of the BiCMOS ESD power clamp needs to be analyzed in conjunction with the circuit elements. BiCMOS ESD power clamps must also establish adequate power supply margin at worst case differential voltages and worst case temperature conditions. In small analog BiCMOS applications, low leakage power clamps are also of interest. BiCMOS ESD power clamps in battery-powered or battery-backup systems, handheld applications, and wireless applications prefer low leakage ESD power clamps. Another difficulty in supporting a wide range of applications is the variety power rail voltage conditions and architectures. BiCMOS applications vary from power amplifiers (PA), variable controlled oscillators (VCO), mixers, hard disk drive circuits, and test equipment. Negative power supply voltages are also present in many BiCMOS applications. As a result, an ESD power clamp strategy must be suitable for CMOS digital blocks, CMOS or BiCMOS analog blocks, and CMOS or BiCMOS RF circuits with a wide variety of voltage conditions as well as negative bias on the substrate. To address this, a BiCMOS ESD design system has both SiGe bipolar-based ESD power clamps and CMOS-based ESD power clamps.

For the BiCMOS analog and RF functional blocks, automated hierarchical ESD power clamps are designed to allow for different voltage trigger conditions and the size of the power clamp. The first ESD power clamp circuit has a fixed trigger voltage based on the BV_{CEO} of the trigger transistor and the output device is a low f_T device with a high BV_{CEO} SiGe HBT *npn* device [9,17,19]. This network is suitable for BiCMOS SiGe chips, bipolar-only implementations, and both zero potential and negative biased substrates. In this implementation, high-, medium-, and low-voltage triggers can be utilized. The circuit diagram of the BiCMOS SiGe ESD power clamp is shown in Figure 9.8.

A hierarchical O[2] *p*-cell of the BiCMOS ESD power clamp circuit is generated allowing for different sizes. In this hierarchical *p*-cell, the design consists of a resistor ballast, two transistors, and bias resistor primitive *p*-cells. In one design method, the bipolar trigger transistor and the bias resistor are fixed in physical size, whereas the output clamp transistor is a repetition group consisting of the ballast resistor and the output transistor stage. The trigger network and the resistor element are not a repeated group reducing the total area of the network elements that are serving only the role of initiating the voltage condition, whereas the output clamp serves the role of providing the

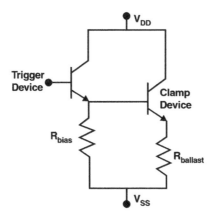

Figure 9.8 ESD Darlington power clamp network

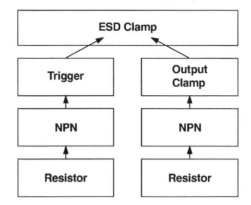

Figure 9.9   Hierarchical structure of parameterized cell ESD power clamp circuit

discharge current between the two power supplies. On the contrary, the output transistor and its associated ballast resistor are formed in a 'quantized' increment. In the quantized repetition group, the bus wiring, the collector interconnects, the emitter interconnects, and the associated vias are repeated with a fixed width bipolar transistor. An illustration of the elements is shown in Figure 9.9.

The formation of the hierarchical parameterized cell can be formed using the graphical methodology, where the output is the repetition group. In the repetition group, stretch lines are parameterized according to the number of quantized steps of the number of parallel power clamp elements to extend the power buses with the size of the physical power clamp (Figure 9.10). The electrical connections, for example, the interconnect between the trigger network and the output clamp, must also be extended as the output clamp size is extended. From the schematic approach, the BiCMOS ESD power clamp network can be generated from the schematic cell view. This circuit can be represented by the full schematic in the semiconductor schematic or a facsimile symbol function.

Symbol function cell views of the BiCMOS power clamp serve as a representation of the ESD network that can be used for identification in verification and checking systems.

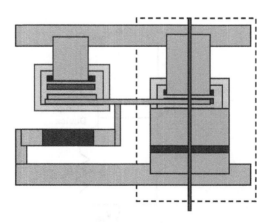

Figure 9.10   Hierarchical ESD power clamp design utilizing a growable ESD power clamp output device using repetition groups and stretch line algorithms utilizing the 'graphical methodology'

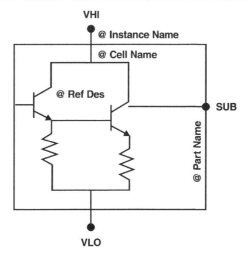

Figure 9.11 An example of a hierarchical parameterized schematic cell 'Symbol' representing the circuit. The symbol contains all inherited parameter and design information

Additionally, symbol functions can serve the purpose of storing the associated information of the ESD power clamp size and the ESD results. Figure 9.11 is an example of a 'symbol' cell view of the BiCMOS ESD power clamp. The 'box' around the symbol of the hierarchical *p*-cell schematic contains all the inherited parameters and circuit information. This is critical for a number of key design purposes:

- stores inherited parameter information;

- stores the circuit *p*-cell hierarchy;

- allows for multiple names to be represented in a common chip design;

- Allows a means of transition from the schematic to the graphical representation of the BiCMOS hierarchical *p*-cell.

An example of the experimental results of the design implementation is shown in Table 9.2. In this implementation, the size of the output transistor is 50 μm. The repetition group repeats

Table 9.2 Hierarchical parameterized cell ESD power clamp ESD results with a 120 GHz/100 GHz f_T/f_{MAX} silicon germanium HBT with carbon incorporation

Power clamp emitter length (μm)	HBM failure voltage (V)	MM failure voltage (V)	TLP failure current (A)
50	2500	240	0.7
100	3100	390	1.25
150	4700	480	1.7
200	5000	600	1.8
250	5900	630	2.1

the output clamp/ballast resistor grouping in a fixed increment size, whereas the trigger element grouping remains fixed. The table shows that the ESD results (HBM, MM, and TLP testing results) increase with the number of repetition groups, demonstrating operability of the design methodology and the circuit implementation [10,11,18,25]. What is evident from the experimental results is that as the size of the power clamp repetition groupings increases, a nonlinearity exists in the results after a set number of repetition groups. This ESD rolloff effect may be because of the voltage distribution or the limited ability for a single fixed trigger element to supply all the base nodes equally and effectively.

Higher order O[3] and O[4] hierarchical parameterized ESD power clamps can be established to address more circuit complexity and to achieve higher function. To address the different power supply conditions, a level-shifting parameterized subdesign p-cell is created that allows the variation in the trigger network. For example, a BiCMOS ESD power clamp can be formed consisting of an O[2] level-shifting p-cell, an O[1] primitive trigger SiGe transistor, a primitive O[1] bias resistor p-cell, and the repetition group of the primitive bipolar transistor clamp output and associated resistor ballast elements. This design has two automation variables; first the trigger condition allows the growth of a string of series SiGe varactors to increase the trigger condition; second, the output clamp and ballast resistor can be increased on the basis of the design area and desired ESD protection level. All interconnects are growable to allow the diode string and clamp size be increased in size automatically. From the schematic methodology, the personalization of the ESD power clamp can be first initiated as a circuit and then as graphical layout.

Figure 9.12 shows a circuit schematic of the variable trigger BiCMOS ESD power clamp [10,11]. The circuit schematics identifies the two growable segments of the power clamp network. First, the 'diode string' can vary in the number of diode elements to adjust the trigger voltage condition. In this implementation, the size of the trigger elements remains fixed in size, but the number of series elements can be varied. As in the last implementation, the trigger network is not repeated as the output clamp is increased in physical size.

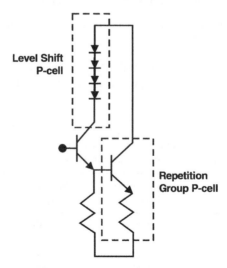

Figure 9.12 Circuit schematic highlighting the ESD power clamp with growable level shift trigger and power clamp output

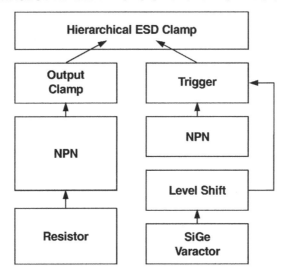

Figure 9.13 ESD power clamp hierarchy highlighting the growable regions of the parameterized ESD network

In the construction of the O[3] hierarchical parameterized cell, the diode string is formed from any physical primitive element that can be utilized as a diode element. The O[2] diode string network must 'grow' as the number of primitive O[1] diode elements is increased. The electrical connections remain connected using the graphical methodology stretch line features.

Figure 9.13 illustrates the hierarchical nature of the BiCMOS ESD power clamp element. In this structure, silicon germanium base–collector varactors are used in a forward biased operation [10,11]. The O[1] primitives of the silicon germanium varactors are integrated to form a growable 'diode string' that can vary in the number of 'diodes' in series forming an O[2] p-cell. This diode string is integrated with the O[1] primitive of the SiGe HBT device in the final circuit network. Moreover, the output clamp is constructed of the O[1] high breakdown SiGe HBT device and the O[1] resistor ballast element.

9.5 ADVANTAGES AND LIMITATIONS OF THE RF ESD DESIGN SYSTEM

In this ESD design system methodology, there are significant advantages and limitations. The methodology developed by Voldman, Strang, and Jordan has been adopted in a foundry environment and has demonstrated significant unanticipated advantages in an analog and mixed signal design team environment.

First, a significant improvement in design and release productivity is evident from implementation of the hierarchical parameterized cell ESD library. Once the family of parameterized cell ESD circuits is compiled and generated, test site development is significantly faster as the inherited parameterized cells allows for rapid generation of the matrix of allowable sizes of the circuit; this provides a productivity improvement in the test site phase.

The second advantage is that the designs are completed at the test site phase; this allows for direct implementation into the design kit release/verification process at the test site phase of the development cycle. This also allows for early customer implementation at an 'alpha' or 'beta' release of a new released technology library. As a result, the initial development evaluation is parallel to the potential of final design release. The ESD p-cell library is defined with full ground rule checking and verification at the test site phase of the development.

The third advantage is that the hierarchical parameterized ESD designs do not need unique d.c. and RF characterization as the designs contain all RF characterized elements; this provides no additional RF characterization workload and will be updated with all design releases. In a customized design environment, where primitive elements are not utilized, d.c. and RF characterization would have to be initiated for each design alteration. In this environment, the d.c. and RF characterizations require no additional work beyond the released design system library O[1] p-cell elements.

The fourth advantage is that customers do not continue to request alternate size structures of different form factors after the initial release. In a custom design environment, the ESD design work would be modified for each form factor in each technology generation or each customer environment. In this environment, the form factors can be modified given the creative implementations of the hierarchical O[n] p-cell utilize the advantages of the design environment.

The fifth advantage of this architecture is that as the design system matures, the number of inherited parameters can be increased to allow increased customer flexibility to address area, form factor, or other issues.

The sixth advantage is the modular nature of the hierarchical ESD designs. The introduction of new ESD networks is possible utilizing the existing implementation and modifying the hierarchy for the modification. This quickly allows for new implementations. Adding new design implementation to the system is also possible as the technology matures. For example, new parameterized cells can be generated and constructed into different input node ESD networks, rail-to-rail networks, or ESD power clamps. The architecture lends itself to this capability. Hence, adding custom implemented salicide-blocked MOSFET p-cells, salicide-blocked resistor elements, and ballasting elements, any concept can be constructed. The subcircuit blocks of these libraries can be used to construct other circuits because of their modular construction. For example, the RC parameterized cell can be used for other implementations and other networks.

The seventh advantage of this methodology is that it is suitable for CMOS, RF CMOS, mixed signal CMOS (MS CMOS), bipolar, BiCMOS, and RF BiCMOS. The RF BiCMOS is a super-set of elements; by removal of the bipolar-based networks, a CMOS-based ESD library exists that is compatible with its BiCMOS counterpart. The design is suitable for a Bipolar technology by removing the CMOS library. In the case of the hybrid O[1] p-cells, these elements require both the CMOS and bipolar design levels. As a result, this strategy can save significant time and cost in the development of an RF CMOS technology; new manuals, new ESD design kits, design release or recharacterization are not necessary when the architecture of this strategy is such that the BiCMOS implementation serves as the ESD 'super-set' of O[1] p-cells elements.

An eighth advantage is the RF-characterization methodology. One concern of the design methodology is that the RF characterization is completed on the base p-cells with no additional measurement of the higher order O[n] p-cell ESD circuit. In actuality, the O[n] p-cell circuits can be measured using additional s-parameter measurements by placement of these circuits into *s*-parameter pad cells. Note that in most cases, the large ESD power clamps would not fit into the *s*-parameter pad sets without additional customized pad sets. Hence, it is necessary to have a methodology that reduces to the primitive O[1] *p*-cells. The RF characterization is established by providing full d.c. and RF models for the primitive O[1] *p*-cells. The interconnects and wiring in the higher order O[*n*] *p*-cells use the RF model for interconnects. In this manner, an RF model of the complete hierarchical O[*n*] *p*-cell circuit is possible, which has the inherited parameters including all the variables of interest for the interconnects and the lower order *p*-cell elements. Choosing this path, a higher level model of the circuit can be generated without additional work and resources.

Another advantage is the ESD characterization. The ESD characterization can be evaluated at any level in the design hierarchy. For ESD technology benchmarking, the base O[1] *p*-cells can be ESD tested and quantified to understand the scaling of the technology and the ESD performance of the released design library of primitive O[1] *p*-cells. This fully characterizes all released elements. The ESD testing is also completed on the full implementation by evaluating the matrix of structure sizes.

In this design process, there are some fundamental limitations as well. Many of these limitations are solvable.

The first limitation is the spatial relation and the number of customer-modified parameters. From the graphical methodology, there are certain directions of growth that can be established as the different sizes are modified. As the number of inherited parameters increases, the design flexibility will also increase. The complexity of the graphical design will become more complex; this is solvable by using the software-based schematic hierarchical *p*-cells instead of graphical methods. Using software-based parameterization, the ability to open more variables will be possible.

The second limitation is the 'quantized' versus 'unquantized' methodology. It is an advantage to have widths and lengths of elements that are not limited by groupings or repetition groups. This is specifically true for RF tuning of elements on input nodes, but also possibly for the ESD rail-to-rail elements and ESD power clamps. Hence, the quantization can influence the performance and design flexibility. From an ESD perspective, the quantization of the metal widths, metal pitch, contact density, and contact spacing will influence the ESD response. With software code generated parameterized cells, many of the ESD design tradeoffs versus functional flexibility can be solved. Although the design method appears to have limitations, these issues are signs of the design strategy success for there is desire to improve its flexibility and modification for further enhancements.

The evolution of our BiCMOS ESD design process has followed the path from a single RF-characterized fixed element to a family of fixed design size ESD networks containing RF-characterized *p*-cells and to a new methodology of variable design size ESD networks containing RF-characterized *p*-cells. The new concept has full RF characterization and models as well as HBM, MM, and TLP characterization; this allows for an automated design change for the cosynthesis of ESD networks and circuit function for digital, analog, and RF circuits. The system allows for autogeneration of the schematic and the layout, as well as mapping from one to the other. The system also allows for placement of multiple implementations of an ESD network with different circuit topologies in a given chip implementation. The system has produced a significant productivity improvement in ESD design, test site development, and customer release. With its evolution, we anticipate an increased opening of the number of inherited parameters, new ESD-optimized *p*-cells, and new ESD circuits. This ESD design concept and architecture improve the current state-of-the-art ESD design and ESD-RF optimization for both RF CMOS and RF BiCMOS technologies.

9.6 GUARD RING *P*-CELL METHODOLOGY

Guard ring design and integration is critical to the RF ESD design methodology. This section will discuss guard ring characteristization and design. Guard rings have been used to lower the CMOS latchup sensitivity and the injection into adjacent circuitry in semiconductor technology. CMOS latchup has been a concern in semiconductor technology in the past and still is a concern today [5–7,33–44]. Guard rings serve the purpose of providing electrical and spatial isolation between adjacent circuit elements and decreasing the risk of intradevice, intracircuit, or intercircuit latchup.

9.6.1 Guard Rings for Internal and External Latchup Phenomena

Guard rings serve the purpose of providing electrical and spatial isolation between adjacent circuit elements. This is achieved by the prevention of minority carriers from within a given injection circuit or minority carriers from entering sensitive circuits. In the first case, the role of the guard ring structure is to prevent the minority carriers from leaving the region of the circuit and influencing the surrounding circuitry. In the second case, the injection is external to the circuit and the objective is to prevent the minority carriers from influencing the circuit of interest. In the discussion of internal latchup, the role of the guard ring is to provide electrical isolation between the *pnp* and the *npn* structure. In this case, the role of the guard ring is to minimize the electrical coupling and prevent regenerative feedback from occurring between the *pnp* and the *npn*; in other words, it is to lower the gain of the feedback loop by reducing the parasitic current gain. In this sense, the guard rings are placed between the *pnp* and the *npn* within the structure. Guard rings within the *pnpn* structures lower the parasitic bipolar gain by the following means:

- increase the base width of the parasitic *pnp* or *npn* structure;
- provide a region of collection of the minority carriers to the substrate or power supply electrodes, 'collecting' the minority carrier via a metallurgical junction or electrical connection;
- provide a region of heavy doping concentration to increase the recombination within the parasitic, 'capturing' the minority carrier via electron–hole pair recombination.

Hence, in the discussion of external latchup, the role of the guard ring is to provide electrical isolation between the first region and the second region. The role of the guard ring is also not to minimize the electrical coupling between the local *pnp* and the *npn* transistor but prevent the traverse of injection mechanisms from the first region to the second region, where the concern lies on one of the two categories:

- Lowering the minority carrier injection from the first region of interest to any exterior region.

- Lowering the minority carrier injection from an exterior region into the circuit region of interest.

In the first case of internal latchup, the focus is on lowering the regenerative feedback between the local *pnp* and the *npn* circuit. Note that this can be within a circuit or between two adjacent circuits. As a result, this can be a 'within' circuit (intracircuit initiated latchup) or a between two circuits (intercircuit initiated latchup) phenomenon. With mixed signal and system-on-chip (SOC) applications, intercircuit is an issue due to lack of coverage of design rule checking (DRC) between adjacent circuits.

In the case of external latchup, there is a higher focus on the transport of minority carrier injection into the region with the initiation of a local. As a result, in this case the issue is not the regenerative feedback within a local *pnpn* but the interaction of an injection source and its electrical stimulus of a *pnpn* parasitic device in both intracircuit and intercircuit cases. Therefore, the focus is on the issue of external initiation of latchup of an intracircuit *pnpn* parasitic structure, where no internal guard rings are present.

Additionally, the difference of whether the concern is the injection from a source to outside the source of minority injection or the opposite view of the injection from entering outside, is a matter of perspective. Hence the quantification of transport from within some region to outside or vice versa is of interest depending on the concern of 'inside' or 'outside' the guard ring structure [5–7,35]. But, in both cases, the role of the guard ring structure serves as a means of lowering the effectiveness of the minority carriers transport via recombination, collection, and spatial separation [6,7,33,35].

In the transport of minority carriers, there is an efficiency at which the carriers transport between the injection region (e.g., the emitter) and the collection region (e.g., the collector). Hence, in the quantification of the guard ring structure, the perspective is that the 'guard rings' serve as a means to mitigate the transport of minority carriers. Thus, the effectiveness of the guard ring to lower the minority carrier transport via collection, recombination, or other means is of interest.

9.6.2 Guard Ring Theory

First, the injection structure, substrate, and collecting structure form a lateral bipolar transistor. From this perspective, the lateral *npn* bipolar characteristics can be quantified. In this manner, the forward and reverse bipolar current gain can be evaluated. The forward and reverse bipolar current gain is evaluated by switching the electrodes. Note that the n^+ injector is enclosed inside the guard rings and *n*-well ring leading to asymmetrical results. In this process, additional structures are placed in the 'base' region between the emitter and the collecting structure. From this perspective, the cumulative structures in between the emitter and collector are evaluated as 'β spoilers' impacting the transport of electron. In this manner, a bipolar-like model can be used to quantify the relationship between the injection structure and the collecting structure. In this experimental test, the 'base region' can be driven to produce I_C versus V_{CE} characteristics as a function of base current I_B.

The second perspective is a more phenomenological formulation that can be generalized. This perspective is formed from a probability view. Minority carriers injected into the substrate are either collected at a junction region or recombined in the bulk or surface.

Hence the probability that an electron is collected plus the probability that an electron is recombined equals unity.

$$P\,(\text{recombine}) + P\,(\text{collected}) = 1$$

Another perspective is that a minority carrier is either trapped or escapes from the guard ring. The sum of the probability that an electron is trapped and the probability that an electron escapes equals unity [6,7,35].

$$P\,(\text{trapping}) + P\,(\text{escape}) = 1$$

In this manner, the probability that an electron is trapped is the probability of recombining or collecting within the guard ring structure (spatial region within the guard ring).

As noted by Troutman [35], the probability of escape is the probability that an electron is collected or recombined outside the guard ring structure. From this perspective, the probability of a guard ring collecting an electron by a double guard ring structure is the current measured at the local p^+ substrate ring and an n-type guard ring normalized to the injection current. The probability that an electron escapes from a guard ring or series of guard rings is the current measured at an additional ring outside the guard rings and the p^+ substrate contact outside the guard ring normalized by the injection current.

The ratio of captured electrons in the guard ring structure to the injected current is a measure of the guard ring effectiveness. When the electron current outside the guard ring is less compared to the current collected on an n-well diffusion, then the escaped collected current normalized to the injected current is also a metric for the evaluation of the effectiveness of the guard ring (or lack of its effectiveness). Let us define a 'guard ring efficiency' injection ratio metric F, where

$$F = \frac{I\,(\text{injector})}{I\,(\text{collector})}$$

If every minority carrier that is injected is collected, the injection ratio would equal unity. As the number of collected minority carriers decreases, the factor F increases. This increase is a measure of the effectiveness of the guard ring to minimize the transport out of the guard ring to a region of interest where the carriers are collected. This can be trivially quantified on a test structure or simulated to quantify the effective transport ratio. As this term is not normalized to the injection current, the inverse relationship is better viewed from a probability perspective.

With normalizing with respect to the injection current, one can evaluate the probability a minority carrier can reach the structure of interest. The inverse term, $1/F$, is the escape probability in the approximation that the collected electrons outside the ring are significantly greater than the electrons that recombine outside the guard ring. This interpretation can change on the basis of the structures between the injecting and the collecting structure. Let us refer to $1/F$ as the guard ring factor (e.g., associated with the collected charge to the outer structure). From a practical perspective of electrical measurement, the collected minority carriers in a metallurgical junction (e.g., an n-well region containing a p-channel transistor) are the structure of interest. Hence, the 'collected current' normalized to the 'injected current' can be quantified using an 'injecting structure' and the 'collecting structure.'

The escape probability is then related to the following design parameters:

- physical dimensions of the injection source (e.g., width, length, and depth);

- semiconductor process of the injection source (e.g., n^+ diffusion, n-well, n-channel transistor);

- physical separation between the injection source and the collecting region;

- physical dimensions of the collecting region (e.g., width length, and depth);

- bias conditions on the collecting region (e.g., n-well to substrate bias);

- current magnitude;

- all structures and guard ring structures between the injection source and the collection source.

9.6.3 Guard Ring Design

One of the key problems in computer aided design (CAD), is the ability to provide checking and verification of the presence of guard rings. Typically, the 'guard rings' were of physical shapes that were not identifiable or distinguishable from other design shapes. Hence, it was difficult to identify the existence of guard rings in a design and that a design team was formed to satisfy the guard ring rules. To address this, guard ring designs were integrated into ESD elements or I/O book as a frame of the elements. In recent years, the lack of verification of the guard rings have lead to both ESD and latchup errors. Hence, in nonparameterized cell design environments, ESD and latchup errors still occur in chip design. Bass *et al.* [45] instituted the design methodology of 'dummy design layers' that was to be placed on the ESD devices and I/O networks. First, this allowed for ESD-specific design rule checking by data compression of all data outside the ESD and I/O regions. Second, it allowed for verification of a one-to-one correspondence, as a reminder to circuit designers, the dummy design layers are placed on the guard rings; this indirectly forced design teams to place the guard rings in chip design with the placement of the 'ESD dummy design level' and 'I/O dummy design level.' But, this still was an inadequate method to provide design rule compliance during design rule checking and verification steps of the existence of the guard ring structure. With the establishment of a hierarchical Cadence^{TM}-based ESD design methodology, the opportunity to integrate guard rings into the design methodology provides built-in compliance, checking, and verification.

In this methodology, the dilemma comes in about the integration of the guard ring structures in the primitive Pcells and the formation of higher level circuits. First, in the design of primitive elements, primitive elements may be internal design elements or external design elements. Hence, in some cases, the primitive element may not require a guard ring around the structure given that it is in the interior or core of the semiconductor chip. Hence, even with the primitive circuit element or device, there are cases of the modeling with or without the guard ring element, and the choice of the guard ring structure may not be suitable for both internal or external chip application. Hence, the integration of a parameterized cell with a defined and fixed guard ring may not be advantageous.

Second, in the case of hierarchical design, the need for multiple guard rings about all the primitive devices may not be necessary; the additional guard rings impacts design area and limits design flexibility. For example, in the formation of a double-diode network, a guard ring is required around the complete double-diode network, but not individually around the p^+/n-well diode and the n-well-to-substrate diode. The integration and layout allow the elimination of some of these rings, and when not doing so, leads to wasted design area. Hence, it is a major design disadvantage without the ability to develop a design system that does not allow the removal of the guard rings in a hierarchical integration.

Additionally, for latchup, it may be desirable to improve the latchup robustness on the basis of the relative placement of certain element adjacent to other circuits (for intercircuit latchup) or injection sources (external latchup).

Additionally, it is desirable to have a guard ring structure whose guard ring efficiency or transmission factor is 'stored' in the guard ring itself, allowing for latchup design evaluation, checking, and verification, as well as quantification of the effectiveness of the guard ring structure associated with intercircuit internal latchup or external latchup verification. In this manner, a latchup guard ring efficiency can be quantified *once* in the technology during technology release and technology library release instead of for each circuit element and circuit to circuit interaction. Additionally, in this manner, the latchup robustness can be adjusted with the introduction of additional guard rings and structures to improve latchup robustness during the floor plan and cosynthesis of floor plan, RF design, ESD, and latchup.

To address design integration of guard rings in a Cadence-based parameterized cell system, Perez and Voldman developed a parameterized guard ring cell, where the guard rings can be integrated with the primitive O[1] device elements or the higher order O[n] hierarchical parameterized cell [5,7]. In this methodology, ESD Pcells are designed such that the guard ring structures can be turned 'off' using switches in the graphical unit interface (GUI). An independent guard ring p-cell is defined as that which contains a plurality of consecutive ring structure types and number. This Pcell allows the designing of a guard ring structure and a choice of guard ring based on guard ring efficiency requirements whose shapes and rings are chosen accordingly. Additionally, it allows for a growable guard ring that expands on the basis of the identification of the element type, the enclosed Pcell device, circuit or function block design input parameters. This concept also allows for designing of a guard ring structure with generated virtual design levels (e.g., guard ring virtual level, I/O virtual level, ESD virtual level), graphical and schematic representation view, guard ring structure symbology, and symbol. Usage of the guard ring Pcell allows for guard ring design checking and verification.

Table 9.3 shows the components of the guard ring Pcell structure. In the guard ring Pcell, there are a plurality of ring structures that can be turned 'on' or 'off.' In the table, a listing of potential cases is shown; note for example, the deep trench (DT) structure can consist of as many as three consecutive rings structures, allowing a large variety of guard ring structures for minimizing latchup, noise or injection issues for RF applications.

In this design methodology, the GUI allows the user to apply the various guard ring structures to the physical design. Figure 9.14 is an example of the GUI interface that initiates the guard ring parameterized cell. In the ESD design system, the 'guard ring' parameterized cell consists of multiple rings.

Table 9.3   Guard ring parameterized cell

Guard ring Pcell components	Examples of guard ring structures
p^+	p^+ (single ring)
n^+	p^+/n^+
n-Well	p^+/NW
Deep trench	$p^+/NW/DT$ inside
	$p^+/NW/DT$ outside
	$p^+/DT/NW/DT$ inside
	$p^+/DT/NW/DT/DT$ out

Figure 9.15 is an example of the graphical implementation of an O[2] RF double-diode network. In the implementation, the guard rings are turned off in the primitive O[1] designs to allow compression of the Pcells for density. In this manner, higher order O[3] Pcell is compiled integrating the ESD element and the guard ring Pcell. The advantage of this method allows significant design flexibility, design checking, and verification of the guard ring existence.

One of the advantage of the guard ring Pcell is the ability to identify the existence of guard ring in the design system for the ESD and the latchup design checking and verification. This can be established by the symbology, design hierarchy, or component review. Figure 9.16 shows the interaction of the ESD network, the circuit, and the guard ring parameterized cells. Another means of design verification is to integrate the guard ring symbol into the Cadence symbol cell view.

9.6.4 Guard Ring Characterization

The guard ring effectiveness can be independently quantified for latchup analysis and design methodology [5,6,7,33,35]. Figure 9.17 is an example of the extraction of the guard ring effectiveness for a Pcell when the n-well guard ring 'switch' is initiated. An injected current is initiated, discharging into the substrate region. The guard ring is placed between the injector and the collector. The ratio of the injected to the collected current, F, is plotted as a

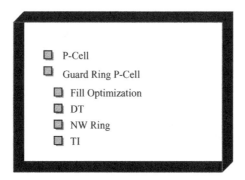

Figure 9.14   Example of the guard ring parameterized cell guard ring automation and graphical unit interface

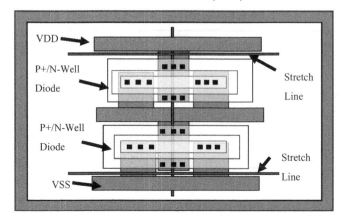

Figure 9.15   RF ESD O[2] hierarchical Pcell with O[1] guard ring *p*-cell forming an O[3] Pcell

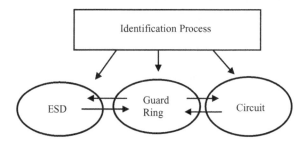

Figure 9.16   Identification process of ESD, guard ring, and circuit parameterized cells

function of the injected current. Different well-bias values are chosen to evaluate the dependence on the guard ring well bias.

Table 9.4 summarizes the structure type for electron injection using n^+ ESD diodes and RF NFETs as the injecting source. In these structures, a series of concentric rings were formed between the injection and collection structure. In essence, the *n*-well collector ring

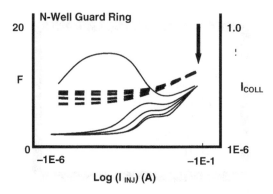

Figure 9.17   An example of the measurement of the extraction of guard ring efficiency (extraction taken at 100 mA current levels)

Table 9.4 Structure type for electron injection

Type	Injector	Guard ring	Collector
A	n^+ Diode	p^+ (single ring)	n-Well
B	n^+ Diode	p^+/n-well (float)	n-Well
C	n^+ Diode	p^+/NW/DT inside	n-Well
D	n^+ Diode	p^+/NW/DT outside	n-Well
E	RF NFET	p^+	n-Well
F	RF NFET	p^+/DT/NW/DT inside	n-Well
G	RF NFET	p^+/DT/NW/DT/DT out	n-Well

Table 9.5 n^+ to n-well current gain and injection efficiency

Type	Guard ring	β_F	β_R	F
A	p^+	0.6	0.22	2.92
B	p^+/n-well	0.130	0.183	11.2
C	p^+/NW/DT	0.050	0.444	19.4
D	p^+/NW/DT/n-well	0.04	0.434	23

also serves as a guard ring structure but is used to serve as a collecting or injecting structure. In these structures, DT is placed within the set of ring structures [6,7].

Table 9.5 is an example of the measurements of the guard ring p-cell where the injector is a n^+ diffusion and the collecting structure is an n-well region. The guard ring structures consist of a plurality of ring cases. Note, the table contains the lateral bipolar gain characteristics and the influence that the guard ring structure has on the forward and reverse gain characteristics.

In Table 9.6, an RF optimized n-channel MOSFET parameterized cell is used as the injection source and enclosed by the guard ring parameterized cell.

To summarize, the hierarchical ESD design methodology was further extended by the development of guard ring parameterized cells that are integrated into the Pcell design hierarchy. The guard ring efficiency of the guard ring p-cell can be quantified independent of the injection source, the primitive circuit element, the ESD Pcell, or I/O network. The information of the physics of the guard ring can also be stored in the design system for evaluation of the guard ring efficiency for internal or external latchup phenomena. The presence of an independent guard ring Pcell allows for the design rule checking and verification, which could not be achieved in other design methodologies. Additionally, the

Table 9.6 RF NFET: n-well bipolar current gain and F factor

Type	Guard ring	β_R	F
E	p^+	0.63	2.69
F	p^+/DT/NW/DT inside	0.898	2.38
G	p^+/DT/p^+/DT/NW/DT	0.093	18.0

flexibility of the design can allow for higher order hierarchical design methods that can optimize the guard ring spacings on the basis of a more complex circuit network and decouple the design from the base primitive cell design. This design methodology was integrated into a 0.13-μm BiCMOS SiGe technology and is suitable for sub-0.13 μm technology for RF CMOS, mixed signal CMOS (MS-CMOS), BiCMOS, BiCMOS SiGe, and system-on-chip (SOC) applications.

9.7 SUMMARY AND CLOSING COMMENTS

In this chapter we have demonstrated new methods of design of ESD elements and hierarchical design methods. With the hierarchical parameterized cell ESD methodology, the size and design of ESD networks can be altered without reevaluation of the RF characteristics; this essential factor enables the ability to cosynthesize ESD networks with the RF characteristics and functional analysis. In recent times, this method is used in RF CMOS, RF BiCMOS, and RF BiCMOS SiGe technology.

In the next chapter, off-chip protection is addressed and the focus will be on on-chip protection networks. We will discuss the concepts of the placement of the ESD protection for the RF circuitry on the board or system level. Moreover, we will discuss the utilization of spark gaps, field emission devices, and conductive polymer ESD protection networks for RF applications.

PROBLEMS

1. What is the problem in the cosynthesis of RF and ESD when primitive elements are not utilized?

2. What are the advantages of quantized unit cells versus nonquantized unit cells (fixed unit cells locally) within a parameterized cells?

3. What is the minimum set of primitive elements (e.g., transistors, resistors, etc.) needed for an RF CMOS ESD library?

4. What is the minimum set of primitive elements required for an RF Bipolar ESD library?

5. What is the minimum set of primitive elements required for an RF BiCMOS ESD library?

6. List the minimum set of hierarchical parameterized ESD cells required to satisfy the ESD design requirements for an RF CMOS technology.

7. List the minimum set of hierarchical parameterized ESD cells required to satisfy the ESD design requirements for an RF BiCMOS technology.

8. Why are new models not required as the hierarchical parameterized ESD elements change physical size? How do the interconnect and wiring not influence the RF response and model results?

9. Explain the problem of integration of guard rings from an RF model perspective and ESD design perspective. How do you integrate primitives into hierarchical designs when the guard rings are present? What is the size impact? What is the modeling impact?

10. What is the minimum number of parameters needed to be open on an RC-triggered MOSFET ESD power clamp? Why?

11. What is the minimum number of parameters needed to be opened on an SiGe Darlington BV_{CEO} triggered ESD power clamp? Why?

12. How many different SiGe HBT Darlington ESD power clamps can be formed with a low and high breakdown transistors? With an additional medium breakdown transistors?

REFERENCES

1. Voldman S, Strang S, Jordan D. An automated electrostatic discharge computer-aided design system with the incorporation of hierarchical parameterized cells in BiCMOS analog and RF technology for mixed signal applications. *Proceedings of the Electrical Overstress/Electrostatic Discharge (EOS/ESD) Symposium*, October 2002. p. 296–305.
2. Voldman S, Strang S, Jordan D. A design system for auto-generation of ESD circuits. *Proceedings of the International Cadence Users Group*, September 2002.
3. Voldman S. Automated hierarchical parameterized ESD network design and checking system. U.S. Patent No. 5,704, 179 (March 9, 2004).
4. Collins D, Jordan D, Strang S, Voldman S. ESD design, verification, and checking system and method of use. U.S. Patent Application No. 20050102644 (May 12, 2005).
5. Perez CN, Voldman S. Method of forming a guard ring parameterized cell structure in a hierarchical parameterized cell design, checking and verification system. U.S. Patent Application No. 20040268284 (December 30, 2004).
6. Watson A, Voldman S, Larsen T. Deep trench guard ring structures and evaluation of the probability of minority carrier escape for ESD and latchup in advanced BiCMOS SiGe technology. *Proceedings of the Taiwan Electrostatic Discharge Conference (T-ESDC)* (November 12–13, 2003). p. 97–103.
7. Voldman S, Perez CN, Watson A. Guard rings: theory, experimental quantification and design. *Proceedings of the Electrical Overstress/Electrostatic Discharge (EOS/ESD) Symposium*, 2005. p. 131–40.
8. Gebreselasie E, Sauter W, St.Onge S, Voldman S. ESD structures and circuits under bond pads for RF BiCMOS silicon germanium and RF CMOS technology. *Proceedings of the Taiwan Electrostatic Discharge Conference (T-ESDC)*, 2005. p. 73–8.
9. Voldman S, Botula A, Hui D, Juliano P. Silicon germanium hetero-junction bipolar transistor ESD power clamps and the Johnson Limit. *Proceedings of the Electrical Overstress/Electrostatic Discharge (EOS/ESD) Symposium*, 2001. p. 326–36.
10. Voldman S. Variable trigger-voltage ESD power clamps for mixed voltage applications using a 120 GHz/100 GHz (f_T/f_{MAX}) silicon germanium hetero-junction bipolar transistor with carbon incorporation. *Proceedings of the Electrical Overstress/Electrostatic Discharge (EOS/ESD) Symposium*, October 2002. p. 52–61.
11. Ronan B, Voldman S, Lanzerotti L, Rascoe J, Sheridan D, Rajendran K. High current transmission line pulse (TLP) and ESD characterization of a silicon germanium hetero-junction bipolar transistor with carbon incorporation. *Proceedings of the International Reliability Physics Symposium (IRPS)*, 2002. p. 175–83.
12. Voldman S, Gebreselasie E. Low-voltage diode-configured SiGe:C HBT triggered ESD power clamps using a raised extrinsic base 200/285 GHz (f_T/f_{MAX}) SiGe:C HBT device. *Proceedings of the Electrical Overstress/Electrostatic Discharge (EOS/ESD) Symposium*, 2004. p. 57–66.
13. Voldman S. ESD robust silicon germanium transistor with emitter NP-block mask extrinsic base ballasting resistor with doped facet region. U.S. Patent No. 6,465,870 (October 15, 2002).

14. Brennan CJ, Voldman S. Internally ballasted silicon germanium transistor. U.S. Patent No. 6,455,919 (September 24, 2002).
15. Brennan CJ, Hershberger DB, Lee M, Schmidt NT, Voldman S. Trench-defined silicon germanium ESD diode network. U.S. Patent No. 6,396,107 (May 28, 2002).
16. Lanzerotti LD, Voldman S. Self-aligned SiGe *npn* with improved ESD robustness using wide emitter polysilicon extensions. U.S. Patent No. 6,441,462 (August 27, 2002).
17. Botula A, Hui DT, Voldman S. Electrostatic discharge power clamp circuit. U.S. Patent No. 6,429,489 (August 6, 2002).
18. Lanzerotti L, Ronan B, Voldman S. Silicon germanium heterojunction bipolar transistor with carbon incorporation. U.S. Patent No. 6,670,654 (December 30, 2003).
19. Voldman S, Botula A, Hui DT. Electrostatic discharge power clamp circuit. U.S. Patent No. 6,549,061 (April 15, 2003).
20. Voldman S. SiGe transistor, varactor, and *p-i-n* velocity saturated ballasting element for BiCMOS peripheral circuits and ESD networks. U.S. Patent No. 6,552,406 (April 22, 2003).
21. Voldman S. Self-aligned silicon germanium heterojunction bipolar transistor device with electrostatic discharge crevice cover for salicide displacement. U.S. Patent No. 6,586,818 (July 1, 2003).
22. Voldman S. Dual emitter transistor with ESD protection. U.S. Patent No. 6,731,488 (May 4, 2004).
23. Voldman S. SiGe transistor, varactor and *p-i-n* velocity saturated ballasting element for BiCMOS peripheral circuits and *ESD* networks. U.S. Patent No. 6,720,637 (April 13, 2004).
24. Voldman S. Electrostatic discharge input and power clamp circuit for high cutoff frequency technology radio frequency (RF) applications. U.S. Patent No. 6,946,707 (September 20, 2005).
25. Coolbaugh D, Voldman S. Carbon-modulated breakdown voltage SiGe transistor for low voltage trigger ESD applications. U.S. Patent No. 6,878,976 (April 12, 2005).
26. Pequignot J, Sloan J, Stout D, Voldman S. Electrostatic discharge protection networks for triple well semiconductor devices. U.S. Patent No. 6,891,207 (May 10, 2005).
27. Voldman S. The effect of deep trench isolation, trench isolation, and sub-collector on the electrostatic discharge (ESD) robustness of radio frequency (RF) ESD STI-bound p^+/n-well diodes in a BiCMOS silicon germanium technology. *Proceedings of the Electrical Overstress/Electrostatic Discharge (EOS/ESD) Symposium*, 2003. p. 214–23.
28. Voldman S. BiCMOS ESD circuit with subcollector/trench-isolated body MOSFET for mixed signal analog/digital RF applications. U.S. Patent No. 6,455,902 (September 24, 2002).
29. Eshun E, Voldman S. High tolerance TCR balanced high current resistor for RF CMOS and RF SiGe BiCMOS applications and Cadence-based hierarchical parameterized cell design kit with tunable TCR and ESD resistor ballasting feature. Patent No. 6,969,903 (November 29, 2005).
30. Voldman S. Tunable semiconductor diodes. U.S. Patent Application No. 200515223 (July 14, 2005).
31. Voldman S, Zierak M. Modulated trigger device. U.S. Patent No. 6,975,015 (December 13, 2005).
32. Voldman S, Johnson R, Lanzerotti LD, St. Onge SA. Deep trench-buried layer array and integrated device structures for noise isolation and latch up immunity. U.S. Patent No. 6,600,199 (July 29, 2003).
33. Troutman R. *Latchup in CMOS technology: the problem and the cure*. New York: Kluwer Publications; 1985.
34. Estreich DB. *The physics and modeling of latchup and CMOS integrated circuits*. Integrated Circuits Laboratory, Stanford University; November 1980.
35. Troutman R. Epitaxial layer enhancement of *n*-well guard rings for CMOS circuits. *IEEE Transactions on Electronic Devices Letters* 1983;**ED-4**:438–40.
36. Hargrove M, Voldman S, Brown J, Duncan K, Craig W. Latchup in CMOS. *Proceedings of the International Reliability Physics Symposium (IRPS)*, April 1998. p. 269–78.
37. Morris W. CMOS latchup. *Proceedings of the International Reliability Physics Symposium (IRPS)*, May 2003. p. 86–92.

38. Voldman S, Watson A. The influence of deep trench and substrate resistance on the latchup robustness in a BiCMOS silicon germanium technology. *Proceedings of the International Reliability Physics Symposium (IRPS)*, 2004. p. 135–42.

39. Voldman S, Watson A. The influence of polysilicon-filled deep trench and sub-collector implants on latchup robustness in RF CMOS and BiCMOS SiGe technology. *Proceedings of the Taiwan Electrostatic Discharge Conference (T-ESDC)*, 2004. p. 15–9.

40. Voldman S. CMOS. Latchup Tutorial. *Tutorial K, Monday Tutorial Notes, Electrical Overstress/Electrostatic Discharge Symposium*, 2004;K1–60.

41. Voldman S. CMOS latchup. *Tutorial Notes of the International Reliability Physics Symposium (IRPS)*, April 17, 2005.

42. Voldman S. A review of CMOS latchup and electrostatic discharge (ESD) in bipolar complimentary MOSFET (BiCMOS) silicon germanium technologies: Part II – Latchup. *Microelectronics Reliability* 2005;**45**:437–55.

43. Voldman S, Gebreselasie E, Lanzerotti L, Larsen T, Feilchenfeld N, St. Onge S, *et al.* The Influence of a silicon dioxide-filled trench isolation structure and implanted sub-collector on latchup robustness. *Proceedings of the International Reliability Physics Symposium (IRPS)*, 2005. p. 112–21.

44. Voldman S, Gebreselasie EG, Hershberger D, Collins DS, Feilchenfeld NB, St. Onge SA, *et al.* Latchup in merged triple well technology. *Proceedings of the International Reliability Physics Symposium (IRPS)*, 2005. p. 129–36.

45. Bass RS, Nickel DJ, Sullivan DC, Voldman S. Method of automated ESD protection level verification. U.S. Patent No. 6,086.627 (July 11, 2000).

10 Alternative ESD Concepts: On-Chip and Off-Chip ESD Protection Solutions

With increasing RF performance objectives, the ability to provide low capacitance ESD protection will increase in importance. When the application frequency increased to 1 GHz, the need to reduce the size of the ESD protection loading effects became an issue for CMOS, silicon-on-insulator (SOI), and BiCMOS technology. As the application frequencies increased from 1 to 5 GHz, the choice of the ESD element based on its ESD robustness versus capacitance loading is evaluated to provide the optimum ESD solution. As the frequency increased above 5 GHz, cosynthesis of ESD and RF performance increased in interest. The question remains, as to what will be the ESD chip and system solution as the frequency of the circuits increase to 10, 100 GHz, and above. In recent times, 100 GHz circuits have been demonstrated.

In this chapter, traditional solutions are abandoned, and new ESD directions are reviewed. On-chip and off-chip ESD solutions that are practiced today and proposed for the future are also discussed. Moreover, the chapter focuses from air breakdown spark gaps [1–7], field emission devices (FEDs) [7], polymer voltage suppression (PVS) devices [8,9], mechanical packaging solutions [10–28], to proximity communication techniques in multichip environments [29].

10.1 SPARK GAPS

With increasing RF performance objectives, the ability to provide low capacitance ESD protection will increase in importance. Spark gaps may play an important role in on-chip or off-chip ESD protection in the future. Spark gaps perform on the concept of air breakdown [1–7]. Spark gaps have been of interest in the systems for circuit boards and on modules in the 1970s [5,6]. Spark gaps have been formed on the printed circuit boards (PCBs), modules, and in multichip environments. Spark gaps can be formed on ceramic substrates, silicon carriers, and other forms of packages that semiconductor chips are mounted upon. Off-chip spark gaps have the advantage of not using area on a semiconductor chip.

Spark gaps and FEDs can be constructed on-chip or off-chip as a form of ESD protection [5–7]. Spark gaps can be formed using metallization patterns formed on the package or substrate material using closely spaced metal lines. Spark gaps formed on the ceramic substrate are limited by the allowed metal line-to-line spacing. A disadvantage of the off-chip implementation of a spark gap is the line-to-line spacing that is significantly larger than what is achievable in an on-chip implementation. Spark gap reliability is critical; this manifests itself in the ability to discharge, as well as repeatability is an important issue in using spark gap as ESD protection. Another limitation of spark gaps is the reaction time. Gas discharges that occur in the spark gaps have a nanosecond reaction time. Another limitation is the breakdown voltages that occur. Hence, spark gaps have the following limitations [7]:

- reliability of electrical discharge initiation;
- repeatability of discharge current and voltage magnitude;
- time constant response of the arc discharge;
- magnitude of the breakdown voltages (e.g., compared to the application device sensitivity).

During discharge events, electrical damage can occur to the spark gap electrodes leading to reliability concerns. Moreover, with damage to the electrode curvature, and material residuals, the repeatability may be inadequate to insure or qualify the off-chip spark gap.

Additionally, for RF applications, the time constant associated with the breakdown, gas ionization, and the reverse recovery time of the ionized gas may be long compared to circuit response of RF circuitry. Spark gap initiation time is a function of the collision and ionization times; the time constant to initiate ionization of gas molecules associated with avalanche multiplication [1–4]. Breakdown in gases is initiated by a feedback induced by the acceleration of carriers leading to secondary carriers. At very high speed, the ability to provide semiconductor devices may be limited.

And lastly, another concern is the magnitude of the trigger voltage of the spark gap. Paschen studied the breakdown physics of gases in planar gap regions [1]. The result of Paschen showed that breakdown process is a function of the product of the gas pressure and the distance between the electrodes. Paschen showed that

$$pd \approx \frac{d}{l}$$

where p is the pressure, d is the distance between the plates, and l is the mean free path of the electrons. From the work of Paschen, a universal curve was established that followed the same characteristics independent on the gas in the gap. The Paschen curve is a plot of the logarithm of the breakdown voltage as a function of the logarithm of the product of the pressure and gap distance.

$$V_{BD} = f(pd)$$

At very low values of the pd product, electrons must accelerate beyond the ionization limit to produce an avalanche process because the likelihood of impacts is very less. In this region, the breakdown voltage decreases with the increasing value of the pressure–gap product. This occurs until a minimum condition is reached. At very high values of the pressure–gap product, the number of inelastic collisions is higher and the breakdown voltage increases. This U-shaped dependence is characteristic of gas phenomenon. Townsend, in 1915, noted that the breakdown occurs at a critical avalanche height [4],

$$H = e^{\alpha d} = \frac{1}{\gamma}$$

In this expression, the avalanche height, H, is equal to the exponential of the product of the probability coefficient of ionization (number of ionizing impacts per electron and unit distance in the direction of the electric field) and electrode spacing. The avalanche height, H can also be expressed as the inverse of the probability coefficient of regeneration (number of new electrons released from the cathode per positive ion).

These devices may also be limited by the resistance of the arc. Toepler, in 1906, established a relationship of the arc resistance in a discharge process [2]. Toepler's law states that the arc resistance at any time is inversely proportional to the charge that has flowed through the arc

$$R(t) = \frac{k_T D}{\displaystyle\int_0^t I(t')\, dt'}$$

where $I(t)$ is the current in the arc discharge at time t and D is the gap between the electrodes. k_T is a constant whose value is 4×10^{-5} V s/cm.

For off-chip spark gaps, the ability to produce a well-defined metal spark gap is difficult because of the lithography and semiconductor process techniques used in the packaging environments.

Experimental work on off-chip spark gaps had limited success for HBM protection. Owing to the nature of the spark gap breakdown, it was found that 'ESD test windows' were observed. Using commercial human body model (HBM) test systems, it was found that off-chip spark gap devices could provide ESD protection at high voltages (e.g., 1000–2000 V HBM). But, with an incremental ESD step-stress at 100 V increments, ESD failure occurred at lower HBM stress levels. Where it was found that the HBM stress provided protection for higher voltages and, semiconductor products were vulnerable at lower HBM levels. Hence, it is important in the application of ESD spark gap elements to demonstrate the operation window and its operability range.

Experimental work on off-chip spark gaps were also found to have had limited success in improving ESD protection. Hyatt utilized a three-stage system protection strategy: the first system level surge protection element, the second-stage on-package spark gap, and the third on-chip ESD protection network used to protect a monolithic microwave semiconductor chip. A very fast-transmission line pulse (VF-TLP) test source, with a 1 ns pulse width demonstrated that the second stage off-chip spark gap (e.g., integrated into the package module) showed no improvement in the three-stage ESD system. The off-chip spark gap had limited response as a result of the lithographic capability to construct on a package (e.g., spark gap width was too large) and limited reaction time for very fast transmission line pulse events.

In the experimental work of Bock [7], on-chip spark gaps were designed for applications of ESD protection of monolithic microwave integrated circuits (MMIC). Bock developed on-chip spark gaps with a gap dimension of 0.4–4.0 μm. At a 1 μm gap dimension, the breakdown voltage of air under normal pressure is in the order of 250 V. He [7] noted that the spark gap breakdown voltage is reduced to 45 V as a result of electric field enhancement factors that are caused by nonplanar surfaces. The primary mode of discharge was associated with field emission in the spark gap, not the actual gas discharge phenomenon. He [7] also observed that for on-chip spark gaps in the micron dimension, the mode of operation was partly by ionization processes as well as field emission between the two electrodes.

The usage of spark gaps as an off-chip protection solution will be more feasible only if the physical dimensions on silicon carriers, ceramic substrates, and other packages are reduced allowing a lower voltage spark gap. To make it suitable for low-voltage semi-conductors, optimization of the field emission properties of the spark gaps must be controlled. Additionally, advanced processing techniques will be required to lower the breakdown voltage. Moreover, as the need for off-chip ESD solutions increases, the need for improved spark gaps with improved reliability will also be increased.

10.2 FIELD EMISSION DEVICES

10.2.1 Field Emission Device (FED) as ESD Protection

Field emission devices as ESD protection have advantages over silicon based devices. Figure 10.1 shows an example of a field emission device constructed in a Gallium Arsenide technology. The advantages are as follows [7]:

- low dielectric constant ($\varepsilon_r = 1$);

- power dissipation;

- power handling capability.

Compared to silicon-based ESD networks, the dielectric constant of air is unity. In semiconductors materials (e.g., silicon, gallium arsenide, germanium, silicon dioxide, and low-k interlevel dielectrics), the dielectric constant is greater than unity, leading to a capacitance advantage in air-based ESD protection networks. In the air gap, the power dissipation and self-heating of the device is not as significant as occurs in silicon-, germanium-, or gallium-based semiconductor devices. Hence, there the ability to handle pulsed power is significant.

From a RF performance and functionality perspective, FED ESD devices have the following advantages [7]:

- low parasitic capacitance (e.g., less than 0.1 pF);

- fast switching times (e.g., less than 1 ps);

- high current densities ($J > 10^8$ A/cm^2).

When the application frequency increases for MMIC applications, under a constraint of a constant reactance, the capacitance of the ESD protection will be scaled. At 100 MHz, 1–10 pF of capacitance is the acceptable loading for ESD protection networks. At 1 GHz,

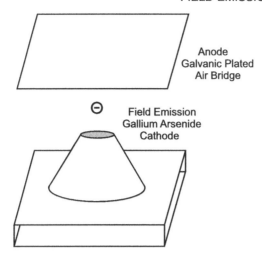

Figure 10.1   Cross section of a field emission device

1 pF of ESD capacitance was acceptable for most CMOS digital logic and RF applications. But, at 10 GHz and above, the ESD capacitance loading must be reduced below 0.1 pF capacitance. In semiconductor-based ESD elements this is achievable by doping optimization or reduction of the ESD network size. Hence, the FED element can achieve the loading capacitance requirement for greater than 10 GHz frequencies.

For switching times, FEDs are limited by the transit time of carrier. As opposed to spark gaps, FEDs are not limited to the ionization processes and avalanche processes, which occurs during air breakdown. As a result, for scaled devices, the potential speed of FEDs can achieve 10^{-12} s time scales. For ESD events, HBM, and machine model specifications, the time constant is 10^{-9} s time scales [7]. The work of Barth shows that the responses can range to 100×10^{-12} s. For charged device model, the rise time is less than 250 ps time. Hence, the FED response time can be as fast or faster than the time scales associated with ESD events. Field emission devices require closely spaced structures and the ability to form a 'point' electrode with a well-defined radius of curvature and emitter-to-collector spacing. Both FEDs and spark gaps utilize air bridge structures to form the physical gap. This requires additional process masks, as well as a process where air exposure is not a reliability issue.

10.2.2 Field Emission Devices in Gallium Arsenide Technology

Bock reported that FEDs can provide higher current densities and have demonstrated higher reliability of discharge repeatability [7]. Using an additional five masks, a air-bridge GaAs FED structure was formed on-wafer with a loading capacitance of less than 0.1 pF [7].

In the structure, the FED was constructed in a gallium arsenide process. The substrate material is a semi-insulating gallium arsenide substrate. An n-epitaxial gallium arsenide layer is formed on the semi-insulating GaAs substrate. The emitter structure is formed using a wet etch process from the n-epitaxial GaAs layer. The etch process forms the emitter points and defines the radius of curvature of the emitter structure; the radius of curvature varies the electric field at the emitter points. The emitter structure is then coated with a low work-function metal film. Using the wet etch process, the emitter tips were designed to 75 nm dimensions [7]. Ohmic contacts are formed near the n-epitaxial region between the emitter points in the emitter structure of the FED.

To provide a well-defined emitter gap, a photo-resist film is deposited over the emitter structure, followed by a second deposition of a conformal metal film. The metal film is masked and etched to form the collector structure. After the removal of the photo-resist, an air-gap is formed between the GaAs emitters and the metal air-bridge collector structure. Hence, control of the design is a function of the ability to form the GaAs emitter points as well as control of the photo-resist film thickness.

10.2.3 Field Emission Device Electronic Blunting Effect

Bock discuss a means of optimization called 'electronic blunting effect' [7]. In a semiconductor film, a saturation effect occurs that leads to a current-limiting phenomenon. As the space charge in the gap increases, the electric field penetrates into the semiconductor emitter structure, where the penetration is a function of the doping concentration. When this occurs, the series resistance of the emitter structure increases; this increase leads to a spatial self-ballasting near the emitter point. As a result, the current distributes more evenly through the emitter region, lowering the peak current density at the emitter tip and provides better current distribution through the emitter structure. Bock and Hartnagel point out that this 'electronic blunting effect' can be utilized to provide a ballasting effect in a multiemitter field effect device. Bock and Hartnagel point out that in the case where a metal film is used in the FED, where there are nonuniform variations in the emitter height, a single emitter may turn-on prior to other parallel emitter structures. But in the case of a semiconductor emitter structure, the 'electronic blunting effect' can lead to simultaneous turn-on of multiple emitter structures, leading to an improved ESD protection device. Hence, although the metal film is superior from a current density perspective for the utilization of an FED ESD element, it is advantageous to use a semiconductor that leads to better current distribution and higher total peak current through the protection element.

10.2.4 Field Emission Device Multiemitter ESD Design

Bock and Hartnagel constructed two different FED ESD protection networks. In the first design, a two emitter structure was implemented with a 20 μm length, and a 250 nm tip radius (Figure 10.2). The emitter regions were coated with a gold film. In this design it was found that degradation was observed in the gold tip emitter structures. In this case, the FED ESD element did demonstrate an improvement in the RF device that was to be protected. This achieved ESD protection levels above 1600 V HBM in GaAs prototypes (e.g., GaAs products typically do not achieve ESD HBM levels above 1000 V HBM) [7].

The second design of the FED ESD network design consisted of a 10-emitter structure, with emitter lengths of 35 μm and a 75 nm emitter tip radius. The emitter-to-emitter design pitch is 14 μm space. The trigger voltage was −7 V with the cathode negatively biased (e.g., emitter pointed wedge has a negative applied voltage relative to the anode air bridge with a

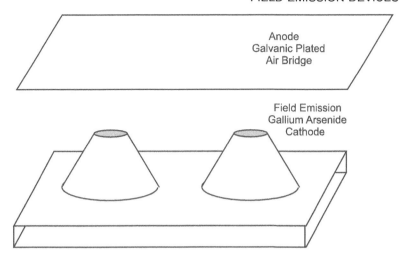

Figure 10.2   Multiemitter GaAs field emission device (FED)

positive bias), and 20 V when the cathode is positively biased (e.g., emitter pointed wedge has a positive applied voltage relative to the anode air bridge with a negative bias). It was pointed out that this implementation provided a bidirectional ESD network operation similar to a ESD 'double-diode' network but with an asymmetric trigger condition [7].

With the utilization of the semiconductor process materials in RF MMIC, the FED ESD element can be integrated into the physical design of a transistor element.

The above design can be improved by adding a control gate structure. Figure 10.3 shows a GaAs field emission triode. This structure uses a n^+ GaAs grid to assist in the control of the field emission process as well as d.c. currents.

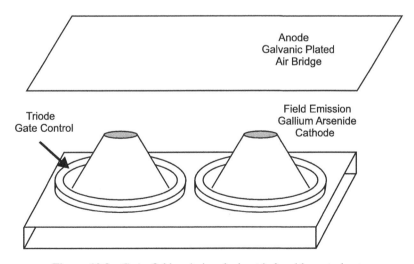

Figure 10.3   GaAs field emission device triode with control gate

10.2.5 Field Emission Device (FED) ESD Design Practices

Key ESD design practices in the implementation of FED ESD networks are as follows:

- FED emitter-to-collector gap control can lead to nonuniform conduction of emitters in multiemitter design.

- FED current uniformity can be improved using a single semiconductor emitter instead of metal emitter.

- FED emitter ballasting can be improved by 'electronic blunting effect' in multiemitter FED structures.

- FED electronic blunting can compensate for emitter-to-collector gap control.

- FED devices can produce a bidirectional (but asymmetric trigger voltage) ESD solution.

In semiconductor chips, the upper surface of the semiconductor chip is passivated with films to avoid corrosion, contamination, and scratching of films and interconnects. In RF semiconductors and with the trend to use micromechanical machines (MEMs), the usage of air bridges and spark gaps are more acceptable than in the past. This will allow opportunities for the implementation and integration of on-chip spark gaps and FEDs. Hence, because of the reliability, the repeatability, the time response, and the breakdown voltage, the usage of on-chip spark gaps and FEDs have provided a greater opportunity and potentially better success for the implementation of the on-chip for the high performance of RF applications. Yet, process cost and other reliability issues may lead to future off-chip implementations.

10.3 OFF-CHIP PROTECTION AND OFF-CHIP TRANSIENT SUPPRESSION DEVICES

With increasing technology performance objectives, the ability to provide on-chip ESD protection will become more difficult. Moreover, with the increase in circuit density and smaller physical geometries, the ability to have on-chip ESD protection only may be a limited perspective. In recent times, the trend toward finding both an on-chip and off-chip protection solution has increased. In today's system, there is an escalating number of signal lines and cable connections, where ESD events occur from the cards, boards, and cables. With dynamically re-configurable systems, cards, boards, and cables, system level solutions have to address environments that are too severe for an 'on-chip ESD only' solution.

At the same time, the speed of the advanced semiconductor may also limit the size of the ESD protection networks. Both the allocated space and the allowed ESD capacitance budget will limit the circuit designers for on-chip ESD protection devices with the increased I/O count. The more stringent capacitance requirements create a potential performance-ESD reliability objectives conflict.

Off-chip voltage suppression devices will be a greater focus for cost and performance reasons. On a system level, circuit elements, and devices used for ESD include off-chip low capacitance air gaps, resistors, capacitors, inductors, varistors, double-diode ESD elements, and combinations thereof. Diodes were commonly used for overshoot and undershoot issues on cards until the need for low capacitance and cost were an issue. System-level diode capacitance in the order of 5 pF are used; these loading effects will be too high for future high-frequency applications. Although diodes are capable of extremely low trigger voltages, the capacitance loading can also cause signal distortion. The need for economical ESD protection of multiple signal lines along with the low capacitance at high frequencies will be required for future high-pin count solutions.

Additionally, there are a number of present and future technologies where on-chip ESD protection is not feasible from a material perspective. For example, magnetic recording industry devices, such as magneto-resistor (MR) heads, giant magneto-resistor (GMR) heads, and tunneling magneto-resistor (TMR) heads are formed on nonsilicon substrates.

The ESD design solution and practice is to incorporate multiple stages of ESD protection in system-level design. In these system-level solutions, the last level of ESD protection will be the on-chip ESD networks. These are further complimented by the addition of spark gaps and surge protection concepts.

10.3.1 Off-Chip Transient Voltage Suppression (TVS) Devices

Off-chip Transient Voltage Suppression (TVS) devices are used to address system-level ESD events. Transient voltage suppressor provides a low-cost solution to protect sensitive integrated circuits connected to high speed data and telecommunications lines from EOS, ESD, and cable discharge events (CDEs) that can enter through external plug ports and cause catastrophic damage. System level engineers are required to improve system-level performance while maintaining the quality and reliability. Electrostatic discharge and electromagnetic emissions (EMI), are a concern in systems. System-level standards and system engineers have long known that charged cables can also introduce system-level concerns. High-voltage and high-current events from charged cables are referred to as cable discharge events (CDE). Charge accumulation on unterminated twisted pair (UTP) cables occur through both tribo-electric charging and induction charging. In the case of tribo-electrification, a UTP cable can be dragged along a floor. A positive charge is established on the outside surface of the insulating film. The positive charge on the outside of the cable attracts negative charge in the twisted pair that leads across the dielectric region. When the negative charge is induced near the outside positive charge, positive charge is induced in the electrical conductor at the ends of the cable. When the cable is plugged into a connector, electrical arcing will occur leading to the charging of the UTP (note: the twisted pair is neutral to this point). If a cable is introduced into a strong electric field, induction charging will occur. When the electric field is removed the cable remains charged until a discharge event from grounding occurs. Given a system that is unpowered, no latchup event can occur. But, if the CDE occurs while a system is powered, the current event can lead to latchup. With the integration of wide area networks (WAN) and local area network (LAN), the Ethernet plays a larger role. When a charged twisted pair cable connects to an Ethernet port with a lower electrical potential, CDEs can occur in LAN systems. In the past, standards (e.g., IEEE 802.3 Section 14.7.2) exhibited the potential for CDE processes in LAN cables. Additionally, the introduction of Category 5 and Category 6 cables have significantly low leakage across the dielectric. As a result, when a tribo-electric charge is established, the conductance of the insulator is so low that the induced charge can maintain for long time scales (e.g., 24-h period). Additional to the system level issues, the latchup robustness of advanced technology are significantly low because of the technology scaling of the latchup critical parameters. Hence, with both system and technology evolution, the reasons for the increased concern for this issue are the following:

- WAN and LAN integration;
- Category 5 and 6 LAN cabling;
- high level incidents of disconnection;
- high level incidents of reconnections.

TVS devices can be used in many environments from one pair of high-speed differential signals (two lines) used in standard 10BaseT (10 Mbits/s), 100BaseT (100 Mbits/s), or 1000BaseTX (1 Gbit/s over copper) Ethernet transceiver interfaces. TVS devices sometimes uses compensation diodes in a full-bridge configuration to reduce the capacitance loading effects, allowing signal integrity to be preserved at the high transmission speeds of GbE interfaces. TVS devices can be below 10 pF capacitance loading and can achieve 100 A (8/20 μs) surge.

ESD design objectives of TVS devices include the following:

- number of protection elements per signal pins;
- low maximum reverse 'standoff voltage';
- low capacitance (1–10 pF);
- low clamping voltages (near peripheral I/O power supply voltage of semiconductor chip levels);
- achieve Telcordia GR-1089 intrabuilding lightning immunity requirements (100 A, 2/10-μs pulse durations);
- peak pulse power maximum (in Watts for 10–100 μs pulse duration time scales);
- minimize board area.

Hence, to address chip-level and system-level electrical overstress events (e.g., EOS, ESD, CDE, and lightning), TVS devices are used as a first line of protection. In future, TVS devices will need to be of low cost, low capacitance, and be able to require a minimum of board area.

10.3.2 Off-Chip Polymer Voltage Suppression (PVS) Devices

The use of filled polymer for circuit protection was initiated in the early 1980s, with a self-resetting polymer fuse that had the capability of resetting after a current surge [9,10]. In these structures, the materials increase in resistance as a function of current. Polymer positive temperature coefficient (PPTC) devices are used across all electronic markets because of their re-set feature and their formation into low profile films [9,10].

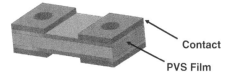

Contact

PVS Film

Figure 10.4   Polymer voltage suppression (PVS) device

Polymer-based voltage suppression (PVS) devices can be used on a package, board, card or system level to suppress high current transient phenomenon [9,10]. To enable ESD protection of electronic products, PVS devices are designed with low capacitance and space-efficient packages for multiple-line ESD protection. PVS devices incorporate system-level ESD protection without interfering with high frequency signals and do not require space on PCBs. The PVS devices can be built in an array or matrix fashion to address multiple parallel pins in a system environment [9]. The array of PVS devices can be incorporated in or in the physical system level connectors.

In PVS components, the polymer provides a low dielectric constant. The dielectric constant of the polymer film can be formulated to meet a wide range of application frequencies. Figure 10.4 shows construction of a single PVS surface-mount device consisting of a PVS film laminated between electrodes. Polymer voltage suppression devices are bidirectional allowing the shunting of both positive and negative polarities from signal lines to ground locations. The PVS device film uses an epoxy polymer, containing uniformly

Figure 10.5   ESD failure of a cell phone that utilizes a GaAs power amplifier

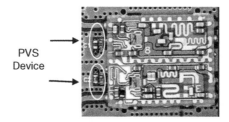

PVS
Device

Figure 10.6   A test board demonstrating integration of polymer voltage suppression (PVS) devices on a board with a GaAs power amplifier. This demonstrates an 'off-chip solution' for RF applications in future cell phone applications

dispersed conductive and nonconductive particles, laminated between electrodes. The laminate is transformed into a surge protection device using process similar to that used for manufacturing PCBs. The overall thickness of the PVS device is in the range less than 10 mils. The PVS material does not require a substrate and consequently, the trigger voltage of the device is set by device dimensions and the polymer formula. These components allow for single- or multiple-array surface mount components and connector arrays. Installing the protection array on the connector itself provides ESD protection at the input traces of ESD sensitive components and semiconductors.

Shrier *et al.* [9] demonstrated PVS elements with approximately 100 fF loading capacitance. Polymer voltage suppression elements were achieved with less than 200 ps rise times and effective clamping time of less than 10 ns [9,10].

Figure 10.5 shows an example of a ESD failure in a cell phone application. Today, gallium arsenide power amplifiers are used in cell phone applications because of the high Johnson Limit of GaAs technology. ESD failure is evident in the passive elements on the cell phone prototype application. ESD damage occurred in a resistor element. Figure 10.6 shows the integration of the PVS devices on the board of the cell phone application. ESD protection is provided off-chip (e.g., on the board) instead of a on-chip GaAs semiconductor chip.

10.4 PACKAGE-LEVEL MECHANICAL ESD SOLUTIONS

In many ESD applications, ESD solutions are best addressed on the package, card, or board level instead of on an on-chip ESD solution. In many applications, the design and application does not lend itself to integration with an ESD element. These can include semiconductor lasers, heterostructures, super-lattices, gallium arsenide power amplifiers, and other low-pin count applications. In high-performance applications, the ability to provide on-chip ESD protection will become more difficult. With the increase in circuit density and smaller physical geometries, the ability to have on-chip ESD protection only may be a limited perspective. ESD solutions on packages have been addressed since the 1960s with electrically connecting spring devices, such as proposed by Wallo [11]. Wallo [11], Kisor [12], Medesha [13], Tolnar and Winyard [14], Dinger *et al.* [15], Bachman and Dimeo [16], and Beecher [17] proposed mechanical shorting solutions to avoid ESD in packages, static charge protective packages, connectors, cartridges, and semiconductor packages. For example, in the 1990s, ESD failure of the semiconductor laser diodes was common. A simple solution developed by Cronin [18], to provide ESD protection between the three leads of a semiconductor diode was to form a helix-like mechanical spring that was inherently part of the packaging structure, forming an electrical short. When the laser element was not inserted, the mechanical spring shorted the leads. When the laser was inserted into the socket, the mechanical packaging spring disengaged by mechanical deflection due to the insertion; this concept provided ESD protection when the element was removed from the application and disengaged during application functional usage. This concept can be extended to other semiconductor packages [18–23], magnetic recording [24–29], and other applications that have an insertion process.

The fundamental off-chip ESD concept, independent of the application, is as follows:

- Mechanical shorting means can be applied between sensitive pins and ground references that is 'shorted' when the application is not functional and 'open' during functional usage.

- Mechanical shorting means can be applied between sensitive pins and ground references when the ESD sensitive element is not inserted into the card, board, or system.

- Mechanical shorting means can be initiated shorting all pins and all references when the application is not functional and disengage when the application is in functional use.

- Mechanical 'disengagement' of the shorting means can be initiated by insertion, power on, or in the functional operation process.

10.5   RF PROXIMITY COMMUNICATIONS CHIP-TO-CHIP ESD DESIGN PRACTICES

In multichip system design, the on-chip bandwidth has significantly improved with semiconductor technology scaling. In system level design, multiple chips and chip sets are placed on a common substrate material. In these multichip systems, the signal lines are required to leave the first chip through wire bonds or solder balls, transfer through a common substrate material, and then re-enter another wire bond or solder ball connection. In these multichip environments, the interchip signals also have ESD protection networks on the transmitting signal pad and the receiving signal pad. In the case of the interchip signal line, the signal is transmitted using an off-chip driver (OCD) of the first chip and received by a receiver network of a second chip. Given that the semiconductor chips were manufactured independently, the first ESD element is placed on the transmitter chip, and the second ESD element on the receiver chip. The total capacitance loading of the solder ball, the MLC interconnect wiring, and the two ESF elements lead to an impact in functional performance. When the on-chip bandwidth increases, the off-chip interconnects and structures do not scale at the same pace; this dichotomy introduces a limitation in multichip systems and a separation between the on-chip and off-chip bandwidths [30].

In a new RF I/O strategy, semiconductor chips can transfer signals through capacitive coupling [30–32]. An insulator film is placed over the pad of the transmitter in the first chip, and a second insulator film is placed over the receiver in the second chip (Figure 10.7). A capacitor is formed between the transmitter signal pad and the receiver signal pad. This allows for capacitive coupling. In this implementation, no solder balls, or wirebonds or

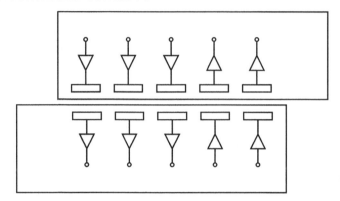

Figure 10.7 Proximity communication

interconnects are used to provide the transfer of the signal from the transmitter to the receiver network. Because an insulator film is placed over the signal pads, the signal pads are not vulnerable to ESD events. The advantage of this implementation allows for the loading effect of the solder balls, the wirebonds, the interconnects, and the ESD networks. As a result, the reduction of the capacitance loading and space provides significant increase in performance. A key issue with this chip-to-chip methodology is the alignment between the transmitter and the receiver structures. This is resolved by subdividing the transmission plate into an array of micro-plates. The transmitter array of micro-plates are formed using an array of 1024 micro-plates. The receiver array of micro-plates consist of 16 micro-plates. In this manner, the misalignment can be corrected by multiplexing of the transmitter and receiver micro-plates [30].

In this capacitive coupling method, the dilemma for the multichip system ESD design is resolved. But there are still few additional ESD questions that remain unresolved. For example, when the first chip is charged and the second chip is grounded, how does the charge transfer across the charged chip and the grounded chip?

ESD design practices to address the interchip signal line issue are the following:

- Place insulator films over the transmitter and receiver films.

- Establish capacitive coupling from the transmitter signal pad to the receiver signal pad via the capacitor formed between the two plates and the insulator films.

- Avoid ESD networks between the interchip signal pads because of the lack of signal wiring, bond pads, and interconnects.

10.6 SUMMARY AND CLOSING COMMENTS

In this chapter we have addressed the issues of nontraditional ESD networks and on-chip and off-chip ESD protection networks. Spark gaps, FEDs, and conductive polymer ESD protection concepts have been demonstrated in gallium arsenide technology and applications with significant success. The transition from nontraditional semiconductor ESD networks to on-chip and off-chip ESD protection may be required as the application frequencies continue to increase and the chip sizes decrease. It is unclear at this time as to which technology type and technology node will require these solutions and when will these become a standard solution for all semiconductor products. Moreover, for these solutions to continue to exist, the scaling of these elements (e.g., in trigger voltages, on-currents, on-resistance, size, and cost) to further improve them for future generations is also an issue.

PROBLEMS

1. For constant reactance scaling, the capacitance loading effect of on-chip protection must be reduced as the application frequency increases. Show a plot of reactance as a function of application frequency. Show a plot of the capacitance as a function of application frequency for a constant reactance ($X = |\omega C|$).

2. Under the assumption of Problem 1, assuming a capacitance per unit area as a constant and the capacitance load as the ESD network, show the scaling of the ESD network size as a function of application frequency. In addition, assuming the power-to-failure as proportional to area, show the reduction in the power-to-failure.

3. Given that a constant power-to-failure is desired, assuming a zero capacitance load off-chip protection, derive the power-to-failure requirement that must be absorbed by the off-chip element as the application frequency increases.

4. Given that the off-chip protection element is a conductive polymer film (resistor), derive the power-to-failure density for a film of thickness t, width W, and length L. Derive the voltage- and current-to-failure according to the Smith–Littau resistor model for the conductive polymer film.

5. Assuming a total power-to-failure condition, whereas the off-chip protection must absorb the residual power (as we scale the on-chip protection network), given a FED of area $W \times W$, calculate the power density. Calculate the current density as the frequency of the application increases.

REFERENCES

1. Paschen F. *Annalen der Physik* 1889;**37**:69.
2. Toepler M. Information on the law of guiding spark formation. *Annalen der Physik* 1906;**21**(12): 193–22.
3. Von Hippel A. Conduction and breakdown. In: *The molecular designing of materials and devices.* Boston: MIT Press; 1965. p. 183–97.
4. Townsend JS. *Electricity in gases.* Oxford: Clarendon Press; 1915.
5. Kleen BG. Printed circuit spark-gap protector. *IBM Technical Disclosure Bulletin* 1972;**14**(2):638.
6. DeBar DE, Francisco H, De La Moneda. Stuby KP and Bertin CL. *IBM Technical Disclosure Bulletin,* "Module Spark Gap", 1975;**18**(7).
7. Bock K. ESD issues in compound semiconductor high-frequency devices and circuits. *Proceedings of the Electrical Overstress/Electrostatic Discharge (EOS/ESD)Symposium,* 1997. p. 1–12.
8. Barth J, Richner J, Verhaege K, Kelly M, Henry L. Correlation Consideration II: Real HBM to HBM testers. *Proceedings of the Electrical Overstress/Electrostatic Discharge (EOS/ESD) Symposium,* 2002. p. 155–62.

9. Shrier K, Truong T, Felps J. Transmission line pulse test methods, test techniques, and characterization of low capacitance voltage suppression device for system level electrostatic discharge compliance. *Proceedings of the Electrical Overstress/Electrostatic Discharge (EOS/ESD) Symposium*, 2004. p. 88–97.

10. Shrier K, Jiaa C. ESD enhancement of power amplifier with polymer voltage suppressor. *Proceedings of the Taiwan Electrostatic Discharge Conference (T-ESDC)*, 2005. p. 110–5.

11. Wallo WH. Electrically connecting spring device. U.S. Patent No. 3,467,930 (September 16, 1967).

12. Kisor TW. Static charge protective packages for electronic devices. U.S. Patent No. 3,653,498 (April 4, 1972).

13. Medesha AL. Package including electrical equipment lead shorting element. U.S. Patent No. 3,774,075 (November 20, 1973).

14. Tolnar Jr EJ, Winyard AH. Connector means having shorting clip. U.S. Patent No. 3,869,191 (March 4, 1975).

15. Dinger ED, Saben DG, VanPatten JR. Static control shorting clip for semiconductor package. U.S. Patent No. 4,019,094 (April 19, 1977).

16. Bachman WJ, Dimeo FR. Plug-in circuit cartridge with electrostatic charge protection. U. S. Patent No. 4,179,178 (December 18, 1979).

17. Beecher R. Cartridge having improved electrostatic discharge protection. U.S. Patent No. 4,531,176 (July 23, 1985).

18. Cronin DV. Electrical connector with attachment for automatically shorting select conductors upon disconnection of connector. U.S. Patent No. 4,971,568 (November 20, 1990).

19. Voldman S. Lightning rods for nanoelectronics. *Scientific American* October 2002. p. 90–8.

20. Cronin DV. Electrostatic discharge protection devices for semiconductor chip packages. U.S. Patent No. 5,108,299 (April 28, 1992).

21. Cronin DV. Electrostatic discharge protection devices for semiconductor chip packages. U.S. Patent No. 5,163,850 (November 17, 1992).

22. Cronin DV. Electrostatic discharge protection device for a printed circuit board. U.S. Patent No. 5,164,880 (November 17, 1992).

23. Johansen AW, Cronin DV. Electrostatic discharge protection device. U.S. Patent No. 5,812,357 (September 22, 1998).

24. Wallash A, Hughbanks T, Voldman S. ESD failure mechanisms of inductive and magnetoresistive recording heads. *Proceedings of the Electrical Overstress/Electrostatic Discharge (EOS/ESD) Symposium*, 1995. p. 322–30.

25. Bajorek CH, Erpelding AD, Garfunkel GA, Pattanaik S, Robertson NL, Wallash AJ. Shorted magnetoresistive head leads for electrical overstress and electrostatic discharge protection during manufacture of a magnetic storage system. U.S. Patent No. 5,465,186 (November 7, 1995).

26. Hughbanks TH, Lee HP, Phipps PB, Robertson NL, Wallash AJ. Shorted magnetoresistive head elements for electrical overstress and electrostatic discharge protection. U.S. Patent No. 5,491,605 (February 13,1996).

27. Voldman S. The impact of technology evolution and revolution in advanced semiconductor technologies on electrostatic discharge (ESD) protection. *Proceedings of the Taiwan Electrostatic Discharge Conference (T-ESDC)*, 2003. p. 2–7.

28. Arya SP, Hughbanks TS, Voldman SH, Wallash AJ. Electrostatic discharge protection system for MR heads. U.S. Patent No. 5,644,454 (July 1, 1997).

29. Arya SP, Hughbanks TS, Voldman SH, Wallash AJ. Electrostatic discharge protection system for MR heads. U.S. Patent No. 5,710,682 (January 20, 1998).

30. Sutherland I. Face to face chips. U.S. Patent No. 6,500,696 (December 31, 2002).

31. Salzman D, Knight T. Capacitively coupled multichip modules. *Proceedings of the Multi-chip Module Conference*, 1994. p. 133–40.

32. Drost RJ, Hopkins RD, Sutherland IE. Proximity Communications, 2003.

Index

ESD: RF Technology and Circuits   Steven H. Voldman
© 2006 John Wiley & Sons, Ltd

Printed and bound by CPI Group (UK) Ltd, Croydon, CR0 4YY

16/04/2025

14658554-0001